The Construction of Optimal Stated Choice Experiments

BICENTENNIAL 1807 WILEY 2007 BICENTENNIAL

The Wiley Bicentennial–Knowledge for Generations

Each generation has its unique needs and aspirations. When Charles Wiley first opened his small printing shop in lower Manhattan in 1807, it was a generation of boundless potential searching for an identity. And we were there, helping to define a new American literary tradition. Over half a century later, in the midst of the Second Industrial Revolution, it was a generation focused on building the future. Once again, we were there, supplying the critical scientific, technical, and engineering knowledge that helped frame the world. Throughout the 20th Century, and into the new millennium, nations began to reach out beyond their own borders and a new international community was born. Wiley was there, expanding its operations around the world to enable a global exchange of ideas, opinions, and know-how.

For 200 years, Wiley has been an integral part of each generation's journey, enabling the flow of information and understanding necessary to meet their needs and fulfill their aspirations. Today, bold new technologies are changing the way we live and learn. Wiley will be there, providing you the must-have knowledge you need to imagine new worlds, new possibilities, and new opportunities.

Generations come and go, but you can always count on Wiley to provide you the knowledge you need, when and where you need it!

William J. Pesce
President and Chief Executive Officer

Peter Booth Wiley
Chairman of the Board

The Construction of Optimal Stated Choice Experiments

Theory and Methods

Deborah J. Street
University of Technology, Sydney

Leonie Burgess
University of Technology, Sydney

BICENTENNIAL 1807 WILEY 2007 BICENTENNIAL

WILEY-INTERSCIENCE
A John Wiley & Sons, Inc., Publication

Published by John Wiley & Sons, Inc., Hoboken, New Jersey.
Published simultaneously in Canada.

Library of Congress Cataloging-in-Publication Data:

Street, Deborah J.
The construction of optimal stated choice experiments : theory and methods / Deborah J. Street, Leonie Burgess.
p. cm.
Includes bibliographical references and index.
ISBN 978-0-470-05332-4
1. Combinatorial designs and configurations. 2. Optimal designs (Statistics) I. Burgess, Leonie. II. Title.
QA166.25.S77 2007
511'.6—dc22 2007011016

CONTENTS

List of Tables xi

Preface xvii

1 Typical Stated Choice Experiments 1

1.1 Definitions 2
1.2 Binary Response Experiments 3
1.3 Forced Choice Experiments 5
1.4 The "None" Option 7
1.5 A Common Base Option 8
1.6 Avoiding Particular Level Combinations 9
1.6.1 Unrealistic Treatment Combinations 9
1.6.2 Dominating Options 10
1.7 Other Issues 11
1.7.1 Other Designs 11
1.7.2 Non-mathematical Issues for Stated Preference Choice Experiments 11
1.7.3 Published Studies 12
1.8 Concluding Remarks 13

2 Factorial Designs 15

2.1 Complete Factorial Designs 16
2.1.1 2^k Designs 16
2.1.2 3^k Designs 19
2.1.3 Asymmetric Designs 24
2.1.4 Exercises 25
2.2 Regular Fractional Factorial Designs 27
2.2.1 Two-Level Fractions 27
2.2.2 Three-Level Fractions 33
2.2.3 A Brief Introduction to Finite Fields 37
2.2.4 Fractions for Prime-Power Levels 39
2.2.5 Exercises 41
2.3 Irregular Fractions 41
2.3.1 Two Constructions for Symmetric OAs 43
2.3.2 Constructing $\text{OA}[2^k; 2^{k_1}, 4^{k_2}; 4]$ 44
2.3.3 Obtaining New Arrays from Old 46
2.3.4 Exercises 52
2.4 Other Useful Designs 53
2.5 Tables of Fractional Factorial Designs and Orthogonal Arrays 55
2.5.1 Exercises 56
2.6 References and Comments 56

3 The MNL Model and Comparing Designs 57

3.1 Utility and Choice Probabilities 58
3.1.1 Utility 58
3.1.2 Choice Probabilities 59
3.2 The Bradley–Terry Model 60
3.2.1 The Likelihood Function 61
3.2.2 Maximum Likelihood Estimation 62
3.2.3 Convergence 65
3.2.4 Properties of the MLEs 67
3.2.5 Representing Options Using k Attributes 70
3.2.6 Exercises 79
3.3 The MNL Model for Choice Sets of Any Size 79
3.3.1 Choice Sets of Any Size 79
3.3.2 Representing Options Using k Attributes 82
3.3.3 The Assumption of Independence from Irrelevant Alternatives 82
3.3.4 Exercises 83
3.4 Comparing Designs 83
3.4.1 Using Variance Properties to Compare Designs 84
3.4.2 Structural Properties 88

3.4.3 Exercises 91
3.5 References and Comments 92

4 Paired Comparison Designs for Binary Attributes 95

4.1 Optimal Pairs from the Complete Factorial 95
4.1.1 The Derivation of the Λ Matrix 97
4.1.2 Calculation of the Relevant Contrast Matrices 99
4.1.3 The Model for Main Effects Only 100
4.1.4 The Model for Main Effects and Two-factor Interactions 105
4.1.5 Exercises 117
4.2 Small Optimal and Near-optimal Designs for Pairs 118
4.2.1 The Derivation of the Λ Matrix 118
4.2.2 The Model for Main Effects Only 119
4.2.3 The Model for Main Effects and Two-Factor Interactions 121
4.2.4 Dominating Options 133
4.2.5 Exercises 134
4.3 References and Comments 134

5 Larger Choice Set Sizes for Binary Attributes 137

5.1 Optimal Designs from the Complete Factorial 138
5.1.1 Difference Vectors 138
5.1.2 The Derivation of the Λ Matrix 143
5.1.3 The Model for Main Effects Only 147
5.1.4 The Model for Main Effects and Two-Factor Interactions 152
5.1.5 Exercises 159
5.2 Small Optimal and Near-Optimal Designs for Larger Choice Set Sizes 159
5.2.1 The Model for Main Effects Only 160
5.2.2 The Model for Main Effects and Two-Factor Interactions 163
5.2.3 Dominating Options 164
5.2.4 Exercises 165
5.3 References and Comments 165

6 Designs for Asymmetric Attributes 167

6.1 Difference Vectors 169
6.1.1 Exercises 173
6.2 The Derivation of the Information Matrix Λ 174
6.2.1 Exercises 180
6.3 The Model for Main Effects Only 180
6.3.1 Exercises 189
6.4 Constructing Optimal Designs for Main Effects Only 189
6.4.1 Exercises 197

6.5 The Model for Main Effects and Two-Factor Interactions 197
6.5.1 Exercises 209
6.6 References and Comments 210
Appendix 211
6. A.1 Optimal Designs for $m = 2$ and $k = 2$ 212
6. A.2 Optimal Designs for $m = 2$ and $k = 3$ 213
6. A.3 Optimal Designs for $m = 2$ and $k = 4$ 217
6. A.4 Optimal Designs for $m = 2$ and $k = 5$ 220
6. A.5 Optimal Designs for $m = 3$ and $k = 2$ 221
6. A.6 Optimal Designs for $m = 4$ and $k = 2$ 223
6. A.7 Optimal Designs for Symmetric Attributes for $m = 2$ 225

7 Various Topics 227

7.1 Optimal Stated Choice Experiments when All Choice Sets Contain a Specific Option 228
7.1.1 Choice Experiments with a None Option 228
7.1.2 Optimal Binary Response Experiments 233
7.1.3 Common Base Option 234
7.1.4 Common Base and None Option 236
7.2 Optimal Choice Set Size 237
7.2.1 Main Effects Only for Asymmetric Attributes 237
7.2.2 Main Effects and Two-Factor Interactions for Binary Attributes 240
7.2.3 Choice Experiments with Choice Sets of Various Sizes 242
7.2.4 Concluding Comments on Choice Set Size 243
7.3 Partial Profiles 243
7.4 Choice Experiments Using Prior Point Estimates 245
7.5 References and Comments 246

8 Practical Techniques for Constructing Choice Experiments 249

8.1 Small Near-Optimal Designs for Main Effects Only 251
8.1.1 Smaller Designs for Examples in Section 6.4 251
8.1.2 Getting a Starting Design 256
8.1.3 More on Choosing Generators 264
8.2 Small Near-Optimal Designs for Main Effects Plus Two-Factor Interactions 269
8.2.1 Getting a Starting Design 269
8.2.2 Designs for Two-Level Attributes 271
8.2.3 Designs for Attributes with More than Two Levels 272
8.2.4 Designs for Main Effects plus Some Two-Factor Interactions 276
8.3 Other Strategies for Constructing Choice Experiments 279

8.4 Comparison of Strategies 291
8.5 References and Comments 293
Appendix 294
8. A.1 Near-Optimal Choice Sets for $\ell_1 = \ell_2 = 2, \ell_3 = \ell_4 = \ell_5 = \ell_6 = 4, \ell_7 = 8, \ell_8 = 36$, and $m = 3$ for Main Effects Only 294
8. A.2 Near-Optimal Choice Sets for $\ell_1 = \ell_2 = \ell_3 = \ell_4 = \ell_5 = 4, \ell_6 = 2, \ell_7 = \ell_8 = 8, \ell_9 = 24$, and $m = 3$ for Main Effects Only 296

Bibliography 301

Index 309

LIST OF TABLES

1.1 Attributes and Levels for the Survey to Enhance Breast Screening Participation 4

1.2 One Option from a Survey about Breast Screening Participation 5

1.3 Six Attributes to be Used in an Experiment to Compare Pizza Outlets 6

1.4 One Choice Set in an Experiment to Compare Pizza Outlets 6

1.5 Attributes and Levels for the Study Examining Preferences for HIV Testing Methods 7

1.6 One Choice Set from the Study Examining Preferences for HIV Testing Methods 8

1.7 Five Attributes to be Used in an Experiment to Investigate Miscarriage Management Preferences 9

1.8 Five Attributes Used to Compare Aspects of Quality of Life 10

2.1 Values of Orthogonal Polynomials for $n = 3$ 22

2.2 A, B, and AB Contrasts for a 3^2 Factorial 23

2.3 A, B, and AB Contrasts for a 3^3 Factorial 24

2.4 Main Effects Contrasts for a $2 \times 3 \times 4$ Factorial 25

2.5 A Regular 2^{4-1} Design 27

2.6 Non-overlapping Regular 2^{5-2} Designs 28

2.7 A 2^{4-1} Design of Resolution 4 29

2.8 Contrasts for the 2^4 design 30

2.9 Smallest Known 2-Level Designs with Resolution at Least 3 32

2.10 Smallest Known 2-Level Designs with Resolution at Least 5 32

2.11 A Design with 6 Binary Factors of Resolution 3 33

2.12 A 3^{4-2} Fractional Factorial Design 34

2.13 Pencils for a 3^2 Factorial Design 35

2.14 Contrasts for a 3^2 Factorial Design 36

2.15 Contrasts for a 3^{4-2} Factorial Design 36

2.16 Smallest Known Regular 3-Level Designs with Resolution at Least 3 37

2.17 Smallest Known 3-Level Designs with Resolution at Least 5 38

2.18 The Finite Field with 4 Elements 39

2.19 A Resolution 3 4^{5-3} Fractional Factorial Design 40

2.20 A Resolution 3 Fractional Factorial Design for Seven 3-Level Factors 44

2.21 Generators for $OA[2^k; 2^{k_1}, 4^{k_2}; 4]$ 45

2.22 The $OA[64;2^3,4^2;4]$ 46

2.23 A Resolution 3 Fractional Factorial Design for Three 2-Level Factors and Four 3-Level Factors 47

2.24 An OA[16,5,4,2] 48

2.25 An OA[4,3,2,2] 48

2.26 An OA[16;2,2,2,4,4,4,4;2] 48

2.27 An OA[16,15,2,2] 49

2.28 An OA[12;2,2,2,3;2] 50

2.29 An OA[16;6,2;2] 51

2.30 An OA[12;2,2,6;2] 51

2.31 An OA[12;2,2,6;2] without 000 52

2.32 An OA[8,6,2,2] 53

2.33 The Blocks a (7,3,1) BIBD 53

2.34 Some Small Difference Sets 54

2.35 Some Small Difference Families 55

3.1	Choice Sets and Choices for Example 3.2.2, with $t = 4$, $s = 5$	63
3.2	Estimates of π_i from All Six Choice Sets	64
3.3	Estimates of π_i from the First Three Choice Sets Only	66
3.4	Unconstrained Representation of Treatment Combinations	71
3.5	Constrained Representation of Treatment Combinations	72
3.6	The Fold-over Pairs with $k = 3$	79
3.7	Triples of Pairs for $k = 2$ and the Corresponding D-, A- and E-Optimum Values	86
3.8	Two Choice Experiments	88
3.9	A Choice Experiment for the Estimation of Main Effects when There Are 3 Attributes with 2, 3, and 6 Levels which Satisfies the Huber-Zwerina Conditions but for which Main Effects Cannot Be Estimated.	90
3.10	An Optimal Choice Experiment for the Estimation of Main Effects when There Are 3 Attributes with 2, 3, and 6 Levels	91
4.1	Six Attributes to Be Used in an Experiment to Compare Pizza Outlets	96
4.2	One Choice Set in an Experiment to Compare Pizza Outlets	96
4.3	The Possible Designs and Corresponding Λ Matrices for $k = 2$	99
4.4	All Possible Pairs when $k = 3$	99
4.5	The 7 Competing Designs for Main Effects Only for Pairs with $k = 3$	104
4.6	The 7 Competing Designs for Main Effects and Two-Factor Interactions for Pairs with $k = 3$	108
4.7	Two Designs with Four Pairs for $k = 4$ Binary Attributes	119
4.8	Non-regular OMEPs of Resolution 3	122
4.9	The Pairs from the Complete Factorial when $k = 3$ and $\mathbf{e} = (0, 1, 1)$	128
4.10	The Pairs from the Fraction in Example 4.2.4 when $\mathbf{e} = (0, 1, 1)$	128
4.11	The Blocks a (7,3,1) BIBD	131
4.12	D-Efficiency and Number of Pairs for Some Constant Difference Choice Pairs	131
5.1	Attributes and Levels for Holiday Packages	138
5.2	One Choice Set for the Possible Holiday Packages	139
5.3	All Possible Triples when $k = 3$	141
5.4	All Possible Triples when $k = 3$ Sorted by Difference Vector	143

5.5 All Possible Choice Experiment Designs for Binary Attributes when $k = 3$ and $m = 3$ 152

5.6 The Efficiency, for the Estimation of Main Effects plus Two-Factor Interactions, of the 7 Competing Designs when $k = 3$ and $m = 3$ 158

5.7 Optimal Choice Sets for Estimating Main Effects Only for $m = 5$ and $k = 9$. 163

5.8 Near-Optimal Choice Sets for Estimating Main Effects and Two-Factor Interactions for $m = 3$ and $k = 4$. 165

6.1 Attributes and Levels for the Study Examining Preferences for HIV Testing Methods 168

6.2 One Choice Set from the Study Examining Preferences for HIV Testing Methods 168

6.3 All Possible Choice Sets when $m = 3$, $k = 2$, $\ell_1 = 2$, and $\ell_2 = 3$ 170

6.4 All Possible Choice Sets when $m = 4$, $k = 2$, $\ell_1 = 2$, and $\ell_2 = 3$ 172

6.5 All Possible Triples when $k = 2$, $\ell_1 = 2$, and $\ell_2 = 3$ Sorted by Difference Vector 173

6.6 All Possible Choice Experiments for $k = 2$, $\ell_1 = 2$, and $\ell_2 = 3$ when $m = 3$ 190

6.7 Choice Sets when $k = 2$, $\ell_1 = 2$, $\ell_2 = 3$, and $m = 3$ 194

6.8 All Possible Choice Experiment Designs for $k = 2$, $\ell_1 = 2$, and $\ell_2 = 3$ when $m = 3$: Main Effects and Two-Factor Interactions 206

6.9 Values of $a_{\mathbf{v}_j}$ for $m = 2$ and $k = 2$ 208

6.10 Values of $\det(C_{MT})$ for $m = 2$ and $k = 2$ 208

7.1 Efficiencies of the Four Designs Discussed in Example 7.1.1 232

7.2 Choice Sets with $k = 3$ Attributes, $\ell_1 = \ell_2 = 2$ and $\ell_3 = 4$ 233

7.3 Efficiency of Different Values of m for Five Attributes, Each with Four Levels for Main Effects Only. 239

7.4 Efficiency of Different Values of m for Four Asymmetric Attributes for Main Effects Only. 241

7.5 Efficiency of Different Values of m for 6 Binary Attributes when Estimating Main Effects and Two-factor Interactions. 241

7.6 Choice Sets of Up to Three Different Sizes for 6 Binary Attributes for Main Effects Only. 243

7.7 Sixteen Attributes Used to Describe Pizzas 244

7.8 (16,20,5,4,1) BIBD 245

7.9	The Four Choice Sets from the First Block of the (16, 20, 5, 4, 1)BIBD	245
7.10	The Treatment Combinations and the Corresponding Assumed π_i	247
8.1	Optimal Choice Sets for $k = 2$, $\ell_1 = 2$, $\ell_2 = 3$ when $m = 3$ for Main Effects Only	252
8.2	Near-Optimal Choice Sets for $k = 2$, $\ell_1 = \ell_2 = 4$ when $m = 6$ for Main Effects Only	254
8.3	Difference Vectors and Sets of Generators for $k = 3$, $\ell_1 = 2$, $\ell_2 = 3$, and $\ell_3 = 6$	254
8.4	$2 \times 3 \times 6//18$ Fractional Factorial Design Obtained from the $3 \times 3 \times 6//18$ by Collapsing One Attribute	255
8.5	Near-Optimal Choice Sets for $k = 3$, $\ell_1 = 2$, $\ell_2 = 3$, and $\ell_3 = 6$ when $m = 3$ for Main Effects Only	256
8.6	Near-Optimal Choice Sets for $k = 4$, $\ell_1 = \ell_2 = 2$, $\ell_3 = 3$, and $\ell_4 = 6$ when $m = 3$ for Main Effects Only	258
8.7	Near-Optimal Choice Sets for $k = 4$, $\ell_1 = \ell_2 = 2$, $\ell_3 = 3$, and $\ell_4 = 6$ when $m = 4$ for Main Effects Only	259
8.8	Starting Design and Near-Optimal Choice Sets for $k = 5$, $\ell_1 = \ell_2 = \ell_3 = 2$, and $\ell_4 = \ell_5 = 4$ when $m = 4$ for Main Effects Only	260
8.9	Common Collapsing/Replacement of Attribute Levels	261
8.10	Choice Sets for $k = 5$, $\ell_1 = \ell_2 = \ell_3 = 2$, and $\ell_4 = \ell_5 = 4$ when $m = 2$ for Main Effects Only	262
8.11	$2^{12} \times 4//16$ Obtained from $4^5//16$ by Expansive Replacement, and Choice Sets for $m = 3$	262
8.12	$2^{12} \times 4//16$ Obtained from $2^{15}//16$ by Contractive Replacement, and Choice Sets for $m = 3$	263
8.13	A Different $2^{12} \times 4//16$ and Choice Sets for $m = 3$	264
8.14	Different Designs for $k = 17$, $\ell_q = 2$, $q = 1, \ldots, 16$, and $\ell_{17} = 13$ when $m = 2, 3, 4, 5$ for Main Effects Only	266
8.15	$4^5 \times 2^2 \times 8//32$ and $4^8 \times 8//32$ OMEPs	268
8.16	$9 \times 4^5 \times 2^2 \times 8//288$ by Adding Another Attribute to the $4^5 \times 2^2 \times 8//32$	269
8.17	$4^5 \times 2 \times 8^3//64$ OMEP Obtained from $8^9//64$ by Collapsing the Levels of 6 Attributes	270
8.18	Fractional Factorial of Resolution 7 for $\ell_q = 2$, $q = 1, \ldots, 7$	272
8.19	Near-Optimal Choice Sets for $\ell_q = 2$, $q = 1, \ldots, 7$ and $m = 2$ for Main Effects and All Two-Factor Interactions	273

8.20 Choice Sets for $k = 5$, $\ell_1 = \ell_2 = \ell_3 = \ell_4 = 2$, $\ell_5 = 4$ when $m = 2$ for Main Effects and All Two-Factor Interactions 275

8.21 Choice Sets for $k = 3$, $\ell_1 = \ell_2 = 3$, and $\ell_3 = 5$ when $m = 3$ for Main Effects and All Two-Factor Interactions 277

8.22 Choice Sets for $k = 4$ Binary Attributes when $m = 2$ for Main Effects and Interactions AB, AD, and BD 278

8.23 Choice Sets for $k = 6$ Attributes, $\ell_1 = 3$, $\ell_2 = \ell_3 = \ell_4 = \ell_5 = 2$, and $\ell_6 = 4$ when $m = 3$ for Main Effects and Interactions AB, AC, AD, AE, and AF 280

8.24 Random Method 1 Choice Sets 281

8.25 Random Method 1 Choice Sets: C_M Matrix 282

8.26 Random Method 2 Choice Sets 283

8.27 Random Method 2 Choice Sets: C_M Matrix 284

8.28 Choice Sets which Satisfy Huber & Zwerina Criteria 285

8.29 Choice Sets which Satisfy Huber & Zwerina Criteria: C_M Matrix 286

8.30 L^{MA} Choice Sets 288

8.31 Choice Sets from SAS Macros 289

8.32 Choice Sets from SAS Macros: C_M^{-1} Matrix 289

8.33 Street–Burgess Choice Sets 290

8.34 Street–Burgess Choice Sets: C_M Matrix 290

8.35 Comparison of Construction Methods for Main Effects Only 292

8.36 Comparison of Construction Methods for Main Effects and Two-Factor Interactions 293

PREFACE

Stated choice experiments are widely used in various areas including marketing, transport, environmental resource economics and public welfare analysis. Many aspects of the design of a stated choice experiment are independent of its area of application, however, and the goal of this book is to present constructions for optimal designs for stated choice experiments. Although we will define "optimal" formally later, informally an optimal design is one which gets as much information as possible from an experiment of a given size.

We assume throughout that all the options in each choice set are described by several attributes, and that each attribute has two or more levels. Usually we will assume that all the choice sets in a particular experiment have the same number of options, although we will relax this constraint in the penultimate chapter. We assume that a multinomial logit model will be used to analyze the results of the stated choice experiment.

In the first chapter we describe typical stated choice experiments and give several examples of published choice experiments. We introduce the terminology that we will use throughout the book.

In Chapter 2 we define and construct factorial designs. These designs are used in various settings to determine the effect of each of several factors, or attributes, on one or more response variables. Factorial designs are appropriate when discussing the design of stated choice experiments, since in most stated choice experiments the options to be considered are described by attributes each of which can take one of several levels. We show how the effects of each factor can be calculated independently of the other factors in the experiment and we show how the joint effects of two or more factors can be determined. We show the relationship between fractional factorial designs and orthogonal arrays, give relevant constructions and discuss how to use some of the tables of such designs that are available.

In Chapter 3 we discuss the use of the multinomial logit (MNL) model to analyze the results of a stated choice experiment. We derive the Bradley–Terry model and extend it to choice sets with more than two options. We show how the attributes that are used to describe the options can be incorporated into the MNL model and hence how to derive the appropriate variance-covariance matrix for the effects of interest. Functions of this matrix are traditionally used to compare designs and we indicate how any two stated choice experiments can be compared so that the "better" design can be determined. We develop the theory for the determination of the optimal stated choice experiment and we show how the optimality value associated with any set of choice sets of any size involving attributes with any number of levels can be calculated. This theory provides a non-subjective way to compare any set of stated choice experiments. We briefly discuss comparing designs using the structural properties of the choice experiments under consideration but as yet there is no firm link between these properties and those of the optimal designs.

In the remaining chapters we discuss some specific choice situations in turn.

In Chapter 4 we discuss the construction of choice experiments in which all of the attributes describing the options have two levels and in which all choice sets have two options. These are often called paired comparison choice experiments. We find the optimal designs for estimating main effects, and main effects plus two-factor interactions. We get the designs from the complete factorial and show how equally good designs can be constructed from fractional factorials, which were constructed in Chapter 2. In both cases the designs always have known efficiency properties.

In Chapter 5 we extend the ideas of the previous chapter to choice sets of any size, although we still retain the restriction that the attributes each have only two levels. We work initially with designs based on the complete factorial and then show how to get smaller designs that are just as good from regular fractional factorial designs.

In Chapter 6 we extend the ideas of the previous two chapters to construct optimal stated choice designs for any number of attributes with any number of levels using choice sets of any size. We derive the upper bound for D-optimal designs and show how to construct small designs that reach this bound for the estimation of main effects. In this case there are no general constructions for optimal designs for the estimation of main effects and two-factor interactions but we give heuristics that give designs that work well in practice. We give tables of optimal designs for some small situations.

In Chapter 7 we briefly consider other important topics in the construction of optimal choice experiments. We look at how to construct optimal designs when there is either a "none of these" option in each choice set or a common base alternative in each choice set. We consider how to design optimal experiments when there are restrictions on the number of attributes that can be different between any two options in a choice set, find the optimal size of the number of options for the choice sets in a choice experiment and look briefly at the use of prior point estimates.

The constructions we have given in the previous chapters are not necessarily the easiest way to construct choice experiments so in Chapter 8 we discuss some techniques that we have used in practice to construct optimal or near-optimal choice experiments. We also compare some commonly used strategies for constructing choice experiments.

For each chapter we provide references to the mathematical and statistical literature for the constructions, to various literature including the marketing literature and the health services literature for examples of applications of the designs, and we provide a number of exercises to help the reader test their understanding of the material presented. Some of these exercises provide interesting extensions to the topics discussed in the chapter.

Software that allows readers to construct choice sets from a starting design by adding sets of generators is available at http://maths.science.uts.edu.au/maths/wiki/SPExpts. For these choice sets, or indeed any set of choice sets, the software will calculate the information matrix and the corresponding variance-covariance matrix.

Our biggest thank you is to Jordan Louviere who introduced the first author to choice experiments a decade ago and who has been a constant source of questions and encouragement ever since. While writing this manuscript we have benefitted greatly from feedback from various people. We would particularly like to thank David Pihlens, Stephen Bush and Amanda Parnis for constructive comments that improved the clarity of the presentation. Each author would like to blame the other author for any mistakes that remain.

DEBORAH J. STREET AND LEONIE BURGESS

Sydney, Australia
March, 2007

CHAPTER 1

TYPICAL STATED CHOICE EXPERIMENTS

People make choices all the time; some of these decisions are of interest to governments and businesses. Governments might want to model demand for health services in the future or to assess the likely electoral impact of a decision to allow logging in a national park. Businesses want to predict the likely market for new goods and services.

Information about choices can be captured from sources like supermarket scanners. However, this "revealed preference" data tells you nothing about products that do not yet exist. Here an experimental approach can help. Such experiments are called "stated preference" or "stated choice" experiments. This book describes the best way to design generic stated preference choice experiments, from a mathematical perspective.

Stated choice experiments are widely used in business although often not published. According to the results of a survey sent out to businesses, there were about 1000 commercial applications in the United States in the 1970s and there were about 400 per year in the early 1980s (Cattin and Wittink (1982); Wittink and Cattin (1989)). Wittink et al. (1994) found less extensive use in Europe in the period 1986–1991, but Hartmann and Sattler (2002) have found that the number of companies using stated choice experiments and the number of experiments conducted had more than doubled in German-speaking Europe by 2001. The range of application areas has also increased and now includes transport (Hensher (1994)), health economics (Bryan and Dolan (2004)), and environmental evaluation (Hanley et al. (2001)), among others.

In the rest of this chapter we will define some concepts that we will use throughout the book. We will use published choice experiments from various application areas to illustrate these concepts. These examples will also illustrate the range of issues that needs to be

The Construction of Optimal Stated Choice Experiments. By D. J. Street and L. Burgess addressed when designing such an experiment. The mathematical and statistical issues raised will be considered in the remainder of this book.

1.1 DEFINITIONS

Stated choice experiments are easy to describe. A *stated choice experiment* consists of a set of choice sets. Each choice set consists of two or more options (or alternatives). Each respondent (also called subject) is shown each choice set in turn and asked to choose one of the options presented in the choice set. The number of options in a choice set is called the *choice set size*. A stated choice or stated preference choice experiment is often called a *discrete choice experiment* and the abbreviations *SP experiment* and *DCE* are very common. We will look at the design of choice experiments for the simplest stated preference situation in this book — the so-called *generic* stated preference choice experiment. In such an experiment, all options in each choice set are described by the same set of attributes, and each of these attributes can take one level from a set of possible levels.

An example involving one choice set might ask members of a group of employees how they will travel to work tomorrow. The five options are {drive, catch a bus, walk, cycle, other} and these five options comprise the choice set. Each respondent will then choose one of these five methods of getting to work.

This simple example illustrates the fact that in many choice experiments people are forced to choose one of the options presented. We call such an experiment a *forced choice* experiment. In this case, being compelled to choose is easy since the respondents were employees (so would be going to work) and every possible way of getting to work was in the choice set since there was an option "other". Thus the list of options presented was *exhaustive*.

Sometimes a forced choice experiment is used even though the list of options presented is not exhaustive. This is done to try to find out how respondents "trade-off" the different characteristics of the options presented. A simple example is to offer a cheap flight with restrictive check-in times or a more expensive flight where there are fewer restrictions on check-in times. In reality, there might be intermediate choices, but these are not offered in the choice set.

However, there are certainly situations where it simply does not make sense to force people to choose. People may well spend several weeks deciding which car to buy and will defer choice on the first few cars that they consider. To allow for this situation, choice experiments include an option variously called "no choice" or "delay choice" or "none of these". We will just talk about having a *none option* to cover all of these situations.

A related situation arises when there is an option which needs to appear in every choice set. This can happen when respondents are being asked to compare a new treatment with an existing, standard treatment for a medical condition, for instance. We speak then of all choice sets having a *common base option*.

Sometimes just one option is described to respondents who are then asked whether or not they would be prepared to use that good or service. Usually several descriptions are shown to each respondent in turn. This is called a *binary response* experiment, and in many ways it is the simplest choice experiment of all. It does not allow for the investigation of trade-offs between levels of different attributes, but it gives an indication of combinations of levels that would be acceptable to respondents.

When constructing choice sets, it is often best to avoid choice sets where one option is going to be preferred by every respondent. In the flight example above, there will be respondents who prefer to save money and so put up with the restrictive check-in conditions, and there will be respondents for whom a more relaxed attitude to the check-in time will be very important. It seems obvious, though, that a cheap flight with the relaxed check-in conditions would be preferred by all respondents. An option which is preferred by all respondents is called a *dominating* or *dominant option*, and it is often important to be able to design choice experiments where it is less likely that there are choice sets in which any option dominates (or where there is an option that is dominated by all others in the choice set). There is a discussion about dominating options with some discussion of earlier work in Huber and Zwerina (1996).

When constructing options which are described by two or more attributes, it can be necessary to avoid unrealistic combinations of attribute levels. For example, when describing a health state and asking respondents whether they think they would want a hip replacement if they were in this health state, it would be unrealistic to describe a state in which the person had constant pain but could easily walk 5 kilometers.

Throughout this book, we will only consider situations where the options in the experiment can be described by several different *attributes*. Each attribute has two or more *levels*. For the flight example we have been describing, the options have two attributes, the cost and the check-in conditions. In general, attributes need to have levels that are plausible and that are varied over a relevant range. For example, health insurance plans can be described by maximum cost to the subscriber per hospital stay, whether or not visits to the dentist are covered, whether or not visits to the physiotherapist are covered, and so on. Although attributes like cost are continuous, in the choice experiment setting we choose a few different costs and use these as discrete levels for the attribute. Thus we do not consider continuous attributes in this book.

Finally, we stress that we will be talking about *generic* stated preference choice experiments throughout this book. We do not consider the construction of optimal designs when the options are labeled, perhaps by brand or perhaps by type of transport, say, and hence the attributes, and the levels, depend on the label. We only consider designs that are analyzed using the MNL model.

1.2 BINARY RESPONSE EXPERIMENTS

As we said above, in binary response experiments the respondents are shown a description of a good or service, and they are asked whether they would be interested in buying or using that good or service. For each option they are shown, they answer "yes" or "no".

One published example of a binary response experiment appears in Gerard et al. (2003). This study was carried out to develop strategies that were likely to increase the participation rates in breast screening programs. The goal of a breast screening program is to achieve a target participation rate across the relevant population since then there should be a reduction in breast cancer mortality across that population. To get this participation rate requires that women participate at the recommended screening rate. The aim of the study described in Gerard et al. (2003) was to "identify attributes of service delivery that eligible screenees value most and over which decision makers have control". The attributes and levels used in the study are given in Table 1.1.

Given these attributes and levels, what does a respondent actually see? The respondent sees a number of options, like the one in Table 1.2, and just has to answer the question. In this particular survey each respondent saw 16 options (invitations) and so answered the question about each of these 16 different possible invitations in turn.

Table 1.1 Attributes and Levels for the Survey to Enhance Breast Screening Participation

Attributes	Attribute Levels
Method of inviting women for screening	Personal reminder letter Personal reminder letter and recommendation by your GP Media campaign Recommendation from family/friends
Information included with invitation	No information sheet Sheet about the procedure, benefits and risks of breast screening
Time to wait for an appointment	1 week 4 weeks
Choice of appointment times	Usual office hours Usual office hours, one evening per week Saturday morning
Time spent traveling	Not more than 20 minutes Between 20 and 40 minutes Between 40 and 60 minutes Between 1 and 2 hours
How staff at the service relate to you	Welcoming manner Reserved manner
Attention paid to privacy	Private changing area Open changing area
Time spent waiting for mammogram	20 minutes 30 minutes 40 minutes 50 minutes
Time to notification of results	8 working days 10 working days 12 working days 14 working days
Level of accuracy of the screening test	70% 80% 90% 100%

Table 1.2 One Option from a Survey about Breast Screening Participation

	Screening Service
How are you informed	Personal reminder letter
Information provided with invitation	No information sheet
Wait for an appointment	4 weeks
Appointment choices	Usual office hours and one evening per week
Time spent traveling (one way)	Not more than 20 minutes
How staff relate to you	Reserved manner
Changing area	Private changing area
Time spent having screen	40 minutes
Time waiting for results	10 working days
Accuracy of the results	90%

Imagine that your next invitation to be screened is approaching.
Would you choose to attend the screening service described above?
(tick one only)

Yes ○ No ○

The statistical question here is: "which options (that is, combinations of attribute levels) should we be showing to respondents so that we can decide which of these attributes, if any, is important, and whether there are any pairs of attribute levels which jointly influence the decision to participate in the breast screening service?" The design of such informative binary response designs is described in Section 7.1.2.

1.3 FORCED CHOICE EXPERIMENTS

In a forced choice experiment, each respondent is shown a number of choice sets in turn and asked to choose the best option from each choice set. There is no opportunity to avoid making a choice in each choice set.

Severin (2000) investigated which attributes made take-out pizza outlets more attractive. In her first experiment, she used the six attributes in Table 1.3 with the levels indicated. A sample choice set for an experiment looking at these six attributes describing take-out pizza outlets is given in Table 1.4. There are three things to observe here.

The first is that all the attributes have two levels; an attribute with only two levels is called a *binary* attribute, and it is often easier to design small, but informative, experiments when all the attributes are binary. We focus on designs for binary attributes in Chapters 4 and 5.

The second is that the question has been phrased so that the respondents are asked to imagine that the two choices presented to them are the last two options that they are considering in their search for a take-out pizza outlet. This assumption means that the respondents are naturally in a setting where it does not make sense not to choose an option, and so they are forced to make a selection even though the options presented are not exhaustive.

Finally, observe that each respondent has been shown only two options and has been asked to state which one is preferred. While it is very common to present only two options in each choice set, it is not necessarily the best choice set size to use; see Section 7.2 for a discussion about the statistically optimal choice set size. Larger choice sets do place more cognitive demands on respondents, and this is discussed in Iyengar and Lepper (2000), Schwartz et al. (2002), and Iyengar et al. (2004).

Table 1.3 Six Attributes to be Used in an Experiment to Compare Pizza Outlets

Attributes	**Attribute Levels**
Pizza type	Traditional Gourmet
Type of Crust	Thick Thin
Ingredients	All fresh Some tinned
Size	Small only Three sizes
Price	$17 $13
Delivery time	30 minutes 45 minutes

Table 1.4 One Choice Set in an Experiment to Compare Pizza Outlets

	Outlet A	**Outlet B**
Pizza type	Traditional	Gourmet
Type of crust	Thick	Thin
Ingredients	All fresh	Some tinned
Size	Small only	Small only
Price	$17	$13
Delivery time	30 minutes	30 minutes

Suppose that you have already narrowed down your choice of take-out pizza outlet to the two alternatives above.
Which of these two would you choose?
(tick one only)

Outlet A ○ Outlet B ○

Most forced choice experiments do not use only binary attributes. Chapter 6 deals with the construction of forced choice experiments for attributes with any number of levels. For example, Maddala et al. (2002) used 6 attributes with 3, 4, 5, 3, 5, and 2 levels, respectively, in a choice experiment examining preferences for HIV testing methods. The attributes, together with the attribute levels, are given in Table 1.5, and one choice set from the study is given in Table 1.6. Each respondent was presented with 11 choice sets and for each of these was asked to choose one of two options. As the respondents were all surveyed at HIV testing locations a forced choice experiment was appropriate.

Table 1.5 Attributes and Levels for the Study Examining Preferences for HIV Testing Methods

Attribute	Attribute Levels
Location	Doctor's office
	Public clinic
	Home
Price	\$0
	\$10
	\$50
	\$100
Sample collection	Draw blood
	Swab mouth
	Urine sample
	Prick finger
Timeliness/accuracy	Results in 1-2 weeks; almost always accurate
	Immediate results; almost always accurate
	Immediate results; less accurate
Privacy/anonymity	Only you know; not linked
	Phones; not linked
	In person; not linked
	Phone; linked
	In person; linked
Counseling	Talk to a counselor
	Read brochure then talk to counselor

Both of the experiments discussed above used only six attributes. Hartmann and Sattler (2002) found that about 75% of commercially conducted stated choice experiments used 6 or fewer attributes and they speculated that this might be because commonly available software often used to generate choice experiments would not allow more than 6 attributes. However, there are choice experiments where many attributes are used; see Section 1.7.2.

1.4 THE "NONE" OPTION

As we said above, sometimes it does not make sense to compel people to choose one of the options in a choice set and so some choice experiments include an option variously called "no choice" or "delay choice" or "none of these" in each choice set.

Often an existing forced choice experiment can be easily modified to include an option not to choose. For instance, in the pizza outlet experiment described in the previous section,

Table 1.6 One Choice Set from the Study Examining Preferences for HIV Testing Methods

Attribute	Option A	Option B
Location	Doctor's office	Public clinic
Price	$100	$10
Sample collection	Swab mouth	Urine sample
Timeliness/accuracy	Results in 1–2 weeks; almost always accurate	Immediate results; less accurate
Privacy/anonymity	In person; not linked	Only you know; not linked
Counselling	Talk to a counselor	Read brochure then talk to counselor
Which of these two testing methods would you choose? *(tick one only)*	Option A ○	Option B ○

we could change the question to ask: "Suppose that you have decided to get a take-out meal. Which of these pizza outlets would you select, or would you go somewhere else?"

Dhar (1997) looks at the situations in which consumers find it hard to choose and so will opt to defer choice if they can. Haaijer et al. (2001) summarize his results by saying "respondents may choose the no-choice when none of the alternatives appears to be attractive, or when the decision-maker expects to find better alternatives by continuing to search. ...adding an attractive alternative to an already attractive choice set increases the preference of the no-choice option and adding an unattractive alternative to the choice set decreases the preference of the no-choice." In Section 7.1.1, we discuss how to design good designs when there is a "none of these" option in each choice set.

1.5 A COMMON BASE OPTION

Some choice experiments have a common (or base) option, sometimes called constant comparator, in each choice set, together with one or more other options. This is often done so that the current situation can be compared to other possibilities. A typical example arises in medicine when the standard treatment option can be compared to a number of possible alternative treatment options.

Ryan and Hughes (1997) questioned women about various possible alternatives to the surgical removal of the product of conception after a miscarriage (note that some such treatment is essential after a miscarriage). They identified the attributes and levels given in Table 1.7 as being appropriate.

In the choice experiment, the common base, which is the current treatment option of surgical management, was described as "having a low level of pain, requiring 1 day and 0 nights in hospital, taking 3–4 days to return to normal activities, costing $350, and there would be complications post-surgery". This raises the question of whether it is sensible to have an option in which it is known with certainty that there will be complications; it might have made more sense to talk of the probability of complications given a particular treatment. This interesting non-mathematical issue will not be addressed in this book.

Table 1.7 Five Attributes to be Used in an Experiment to Investigate Miscarriage Management Preferences

Attributes	Attribute Levels
Level of pain	Low Moderate Severe
Time in hospital	1 day and 0 nights 2 days and 1 night 3 days and 2 nights 4 days and 3 nights
Time taken to return to normal activities	1–2 days 3–4 days 5–6 days More than 7 days
Cost to you of treatment	\$100 \$200 \$350 \$500
Complications following treatment	Yes No

Having a common option in all choice sets is not as good as allowing all the options to be different from one choice set to another but when a common base is appropriate we show how to design as well as possible for this setting in Section 7.1.3.

1.6 AVOIDING PARTICULAR LEVEL COMBINATIONS

Sometimes a set of level combinations of at least two of the attributes is unrealistic and sometimes a set of level combinations is clearly the best for all respondents and so will always be chosen.

We discuss examples of each of these situations here. We give some ideas for how to design choice experiments when these circumstances pertain in Chapters 4, 5, and 8.

1.6.1 Unrealistic Treatment Combinations

To illustrate this idea consider the descriptions of 5 attributes describing health states devised by EuroQol; see EuroQoL (2006). These attributes and levels are given in Table 1.8 and are used to describe health states for various purposes. In the context of a stated preference choice experiment we might describe two health states and ask respondents which one they prefer.

But even a quick look at the levels shows that some combinations of attribute levels do not make sense. A health state in which a person is "Confined to bed" in the mobility attribute is not going to be able to be linked with "No problems with self-care" in the self-care attribute. Thus it is necessary to determine the level combinations that are *unrealistic* before using choice sets that are generated for these attributes using the techniques developed in later chapters.

Table 1.8 Five Attributes Used to Compare Aspects of Quality of Life

Attributes	Attribute Levels
Mobility	No problems in walking about Some problems in walking about Confined to bed
Self-Care	No problems with self-care Some problems with self-care Unable to wash or dress one's self
Usual Activities (e.g., work, study housework)	No problems with performing one's usual activities Some problems with performing one's usual activities Unable to perform one's usual activities
Pain/Discomfort	No pain or discomfort Moderate pain or discomfort Extreme pain or discomfort
Anxiety/Depression	Not anxious or depressed Moderately anxious or depressed Extremely anxious or depressed

1.6.2 Dominating Options

Many attributes can have ordered levels in the sense that all respondents agree on the same ordering of the levels for the attribute. In the levels presented in Table 1.8 it is clear that in every attribute the levels go from the best to the worst. Thus a choice set that asks people to choose between the health state (No problems in walking about, No problems with self-care, No problems with performing one's usual activities, No pain or discomfort, Not anxious or depressed) and any other health state is not going to give any information - all respondents will choose the first health state. We say that the first health state *dominates* the other possible health states.

It is possible to have a choice set in which one option dominates the other options in the choice set even though the option is not one that dominates all others in the complete set of level combinations. So if we use 0, 1, and 2 to represent the levels for each of the attributes in Table 1.8 then the best health state overall is 00000. But a choice set that contains (00111, 01222, 12211), for example, has a *dominating option* since 00111 has at least as good a level on every attribute as the other two options in the choice set. It is not clear, however, whether 01222 or 12211 would be preferred since it is not necessarily true that the same utility values apply to the same levels of different attributes.

We discuss ways of avoiding choice sets with dominating options in Chapters 4, 5, and 8.

1.7 OTHER ISSUES

In this chapter we have discussed a number of different types of choice experiments that have been published in the literature. We have indicated that, in later chapters, we will describe how to construct optimal designs for binary responses, for forced choice stated preference experiments, for stated choice experiments where a "none" option is included in every choice set, and for stated choice experiments where a common base option is included in every choice set. In all cases we consider only generic options analyzed using the MNL model. In this section we want to mention a couple of designs that we will not be constructing and to discuss briefly some non-mathematical issues that need to be considered when designing choice experiments.

1.7.1 Other Designs

In some experiments, options are described not only by attribute levels but also by a brand name or label; for example, the name of the airline that is providing the flight. Such experiments are said to have *branded* or *labeled* options (or alternatives), and these have *alternative specific* attributes.

In many situations, people choose from the options that are available at the time they make their choice rather than deferring choosing until some other option is available. Experimentally, we can model this by having a two-stage design process. We have a design which says what options are available and another design that determines the specific options to present given what is to be available. Such designs are termed *availability designs*.

We will not be discussing the construction of designs for branded alternatives or for availability experiments in this book. The interested reader is referred to Louviere et al. (2000).

1.7.2 Non-mathematical Issues for Stated Preference Choice Experiments

The first issue, and one that we have alluded to in Sections 1.3 and 1.4, is the question of task complexity and thus of respondent efficiency. If a task is too complicated (perhaps because there are too many attributes being used to describe the options in a choice set or because there are too many options in each choice set), then the results from a choice experiment are likely to be more variable than expected. Aspects of this problem have been investigated by various authors, and several relevant references are discussed in Iyengar and Lepper (2000). Louviere et al. (2007) found that completion rates are high even for what would be considered large choice experiments in terms of the number of attributes and the number of options.

A related issue concerns the number of choice sets that respondents can reasonably be expected to complete. If there are too many choice sets, then respondents may well become tired and give more variable results over the course of the experiment. This situation has been investigated by Brazell and Louviere (1995) and Holling et al. (1998).

The discrete choice experiment task needs to be thought about in context. Are the choices one time only or are they repeated? How important is the outcome of the decision? (In a medical setting you could be asking people to think about life-or-death decisions.) How familiar is the context? The choice of a health insurance provider may be familiar while other choices, such as for liver transplant services, may require the provision of detailed information in the experiment so that respondents can make an informed choice.

The attributes that are used to describe the options in a choice experiment need to be appropriate and plausible, and the combinations that are presented in the experiment need to be realistic; otherwise, respondents may take the task less seriously or be confused by it.

The mechanics of setting up a stated preference experiment are outlined in Chapter 9 of Louviere et al. (2000). Dillman and Bowker (2001) discuss many aspects of both mail and internet surveys.

There are various sources of bias that have been identified as sometimes occurring in choice experiments. One is what has been termed *affirmation bias* when respondents choose responses to be consistent with what they feel the study objectives are. This is why experimenters sometimes include attributes that are not of immediate interest to mask the main attributes that are under investigation.

A second possible source of bias is called *rationalization bias*, where responses are given that justify the actual behavior. This serves to reduce cognitive dissonance for the respondents. A third possible source of bias results from the fact that there are no transaction costs associated with choices in a stated preference study. Some respondents try to respond in a way that they believe will influence the chance of, or the magnitude of, changes in the real world. This is termed *strategic* or *policy response bias*. Finally, people may not be prepared to indicate preferences which they feel are socially unacceptable or politically incorrect. These terms are defined and used in Walker et al. (2002).

Other sources of bias can be related to the actual topic under investigation. Carlsson (2003) was investigating business passengers preferences, and some of his options were more environmentally friendly, but more expensive, than other options, and he spoke of respondents perhaps aiming to have a "warm glow" or "purchasing moral satisfaction" when making choices.

Severin (2000) has shown, in a paired comparison experiment, that if there are a large number of attributes describing each option, then respondents find the task more difficult and she has suggested that about 8 or 9 attributes seem to be effectively processed; see her thesis for more details. For a discussion of the role of cognitive complexity in the design of choice experiments, readers are referred to DeShazo and Fermo (2002) and Arentze et al. (2003) and references cited therein.

These and other psychological and cognitive issues are beyond the scope of this book, and we refer the reader interested in such issues elsewhere. The papers by Iyengar and Lepper (2000) and Schwartz et al. (2002) and the references cited therein provide a good starting point to find out more about these issues.

1.7.3 Published Studies

As we have said before, stated preference choice experiments are used in many areas. Here we give references to a few published studies, together with a very brief indication of the question being investigated and the type of design being used.

Chakraborty et al. (1994) describe a choice experiment to investigate how consumers choose health insurance. As well as the actual company offering the insurance, 23 other attributes were used to describe health insurance plans. Respondents were presented with choice sets with 4 options in each and asked to indicate their preferred plan.

Hanley et al. (2001) describe a stated preference study to investigate demand for climbing in Scotland. Each choice set contained two possible climbs and a "neither of these" option. They also give a table with details of about 10 other studies that used DCEs to investigate questions in environmental evaluation.

Kemperman et al. (2000) used an availability design to decide which of four types of theme parks would be available to respondents in each of spring and summer. Each choice set contained four theme parks and a "none of these" option.

McKenzie et al. (2001) describe a study in which five attributes that are commonly used to describe the symptoms of asthma were included and three levels for each of these chosen. Patients with moderate to severe asthma were shown "pairs of scenarios characterized by different combinations of asthma symptoms", and were asked which of the two scenarios they thought would be better to have or whether they felt there was no difference.

Ryan and Gerard (2003) and Bryan and Dolan (2004) give examples of the use of DCEs in the health economics context.

Scarpa et al. (2004) discuss experiments to "characterise the preference for fifteen different attributes related to water provision".

An example of a choice experiment that involved labeled options is given in Tayyaran et al. (2003). The authors were interested in investigating whether telecommuting and intelligent transportation systems had an impact on residential location. Each choice set contained three residential options as well as a "none of these" option. The residential options were "branded" as *central cities*, *first-tier satellite nodes*, and *second-tier satellite nodes*. Each residential option was described by 7 attributes, 4 of which were the same for all of the locations.

Walker et al. (2002) used two stated choice experiments to model tenants' choices in the public rental market.

1.8 CONCLUDING REMARKS

In this chapter we have seen that there are a number of areas where stated choice experiments have been applied and that there are a number of issues, both mathematical and non-mathematical, which need to be considered in the construction of the best possible experiments for a given situation. In the remainder of this book we look at the mathematical issues that need to be considered to design a good generic stated choice experiment to be analyzed using the MNL model. In the next chapter, we collect a number of results about factorial designs which are intimately connected with the representation of options by attributes. We follow this with a discussion about parameter estimators and their distribution. Over the following three chapters we show how to get the best designs for any number of attributes, each attribute being allowed to have any number of levels and with choice sets of any size. In the penultimate chapter we consider how to construct good designs for other situations, such as the inclusion of a "none" option in every choice set. The final chapter illustrates the application of the results in the book to the construction of a number of experiments that we have designed in the last five years.

CHAPTER 2

FACTORIAL DESIGNS

Comparative experiments are carried out to compare the effects of two or more treatments on some response variable (or variables). They were developed in an agricultural setting, and often the effects of several factors on the yield of a crop, say, were investigated simultaneously. Such experiments are called *factorial experiments* and were introduced by Sir Ronald Fisher in the 1920s.

Suppose that an experiment is designed to investigate the effect of k factors on the yield of a crop. Then the treatments of interest are in fact combinations of levels of each of the k factors under investigation. We refer to these as *treatment combinations*. If the qth factor has ℓ_q levels, then there are $L = \prod_{q=1}^{k} \ell_q$ possible treatment combinations, but often only a subset of these are actually used in the experiment.

In this chapter we will show the link between the treatment combinations in a factorial design and the options used in a stated preference choice experiment. We will review the constructions that exist for obtaining subsets of the treatment combinations from a factorial design so that specific effects of interest can still be estimated. While these constructions were developed to provide good designs when the results of the experiment are analyzed using a linear model with normal errors, it turns out that these constructions are still useful when the multinomial logit model is used. Thus the results developed here form the basis for much of the rest of the book.

Throughout this chapter, we will assume that $y_{i_1,i_2,\ldots,i_k}$ is the response with the first factor at level i, the second factor at level j, and so on, until the final factor is at level q in a traditional comparative experiment. (In choice experiments the word *attribute* is used instead of the word *factor*. In this chapter we use the word "factor," since we are summarizing constructions from the statistical literature. In later chapters we will use the word "attribute" instead, to make the results more consistent with the use in various application areas.)

2.1 COMPLETE FACTORIAL DESIGNS

A *complete factorial design* is one in which each of the possible level combinations appears at least once. If all of the factors have the same number of levels, then the factorial design is said to be *symmetric*; otherwise it is *asymmetric*.

Suppose that there are k factors and that each of the factors has ℓ levels. Then the factorial design is said to be a symmetric design with ℓ levels, and we talk about an ℓ^k factorial design. The smallest symmetric factorial design is a design in which two factors each have two levels, a 2^2 design.

We will represent the levels of an ℓ level factor by $0, 1, \ldots, \ell - 1$.

■ EXAMPLE 2.1.1.
The combinations of factor levels in a 2^3 factorial design are 000, 001, 010, 011, 100, 101, 110, and 111. □

In an asymmetric factorial design, the factors may have different numbers of levels. If factor q has ℓ_q levels, then we speak of an $\ell_1 \times \ell_2 \times \ldots \times \ell_k$ factorial design.

■ EXAMPLE 2.1.2.
The combinations of factor levels in a 3×4 factorial design are 00, 01, 02, 03, 10, 11, 12, 13, 20, 21, 22, and 23. □

2.1.1 2^k Designs

Designs in which all factors have only two levels are the most commonly used factorial designs. Typically, the two levels are chosen to be the largest and smallest levels that are deemed to be plausible for that attribute (factor). For example, in an experiment to investigate treatment preferences for asthma, one of the attributes might be *sleep disturbance* with levels *no disturbance* and *woken more than 5 times per night*. In an experiment describing employment conditions, one of the attributes might be *amount of annual leave* with levels *2 weeks per year* and *6 weeks paid leave and up to 4 weeks unpaid leave per year*. In an experiment to compare plane flights, one of the attributes might be *in-flight service* with levels *beverages* and *hot meals*.

2.1.1.1 Main Effects As a result of conducting the experiment, we would like to be able to estimate the effect of each of the factors, individually, on the response. This is termed the *main effect* of that factor.

If we fix the levels of all factors except one, and then look at the difference in response between the high level and the low level of this one factor, this difference in response is called the *simple effect* of the factor at the particular levels of the other factors. The *main effect* of a factor is the average of all the simple effects for that factor.

■ EXAMPLE 2.1.3.
Suppose that there are $k = 3$ factors, A, B, and C, each with two levels. Suppose that a complete experiment has been carried out. Then there are four simple effects to calculate

for each of the three factors. Consider factor A. Then the simple effect of A at level 0 for B and level 0 for C is given by

$$y_{100} - y_{000}.$$

The simple effect of A at level 0 for B and level 1 for C is given by

$$y_{101} - y_{001}.$$

The simple effect of A at level 1 for B and level 0 for C is given by

$$y_{110} - y_{010}.$$

Finally, the simple effect of A at level 1 for B and level 1 for C is given by

$$y_{111} - y_{011}.$$

The main effect of A is the average of these four simple effects and is given by

$$\frac{1}{4}(y_{100} - y_{000} + y_{101} - y_{001} + y_{110} - y_{010} + y_{111} - y_{011}). \qquad \square$$

It is often easier to describe the main effect of a factor as the difference between the average response to that factor when it is at level 1 and the average response when it is at level 0. To write this mathematically, we let the treatment combinations be represented by binary k-tuples $(x_1 x_2 \ldots x_k)$, where $x_q = 0$ or 1, $q = 1, \ldots, k$. Then we could write the main effect of the qth factor as

$$\frac{1}{2^{k-1}}\left(\sum_{x_q=1} y_{x_1 x_2 \ldots x_k} - \sum_{x_q=0} y_{x_1 x_2 \ldots x_k}\right).$$

2.1.1.2 Interaction Effects Two factors are said to *interact* if the effect of one of the factors on the response depends on the level of the other factor.

Formally the *two-factor interaction effect* of factors A and B is defined as the average difference between the simple effect of A at level 1 of B and the simple effect of A at level 0 of B.

■ EXAMPLE 2.1.4.

Suppose that $k = 2$. Then the simple effect of A at level 1 of B is $(y_{11} - y_{01})$ and the simple effect of A at level 0 of B is $(y_{10} - y_{00})$. The interaction effect of A and B, denoted by AB, is the average difference between these two simple effects, that is,

$$AB = \frac{1}{2}\left((y_{11} - y_{01}) - (y_{10} - y_{00})\right) = \frac{1}{2}\left(y_{11} + y_{00} - (y_{01} + y_{10})\right). \qquad \square$$

In general, a two-factor interaction is

$$\frac{1}{2}\left\{\frac{1}{2^{k-2}}\left(\sum_{x_{q_1}=1, x_{q_2}=1} y_{x_1 x_2 \ldots x_k} - \sum_{x_{q_1}=0, x_{q_2}=1} y_{x_1 x_2 \ldots x_k}\right)\right.$$
$$\left. - \frac{1}{2^{k-2}}\left(\sum_{x_{q_1}=1, x_{q_2}=0} y_{x_1 x_2 \ldots x_k} - \sum_{x_{q_1}=0, x_{q_2}=0} y_{x_1 x_2 \ldots x_k}\right)\right\}$$

$$= \frac{1}{2^{k-1}} \left(\sum_{x_{q_1}+x_{q_2}=0} y_{x_1x_2\ldots x_k} - \sum_{x_{q_1}+x_{q_2}=1} y_{x_1x_2\ldots x_k} \right).$$

Higher-order interactions are defined recursively. Thus the three-factor interaction between factors A, B, and C is defined as the average difference between the AB interaction at level 1 of C and the AB interaction at level 0 of C.

■ EXAMPLE 2.1.5.
Suppose that $k = 4$. Then the AB interaction at level 1 of C is

$$\frac{1}{4}\left(y_{1111} + y_{0011} - (y_{0111} + y_{1011}) + y_{1110} + y_{0010} - (y_{0110} + y_{1010})\right).$$

The AB interaction at level 0 of C is

$$\frac{1}{4}(y_{1101} + y_{0001} - (y_{0101} + y_{1001}) + y_{1100} + y_{0000} - (y_{0100} + y_{1000})).$$

The average difference between these is

$$\begin{aligned} ABC = &\frac{1}{2}\left\{\left[\frac{1}{4}\left(y_{1111} + y_{0011} - (y_{0111} + y_{1011}) + y_{1110} + y_{0010} - (y_{0110} + y_{1010})\right)\right]\right. \\ &\left. - \left[\frac{1}{4}\left(y_{1101} + y_{0001} - (y_{0101} + y_{1001}) + y_{1100} + y_{0000} - (y_{0100} + y_{1000})\right)\right]\right\} \\ = &\frac{1}{8}\left((y_{1111} + y_{0011} + y_{1110} + y_{0010} + y_{0101} + y_{1001} + y_{0100} + y_{1000})\right. \\ &\left. - (y_{0111} + y_{1011} + y_{0110} + y_{1010} + y_{1101} + y_{0001} + y_{1100} + y_{0000})\right). \end{aligned}$$ □

This is a specific case of the following general result.

■ THEOREM 2.1.1.
The interaction effect of factors $q_1, q_2, \ldots, q_t$ in a 2^k factorial experiment is estimated by $1/2^{k-1}$ times the difference between the sum of the treatment combinations with

$$x_{q_1} + x_{q_2} + \ldots + x_{q_t} = t \pmod 2$$

and the sum of the treatment combinations with

$$x_{q_1} + x_{q_2} + \ldots + x_{q_t} = (t+1) \pmod 2.$$

Proof. We will prove the result by induction on t.

The main effect of factor q_1 is the difference between the sum of the treatment combinations receiving factor q_1 at the high level (and so with $x_{q_1} = 1$) and the sum of the treatment combinations receiving factor q_1 at the low level (and so with $x_{q_1} = 0$). Thus the result holds for $t = 1$.

The interaction between the factors q_1 and q_2 is the difference between the main effect of q_1 at the high level of q_2 and the main effect of q_1 at the low level of q_2. Thus we get

$$\frac{1}{2}\left\{\frac{1}{2^{k-2}}\left(\sum_{x_{q_1}=1, x_{q_2}=1} y_{x_1x_2\ldots x_k} - \sum_{x_{q_1}=0, x_{q_2}=1} y_{x_1x_2\ldots x_k}\right)\right.$$

$$- \frac{1}{2^{k-2}} \left(\sum_{x_{q_1}=1, x_{q_2}=0} y_{x_1 x_2 \ldots x_k} - \sum_{x_{q_1}=0, x_{q_2}=0} y_{x_1 x_2 \ldots x_k} \right) \Bigg\}$$

$$= \frac{1}{2^{k-1}} \left(\sum_{x_{q_1}+x_{q_2}=0} y_{x_1 x_2 \ldots x_k} - \sum_{x_{q_1}+x_{q_2}=1} y_{x_1 x_2 \ldots x_k} \right).$$

Hence the result follows by induction. □

■ EXAMPLE 2.1.6.

Let $k = 4$. Then the ABC interaction is estimated by 1/8 of the difference between the sum of the treatment combinations with $x_1 + x_2 + x_3 = 1 \pmod 2$ and the treatment combinations with $x_1 + x_2 + x_3 = 0 \pmod 2$; this is consistent with the result of Example 2.1.5. □

2.1.2 3^k Designs

In a 3^k design there are k factors and all of the factors have three levels each. The factor might be *quantitative*, such as the *temperature* with levels 40°, 50°, and 60°, or the factor might be *qualitative*, such as *severity of symptoms* with levels *mild*, *moderate*, and *severe*.

In either case we are again interested in the effect of each factor, independently, on the response and on the joint effect of two or more factors on the response.

2.1.2.1 Main Effects

The *main effect* of a factor is the effect of that factor, independent of any other factor, on the response. So it seems natural, for each factor, to group the 3^k responses to the treatment combinations into three sets. For factor q, these sets are $\{y_{x_1 x_2 \ldots x_k} | x_q = 0\}$, $\{y_{x_1 x_2 \ldots x_k} | x_q = 1\}$ and $\{y_{x_1 x_2 \ldots x_k} | x_q = 2\}$.

When a factor had two levels, we compared the average response in each of the two sets of responses. When there are three sets, there are a number of comparisons that might be of interest. We might be interested in comparing the average response of the three sets pairwise, giving:

$$\frac{1}{3^{k-1}} \left(\sum_{x_q=1} y_{x_1 x_2 \ldots x_k} - \sum_{x_q=0} y_{x_1 x_2 \ldots x_k} \right),$$

which compares responses to levels 1 and 0;

$$\frac{1}{3^{k-1}} \left(\sum_{x_q=2} y_{x_1 x_2 \ldots x_k} - \sum_{x_q=0} y_{x_1 x_2 \ldots x_k} \right),$$

which compares responses to levels 2 and 0; and

$$\frac{1}{3^{k-1}} \left(\sum_{x_q=2} y_{x_1 x_2 \ldots x_k} - \sum_{x_q=1} y_{x_1 x_2 \ldots x_k} \right),$$

which compares responses to levels 2 and 1. But these three comparisons are not independent, since we can calculate the third comparison if we know the other two. How can we get a set of independent comparisons?

We want each comparison to be independent of the mean of all the responses and of every other comparison so that each comparison can be tested independently of every other comparison and of the mean. (Recall that if $\mathbf{y} \sim N(\boldsymbol{\mu}, \Sigma)$ then $\mathbf{a}'\mathbf{y} \sim N(\mathbf{a}'\boldsymbol{\mu}, \mathbf{a}'\Sigma\mathbf{a})$. By definition, two normally distributed random variables, x and z are independent if $\text{Cov}(x, z) = 0$. If A is a matrix and if $\mathbf{y} \sim N(\boldsymbol{\mu}, \Sigma)$, then $A\mathbf{y} \sim N(A\boldsymbol{\mu}, A\Sigma A')$. These results can be used to establish when comparisons are independent; see below and Exercise 2.1.4.10.)

We now give some definitions to formalize the ideas that we need.

A *comparison* is any linear combination of the responses. We will write a comparison as

$$\sum_{x_1 x_2 \ldots x_k} \lambda_{x_1 x_2 \ldots x_k} y_{x_1 x_2 \ldots x_k}.$$

Thus the mean is a comparison with $\lambda_{x_1 x_2 \ldots x_k} = \frac{1}{3^k}$. Two comparisons with coefficients $\lambda_{x_1 x_2 \ldots x_k}$ and $\nu_{x_1 x_2 \ldots x_k}$ are *orthogonal* or *independent* if

$$\sum_{x_1 x_2 \ldots x_k} \lambda_{x_1 x_2 \ldots x_k} \nu_{x_1 x_2 \ldots x_k} = 0$$

(see Exercise 2.1.4.10). Thus a comparison is independent of the mean if

$$\sum_{x_1 x_2 \ldots x_k} \lambda_{x_1 x_2 \ldots x_k} = 0.$$

Any such comparison is called a *contrast*.

We want to find two contrasts that are also orthogonal (equivalently pairwise independent). Since all the treatment combinations in each set will have the same coefficient, there are only three distinct coefficients that we need to find. We let λ_0 be the coefficient for all the treatment combinations in the set $\{y_{x_1 x_2 \ldots x_k} | x_q = 0\}$, λ_1 be the coefficient for all the treatment combinations in the set $\{y_{x_1 x_2 \ldots x_k} | x_q = 1\}$, and we let λ_2 be the coefficient for all the treatment combinations in the set $\{y_{x_1 x_2 \ldots x_k} | x_q = 2\}$. We define ν_0, ν_1 and ν_2 similarly. So we need to find $\lambda_0, \lambda_1, \lambda_2, \nu_0, \nu_1$ and ν_2 such that $\lambda_0 + \lambda_1 + \lambda_2 = 0$, $\nu_0 + \nu_1 + \nu_2 = 0$ and $\lambda_0\nu_0 + \lambda_1\nu_1 + \lambda_2\nu_2 = 0$.

While there are infinitely many solutions to these equations, there is one solution that is commonly used. Let $\lambda_0 = -1$, $\lambda_1 = 0$ and $\lambda_2 = 1$, and let $\nu_0 = \nu_2 = 1$ and $\nu_1 = -2$. Then these contrasts give the linear and quadratic components of the response to a quantitative factor with equally spaced levels; we prove this below.

Any two independent sets of contrasts will give the same sum of squares for a main effect for a normally distributed random vector (see Exercise 2.1.4.10).

■ **EXAMPLE 2.1.7.**

Suppose that $k = 3$. Then the three sets of responses for the first factor are

$$\{y_{000}, y_{001}, y_{002}, y_{010}, y_{011}, y_{012}, y_{020}, y_{021}, y_{022}\},$$

$$\{y_{100}, y_{101}, y_{102}, y_{110}, y_{111}, y_{112}, y_{120}, y_{121}, y_{122}\},$$

$$\{y_{200}, y_{201}, y_{202}, y_{210}, y_{211}, y_{212}, y_{220}, y_{221}, y_{222}\}.$$

The contrast corresponding to $\lambda_0 = -1$, $\lambda_1 = 0$ and $\lambda_2 = 1$ is

$$\frac{1}{3^{k-1}}((y_{200} + y_{201} + y_{202} + y_{210} + y_{211} + y_{212} + y_{220} + y_{221} + y_{222})$$
$$- (y_{000} + y_{001} + y_{002} + y_{010} + y_{011} + y_{012} + y_{020} + y_{021} + y_{022})). \quad \square$$

2.1.2.2 Orthogonal Polynomial Contrasts Suppose that we have a set of pairs (x_i, Y_i), $i = 1, 2, \ldots, \ell$ and that we want to predict $\mathbf{Y}$ as a polynomial function of x of order $n = 0, 1, \ldots, \ell - 1$. We usually do this by constructing a matrix X with the first column with entries of 1 (for the constant term), second column with the values of the x_i, the third column with the values of the x_i^2 and so on until the final column of X contains the values of x_i^n. Thus we can write $E(\mathbf{Y}) = X\boldsymbol{\beta}$ and $\hat{\boldsymbol{\beta}} = (X'X)^{-1}X'\mathbf{Y}$. In general, nothing can be said about the form of the matrix $X'X$. However, we could just as readily use other polynomials to determine the entries in the X matrix so that the resulting $X'X$ is diagonal (and hence easy to invert).

To do this we proceed as follows.

We define

$$P_n(x) = x^n + \alpha_{n,n-1}x^{n-1} + \alpha_{n,n-2}x^{n-2} + \ldots + \alpha_{n,1}x + \alpha_{n,0}$$

for $n = 1, 2, \ldots$ and we let $P_0(x) = 1$. We will then do a regression on $P_1(x)$, $P_2(x)$ and so on, and we want the resulting X matrix to be such that $X'X$ is diagonal.

The off-diagonal entries in $X'X$ are of the form

$$\sum_{i=1}^{\ell} P_n(x_i)P_{n-j}(x_i), j = 1, \ldots, n; \tag{2.1}$$

and so we require this sum to equal 0.

We will assume that the values of the x_i are equally spaced; thus, $x_{i+1} - x_i = d$ for all i.

The values of the coefficients in $P_n(x)$ depend on the value of ℓ. We illustrate the calculations using $\ell = 3$.

As $\ell = 3$, we want to evaluate $P_0(x)$, $P_1(x)$ and $P_2(x)$. But we already know that $P_0(x) = 1$ (by assumption, since $P_0(x)$ is some constant function). Using Equation (2.1) with $n = 1$ gives

$$\sum_{i=1}^{3} P_1(x_i)P_0(x_i) = \sum_{i=1}^{3} P_1(x_i) = 0.$$

Now substituting for P_1 we get

$$\begin{aligned}\sum_{i=1}^{3} P_1(x_i) &= \sum_{i=1}^{3}(x_i + \alpha_{1,0}) \\ &= \sum_{i=1}^{3} x_i + 3\alpha_{1,0} \\ &= 0.\end{aligned}$$

So we see that $\alpha_{1,0} = -\overline{x}$ and hence $P_1(x) = x - \overline{x}$.

Using Equation (2.1) with $n = 2$ and $j = 1$ gives

$$\begin{aligned}\sum_{i=1}^{3} P_2(x_i)P_1(x_i) &= \sum_{i=1}^{3}(x_i^2 + \alpha_{2,1}x_i + \alpha_{2,0})(x_i - \overline{x}) \\ &= 0.\end{aligned} \tag{2.2}$$

Using Equation (2.1) with $n = 2$ and $j = 2$ gives

$$\begin{aligned}\sum_{i=1}^{3} P_2(x_i)P_0(x_i) &= \sum_{i=1}^{3}(x_i^2 + \alpha_{2,1}x_i + \alpha_{2,0}) \\ &= 0. \end{aligned} \tag{2.3}$$

We now use these two equations to solve for $\alpha_{2,1}$ and $\alpha_{2,0}$. From Equation (2.2) we get (collecting terms)

$$\alpha_{2,1}\left(\sum_i x_i^2 - 3\overline{x}^2\right) + \alpha_{2,0}\left(3\overline{x} - 3\overline{x}\right) + \sum_i x_i^3 - \overline{x}\sum_i x_i^2 = 0,$$

from which we see that

$$\alpha_{2,1} = \frac{\overline{x}\sum_i x_i^2 - \sum_i x_i^3}{\sum_i x_i^2 - 3\overline{x}^2}.$$

Using Equation (2.3), we get that

$$3\alpha_{2,0} = -\sum_i x_i^2 - 3\alpha_{2,1}\overline{x}.$$

These are general expressions and we now simplify them by considering the case when $x_1 = 0$, $x_2 = 1$ and $x_3 = 2$. Then

$$\overline{x} = (0 + 1 + 2)/3 = 1,$$

$$\sum_i x_i^2 = 0 + 1 + 4 = 5,$$

and

$$\sum_i x_i^3 = 0 + 1 + 8 = 9.$$

So we get

$$\alpha_{2,1} = (1 \times 5 - 9)/(5 - 3 \times 1) = -2$$

and

$$\alpha_{2,0} = (-5 - 3 \times (-2) \times 1)/3 = 1/3.$$

We usually record the values of the orthogonal polynomials for a given value of n rather than the actual polynomials. Hence we would get a table like Table 2.1 for $n = 3$.

Table 2.1 Values of Orthogonal Polynomials for $n = 3$

x	0	1	2	
$P_1(x)$	−1	0	1	Linear component
$3P_2(x)$	1	−2	1	Quadratic component

We can find the values of orthogonal polynomials for any value of n in a similar way. Tables of orthogonal polynomials may be found in Kuehl (1999) and Montgomery (2001).

It is possible to evaluate orthogonal polynomials for unequally spaced levels; see Addelman (1962) and Narula (1978).

For a factor with ℓ levels, the values of $P_1(x)$ are the coefficients of the linear contrast and the values of $P_2(x)$ are the coefficients of the quadratic contrast. Higher-order contrasts are defined similarly.

2.1.2.3 Interaction Effects We start with two-factor interactions. The two-factor interaction effect of factors A and B is the joint effect of the two factors on the response and is again called AB. Now, however, there are four independent contrasts associated with the interaction. This is because there are nine distinct pairs of levels for factors A and B and so 8 independent contrasts possible between these sets. Four of these contrasts have been used to calculate the main effects of the factors, leaving 4 contrasts for the interaction effect.

The easiest way to determine these contrasts is to divide the responses to the treatment combinations into nine sets, based jointly on the levels of A and B. So we get sets

$$AB_{\theta\psi} = \{y_{x_1x_2\ldots x_k} | x_A = \theta, x_B = \psi\}, \text{ for } \theta = 0, 1, 2; \ \psi = 0, 1, 2;$$

for any two factors A and B. There are 8 independent contrasts possible between 9 sets. However, these sets are subsets of the sets that we used to define the main effects of factors A and B. To ensure that the main effects and the interaction effect are independent, we have to use those four contrasts (two from each main effect) to be 4 of the 8 contrasts. Then any other four independent contrasts can be chosen to represent the interaction effect. One common way to get the final four independent contrasts is to take the component-wise product of the contrasts for main effects. This gives contrasts that are interpretable as the *linear* × *linear* interaction, the *linear* × *quadratic* interaction, the *quadratic* × *linear* interaction and the *quadratic* × *quadratic* interaction. It is easy to verify that these contrasts are independent of the contrasts for main effects.

■ **EXAMPLE 2.1.8.**

Suppose that $k = 2$. Then the contrasts for main effects and for the linear × linear contrast, the linear × quadratic contrast, the quadratic × linear contrast and the quadratic × quadratic contrast are given in Table 2.2. □

Table 2.2 A, B, and AB Contrasts for a 3^2 Factorial

00	01	02	10	11	12	20	21	22	Treatment combinations
1	1	1	1	1	1	1	1	1	Mean
−1	−1	−1	0	0	0	1	1	1	A Linear $= A_L$
1	1	1	−2	−2	−2	1	1	1	A Quadratic $= A_Q$
−1	0	1	−1	0	1	−1	0	1	B Linear $= B_L$
1	−2	1	1	−2	1	1	−2	1	B Quadratic $= B_Q$
1	0	−1	0	0	0	−1	0	1	$A_L \times B_L$
−1	2	−1	0	0	0	1	−2	1	$A_L \times B_Q$
−1	0	1	2	0	−2	−1	0	1	$A_Q \times B_L$
1	−2	1	−2	4	−2	1	−2	1	$A_Q \times B_Q$

Another way to get the four contrasts that correspond to the two-factor interaction effects is to use the sets

$$\{y_{x_1x_2\ldots x_k} | x_{q_1} + x_{q_2} = \theta\}, \ \theta = 0, 1, 2$$

and

$$\{y_{x_1x_2\ldots x_k} | x_{q_1} + 2x_{q_2} = \theta\}, \ \theta = 0, 1, 2.$$

Then the linear and quadratic contrasts can be used on these sets, although it is not clear what interpretation might be put on them, even for quantitative factors. Note, though, that these sets are the pencils of an affine plane and that this link has been exploited in proving results about fractional factorial designs; see Section 2.2 and Raghavarao (1971) for more details.

■ **EXAMPLE 2.1.9.**
Suppose that $k = 3$. The set given by $x_1 + x_2 = 0$ is

$$\{y_{000}, y_{001}, y_{002}, y_{120}, y_{121}, y_{122}, y_{210}, y_{211}, y_{212}\}.$$

If we let

$$AB_{00} = \{x_1x_2x_3 | x_1 = 0, x_2 = 0\}, \quad AB_{01} = \{x_1x_2x_3 | x_1 = 0, x_2 = 1\},$$

and so on, then the set given by $x_1 + x_2 = 0$ is the union of AB_{00}, AB_{12}, and AB_{21}. The other 5 sets can be defined similarly. The 9 orthogonal contrasts are then as given in Table 2.3. □

Table 2.3 A, B, and AB Contrasts for a 3^3 Factorial

AB_{00}	AB_{01}	AB_{02}	AB_{10}	AB_{11}	AB_{12}	AB_{20}	AB_{21}	AB_{22}	
1	1	1	1	1	1	1	1	1	Mean
−1	−1	−1	0	0	0	1	1	1	A Effect
1	1	1	−2	−2	−2	1	1	1	
−1	0	1	−1	0	1	−1	0	1	B Effect
1	−2	1	1	−2	1	1	−2	1	
−1	0	1	0	1	−1	1	−1	0	AB Effect
1	−2	1	−2	1	1	1	1	−2	
−1	1	0	0	−1	1	1	0	−1	
1	1	−2	−2	1	1	1	−2	1	

The definition of the sets of treatment combinations for the determination of higher-order interaction terms is similar. We can either take the orthogonal polynomial approach with linear × linear × linear and so on, or we can use the second, "geometric", approach. With the geometric approach, the three-factor interaction between factors A, B, and C requires four sets to define the eight contrasts; these are

$$x_1 + x_2 + x_3 = i \pmod 3, \quad x_1 + x_2 + 2x_3 = i \pmod 3,$$

$$x_1 + 2x_2 + x_3 = i \pmod 3, \quad x_1 + 2x_2 + 2x_3 = i \pmod 3.$$

We will take the geometric approach when we come to construct fractional factorial designs in Section 2.2.

2.1.3 Asymmetric Designs

As we said earlier, we use $\ell_1 \times \ell_2 \times \ldots \times \ell_k$ to refer to a general factorial design in which there are no restrictions on the number of levels for any of the factors.

2.1.3.1 Main Effects

As before, we want to be able to estimate the effect of each of the factors, individually, on the response. For factor q, we calculate ℓ_q sets associated with that factor and use contrasts between these sets to calculate the main effects for that factor.

■ EXAMPLE 2.1.10.

Consider a $2 \times 3 \times 4$ complete factorial design. The two sets for the first factor are

$$\{000, 001, 002, 003, 010, 011, 012, 013, 020, 021, 022, 023\},$$

$$\{100, 101, 102, 103, 110, 111, 112, 113, 120, 121, 122, 123\}.$$

The second factor has three sets associated with it:

$$\{000, 001, 002, 003, 100, 101, 102, 103\}, \{010, 011, 012, 013, 110, 111, 112, 113\},$$

$$\{020, 021, 022, 023, 120, 121, 122, 123\}.$$

Similarly, there are four sets associated with the third factor. The corresponding matrix of contrasts for main effects for the three factors is given in Table 2.4. □

Table 2.4 Main Effects Contrasts for a $2 \times 3 \times 4$ Factorial

Treatment Combinations																								
0	0	0	0	0	0	0	0	0	0	0	0	1	1	1	1	1	1	1	1	1	1	1	1	
0	0	0	0	1	1	1	1	2	2	2	2	0	0	0	0	1	1	1	1	2	2	2	2	
0	1	2	3	0	1	2	3	0	1	2	3	0	1	2	3	0	1	2	3	0	1	2	3	
−1	−1	−1	−1	−1	−1	−1	−1	−1	−1	−1	−1	1	1	1	1	1	1	1	1	1	1	1	1	A
−1	−1	−1	−1	0	0	0	0	1	1	1	1	−1	−1	−1	−1	0	0	0	0	1	1	1	1	B_L
1	1	1	1	−2	−2	−2	−2	1	1	1	1	1	1	1	1	−2	−2	−2	−2	1	1	1	1	B_Q
−3	−1	1	3	−3	−1	1	3	−3	−1	1	3	−3	−1	1	3	−3	−1	1	3	−3	−1	1	3	C_L
1	−1	−1	1	1	−1	−1	1	1	−1	−1	1	1	−1	−1	1	1	−1	−1	1	1	−1	−1	1	C_Q
−1	3	−3	1	−1	3	−3	1	−1	3	−3	1	−1	3	−3	1	−1	3	−3	1	−1	3	−3	1	C_C

2.1.3.2 Interaction Effects

For interaction effects, we again use the polynomial contrasts for each factor and take all possible component-wise products.

■ EXAMPLE 2.1.11.

Consider the 2×3 factorial. Then the six treatment combinations are 00, 01, 02, 10, 11, and 12.

The contrast for the main effect of the first factor is $-1, -1, -1, 1, 1, 1$.

For the second factor the main-effect contrasts are $-1, 0, 1, -1, 0, 1$ and $1, -2, 1, 1, -2, 1$.

Thus the two contrasts for the interaction are $1, 0, -1, -1, 0, 1$ and $-1, 2, -1, 1, -2, 1$. □

2.1.4 Exercises

1. Give all the level combinations in a $2 \times 2 \times 3$ factorial.
2. Give the B and C main effects for Example 2.1.3.
3. Use $\frac{1}{2^{k-1}}(\sum_{x_q=1} y_{x_1x_2\ldots x_k} - \sum_{x_q=0} y_{x_1x_2\ldots x_k})$ to confirm the results in Example 2.1.3 and the previous exercise.

4. Consider factors with two levels and let $k = 2$ Confirm that the AB interaction and the BA interaction are equal. Repeat with $k = 3$.

5. Consider factors with two levels and let $k = 3$. Show that the ABC interaction is also equal to the average difference between the AC interaction at level 1 of B and the AC interaction at level 0 of B. Are there any other ways that you could describe this interaction?

6. If $\ell = 4$, verify that the following components are correct.

-3	-1	1	3	*Linear*
1	-1	-1	1	*Quadratic*
-1	3	-3	1	*Cubic*

7. Suppose that $k = 2$ and that both attributes have 4 levels. Give possible entries for the 6 polynomial contrasts for main effects and the 9 polynomial contrasts for the two-factor interaction.

8. Suppose that $\ell = 3$.
   (a) Let $\mathbf{x} = (0, 1, 2)'$. Let $X = [\mathbf{1}, \mathbf{x}, \mathbf{x}^2]$. Calculate $X'X$.
   (b) Now let $X = [\mathbf{1}, P_1(x), P_2(x)]$ and calculate $X'X$.
   (c) Comment.
   (d) Suppose that you use the representations in both of the first two parts to fit a quadratic polynomial. What is the relationship between the estimates?

9. (a) List the 6 sets of treatment combinations in a 3^2 factorial experiment corresponding to the main effects.
   (b) Give two independent contrasts for the main effects for each of the factors.
   (c) List the six sets that correspond to the two-factor interaction.
   (d) Give four independent contrasts corresponding to the two-factor interaction.
   (e) Verify that all the eight contrasts (for main effects and the two-factor interaction) that you get are orthogonal (that is, independent) in this case.

10. Suppose that $y_1, y_2, \ldots, y_n$ are independently identically distributed $N(\mu, \sigma^2)$ random variables.
   (a) Let $\mathbf{y}' = (y_1, y_2, \ldots, y_n)$ and let $\mathbf{a}' = (a_1, a_2, \ldots, a_n)$. Then
   $$\mathbf{a}'\mathbf{y} \sim N(\mu \sum_i a_i, \sigma^2 \mathbf{a}'\mathbf{a}).$$
   Hence deduce that the sum of squares for testing $H_0 : \mu \sum_i a_i = 0$ is given by $(\mu \sum_i a_i)^2 n/s^2$ where s^2 is an unbiased estimate of σ^2.
   (b) By definition, two normally distributed random variables, x and z are independent if $Cov(x, z) = 0$. If A is a matrix, then $A\mathbf{y} \sim N(A\boldsymbol{\mu}, \sigma^2 AA')$, where $\boldsymbol{\mu} = (\mu, \mu, \ldots, \mu)$. Suppose that the rows of A form a set of independent contrasts. Then AA' is a diagonal matrix. Derive the sum of squares for testing $H_0 : A\boldsymbol{\mu} = 0$.
   (c) Hence show that, in an ordinary least squares model, the sum of squares for a factor is independent of which set of independent contrasts is chosen.

11. Verify that the five contrasts in Example 2.1.11 are all mutually orthogonal.

2.2 REGULAR FRACTIONAL FACTORIAL DESIGNS

A *fractional factorial design* is one in which only a subset of the level combinations appears.

Fractional factorial designs are used when the number of treatment combinations in the complete factorial is just too large to be practical, either because it will take too long to complete the experiment or it will cost too much. So a fractional factorial design is faster and cheaper and, with a suitably chosen fraction, you can still get all the information that you want from the experiment. In this section we will look at the ways of getting fractional factorials that allow you to estimate all the effects you are interested in.

2.2.1 Two-Level Fractions

Recall that, if each of the k factors has 2 levels, then we talk about a 2^k design. We use 2^{k-p} to denote a fractional factorial design in which only 2^{k-p} treatment combinations appear. A *regular fraction* of a 2^k factorial is a fraction in which the treatment combinations can be described by the solution to a set of binary equations. Equivalently, a regular fraction is one in which there are some generator factors and all other factors can be defined in terms of these generators. The binary equations are called the *defining equations* or the *defining contrasts* of the fractional factorial design. A regular 2^{k-p} fraction is defined by p independent binary equations or, equivalently, by $k-p$ generators.

An *irregular fraction* is a subset of the treatment combinations from the complete factorial, but the subset is determined in some ad-hoc fashion. Usually all treatments combinations in an irregular fraction have to be listed explicitly.

■ **EXAMPLE 2.2.1.**

A regular 2^{4-1} has 8 treatment combinations and these 8 treatment combinations are the solutions to one binary equation. The solutions to the binary equation

$$x_1 + x_2 + x_3 = 0 \pmod{2}$$

are the treatment combinations in Table 2.5. If $x_1 = 1$ and $x_2 = 0$ then $1 + 0 + x_3 = 0 \pmod{2}$; so $x_3 = 1$. The defining equation places no restrictions on x_4; so both 1010 and 1011 are solutions to the defining equation. This fraction may also be written as $I \equiv ABC$. (Recall that, when working modulo 2, $0+0 = 1+1 = 0$ and $0+1 = 1+0 = 1$.) □

Table 2.5 A Regular 2^{4-1} Design

0	0	0	0
0	0	0	1
0	1	1	0
0	1	1	1
1	0	1	0
1	0	1	1
1	1	0	0
1	1	0	1

As we have said, for a regular 2^{k-p} design, the treatment combinations will satisfy p independent binary equations. A set of equations is said to be *independent* if no non-zero linear combination of the equations is identically 0. We now consider an example of this.

■ EXAMPLE 2.2.2.
Let $k = 6$. Consider the three binary equations

$$\begin{aligned} x_1 + x_2 + x_3 + x_4 &= 0, \\ x_1 + x_2 + x_5 + x_6 &= 0, \\ x_1 + x_3 + x_5 &= 0. \end{aligned}$$

To show that these equations are independent, we need to consider all the sums of these equations; thus we consider the three sums of two equations and the sum of all three equations. We get

$$\begin{aligned} x_1 + x_2 + x_3 + x_4 + x_1 + x_2 + x_5 + x_6 &= x_3 + x_4 + x_5 + x_6, \\ x_1 + x_2 + x_3 + x_4 + x_1 + x_3 + x_5 &= x_2 + x_4 + x_5, \\ x_1 + x_2 + x_5 + x_6 + x_1 + x_3 + x_5 &= x_2 + x_3 + x_6, \\ x_1 + x_2 + x_3 + x_4 + x_1 + x_2 + x_5 + x_6 + x_1 + x_3 + x_5 &= x_1 + x_4 + x_6, \end{aligned}$$

and these are the only linear combinations possible for binary equations. There are 8 solutions to these equations: 000000, 001111, 010101, 011010, 100110, 101001, 110011, 111100.

On the other hand, the equations $x_1 + x_2 + x_3 + x_4 = 0$, $x_1 + x_2 + x_5 + x_6 = 0$ and $x_3 + x_4 + x_5 + x_6 = 0$ have 16 solutions since the third equation is the sum of the first two and so there are only two *independent* equations. □

We have defined regular fractions by setting linear combinations of the x_i equal to 0. This is called the *principal fraction*. Other fractions are obtained by equating some (or all) of the linear combinations to 1. Indeed if we use the same linear combinations with all possible solutions we obtain a partition of the complete factorial into regular fractions, as the following example shows.

■ EXAMPLE 2.2.3.
Let $k = 5$ and consider the linear combinations $x_1 + x_2 + x_3$ and $x_3 + x_4 + x_5$. The corresponding partition of the complete 2^5 factorial into four regular fractions is given in Table 2.6. □

Table 2.6 Non-overlapping Regular 2^{5-2} Designs

$x_1+x_2+x_3=0$, $x_3+x_4+x_5=0$	$x_1+x_2+x_3=0$, $x_3+x_4+x_5=1$	$x_1+x_2+x_3=1$, $x_3+x_4+x_5=0$	$x_1+x_2+x_3=1$, $x_3+x_4+x_5=1$
0 0 0 0 0	0 0 0 0 1	0 1 0 0 0	0 1 0 0 1
0 0 0 1 1	0 0 0 1 0	0 1 0 1 1	0 1 0 1 0
0 1 1 0 1	0 1 1 0 0	0 0 1 0 1	0 0 1 0 0
0 1 1 1 0	0 1 1 1 1	0 0 1 1 0	0 0 1 1 1
1 0 1 0 1	1 0 1 0 0	1 1 1 0 1	1 1 1 0 0
1 0 1 1 0	1 0 1 1 1	1 1 1 1 0	1 1 1 1 1
1 1 0 0 0	1 1 0 0 1	1 0 0 0 0	1 0 0 0 1
1 1 0 1 1	1 1 0 1 0	1 0 0 1 1	1 0 0 1 0

For any fractional factorial design, we will estimate an effect by using the same coefficients as we would have used for that effect in the complete factorial design. This is sometimes called *definition by restriction*.

■ **EXAMPLE 2.2.4.**
Let $k = 4$ and consider the 2^{4-1} fraction given by $x_1 + x_2 + x_3 + x_4 = 0$; see Table 2.7. There are 8 treatment combinations and so there will be 8-1=7 orthogonal contrasts possible. The 15 contrasts from a 2^4 design are given in Table 2.8. The contrasts that we use for the fraction are obtained from the contrasts in Table 2.8 by considering only the eight columns that correspond to the eight treatments in the fractional factorial design. These are indicated in bold. Doing this, we see that $ABCD$ is indistinguishable from the mean. This is consistent with Theorem 2.1.1 since all the treatment combinations with $x_1 + x_2 + x_3 + x_4 = 0$ have the same sign in the $ABCD$ contrast. All the other contrasts come in pairs with, say, the effects of A and BCD being indistinguishable. This is because the two sets with different coefficients in the A contrast are given by $x_1 = 0$ and $x_1 = 1$. For the BCD contrast these sets are $x_2 + x_3 + x_4 = 1$ and $x_2 + x_3 + x_4 = 0$. If we know that $x_1 + x_2 + x_3 + x_4 = 0$, then $x_1 = x_2 + x_3 + x_4$ and so the sets for the A contrast and the BCD contrast coincide. Similarly, $x_2 = x_1 + x_3 + x_4$, $x_3 = x_1 + x_2 + x_4$, $x_4 = x_1 + x_2 + x_2$, $x_1 + x_2 = x_3 + x_4$, $x_1 + x_3 = x_2 + x_4$, and $x_1 + x_4 = x_2 + x_3$. So A and BCD, B and ACD, C and ABD, D and ABC, AB and CD, AC and BD, and AD and BC form pairs of effects with contrasts that are the same in the 2^{4-1} fraction given by $x_1 + x_2 + x_3 + x_4 = 0$. □

Table 2.7 A 2^{4-1} Design of Resolution 4

0	0	0	0
0	0	1	1
0	1	0	1
0	1	1	0
1	0	0	1
1	0	1	0
1	1	0	0
1	1	1	1

We say that effects with the same contrast in a fraction are *aliased* or *confounded* effects. The effects which have the same coefficient for every treatment combination in the fraction are called the *defining effects* or the *defining contrasts*. The list of all the aliases for each effect in a design is called the *alias structure* of the design. Given the defining contrasts, we can calculate the alias structure (and conversely).

If we know nothing about the factors and possible interactions, then it is usually best to be able to estimate low order interactions (main effects and two-factor interactions) independently of each other in a fractional factorial design. So fractional factorial designs are classified by the alias structure of the design. If no main effect is confounded with any other main effect, but at least one main effect is confounded with a two-factor interaction, then the design is said to be of *resolution 3*. If at least one main effect is confounded with a three-factor interaction but no two main effects are confounded with each other and no main effect is confounded with a two-factor interaction, then the design is said to be of *resolution 4*. So the design in Table 2.7 is of resolution 4 since each main effect is

Table 2.8 Contrasts for the 2^4 design

	0	0	0	0	0	0	0	0	1	1	1	1	1	1	1	1
	0	0	0	0	1	1	1	1	0	0	0	0	1	1	1	1
	0	0	1	1	0	0	1	1	0	0	1	1	0	0	1	1
	0	1	0	1	0	1	0	1	0	1	0	1	0	1	0	1
A	−1	−1	−1	−1	−1	−1	−1	−1	1	1	1	1	1	1	1	1
B	−1	−1	−1	−1	1	1	1	1	−1	−1	−1	−1	1	1	1	1
C	−1	−1	1	1	−1	−1	1	1	−1	−1	1	1	−1	−1	1	1
D	−1	1	−1	1	−1	1	−1	1	−1	1	−1	1	−1	1	−1	1
AB	1	1	1	1	−1	−1	−1	−1	−1	−1	−1	−1	1	1	1	1
AC	1	1	−1	−1	1	1	−1	−1	−1	−1	1	1	−1	−1	1	1
AD	1	−1	1	−1	1	−1	1	−1	−1	1	−1	1	−1	1	−1	1
BC	1	1	−1	−1	−1	−1	1	1	1	1	−1	−1	−1	−1	1	1
BD	1	−1	1	−1	−1	1	−1	1	1	−1	1	−1	−1	1	−1	1
CD	1	−1	−1	1	1	−1	−1	1	1	−1	−1	1	1	−1	−1	1
ABC	−1	−1	1	1	1	1	−1	−1	1	1	−1	−1	−1	−1	1	1
ABD	−1	1	−1	1	1	−1	1	−1	1	−1	1	−1	−1	1	−1	1
ACD	−1	1	1	−1	−1	1	1	−1	1	−1	−1	1	1	−1	−1	1
BCD	−1	1	1	−1	1	−1	−1	1	−1	1	1	−1	1	−1	−1	1
$ABCD$	1	−1	−1	1	−1	1	1	−1	−1	1	1	−1	1	−1	−1	1

confounded with a three-factor interaction. If at least one main effect is confounded with a four-factor interaction and no main effect is confounded with anything smaller, or if at least one two-factor interaction is confounded with a three-factor interaction but no pair of two-factor interactions are confounded, then the design is said to be of *resolution 5*.

We can determine the resolution of a design directly from the defining equations. If there are r non-zero coefficients in a defining equation, then main effects corresponding to the x_q with non-zero coefficients in the equation are confounded with interactions of $r-1$ factors, two-factor interactions corresponding to pairs of x_q with non-zero coefficients are confounded with interactions with $r-2$ factors, and so on.

■ **EXAMPLE 2.2.5.**
Let $k = 5$ and consider the fraction given by the defining equations

$$x_1 + x_2 + x_3 = 0 \text{ and } x_1 + x_2 + x_4 + x_5 = 0.$$

The aliasing structure is determined by noting, first, that

$$x_1 + x_2 + x_3 + x_1 + x_2 + x_4 + x_5 = x_3 + x_4 + x_5 = 0.$$

Thus there are in total three defining equations; these are

$$x_1 + x_2 + x_3 = x_1 + x_2 + x_4 + x_5 = x_3 + x_4 + x_5 = 0.$$

From these we can now establish the sets of equations that have the same solutions. We get

$$\begin{array}{rclclcl}
x_1 &=& x_2 + x_3 &=& x_2 + x_4 + x_5 &=& x_1 + x_3 + x_4 + x_5, \\
x_2 &=& x_1 + x_3 &=& x_1 + x_4 + x_5 &=& x_2 + x_3 + x_4 + x_5, \\
x_3 &=& x_1 + x_2 &=& x_1 + x_2 + x_3 + x_4 + x_5 &=& x_4 + x_5, \\
x_4 &=& x_1 + x_2 + x_3 + x_4 &=& x_1 + x_2 + x_5 &=& x_3 + x_5, \\
x_5 &=& x_1 + x_2 + x_3 + x_5 &=& x_1 + x_2 + x_4 &=& x_3 + x_4.
\end{array}$$

Thus we see that the main effect of A, for instance, is confounded with BC, BDE and $ACDE$. As there is at least one main effect that is confounded with a two-factor interaction (and no pair of main effects that are confounded), we have confirmed that the design is of resolution 3. □

■ **EXAMPLE 2.2.6.**
Let $k = 6$. The defining equations

$$x_1 + x_2 + x_3 + x_4 = 0 \text{ and } x_3 + x_4 + x_5 + x_6 = 0$$

give a design of resolution 4. We can see this because

$$x_1 + x_2 + x_3 + x_4 + x_3 + x_4 + x_5 + x_6 = x_1 + x_2 + x_5 + x_6 = 0.$$

Thus each main effect is confounded with two 3-factor interactions and a 5-factor interaction. Each 2-factor interaction is confounded with at least one other 2-factor interaction. □

If there are more than three factors ($k > 3$), then we can find a resolution 3 design that uses fewer treatment combinations than the complete factorial.

Designs of resolution 5 can be used to estimate main effects and two-factor interactions. If there are four or fewer factors (so $k \leq 4$), then the only resolution 5 design is the complete factorial, independent of the number of levels each of the factors have. If there are more than four factors ($k > 4$), then we can find a resolution 5 design that uses fewer treatment combinations than the complete factorial. We do this by finding a set of defining equations such that each defining equation has at least 5 non-zero coefficients and such that each linear combination of the equations has at least 5 non-zero coefficients just as we did for resolution 4 in Example 2.2.6.

2.2.1.1 Regular Designs with Factors at 2 Levels Regular fractions can be constructed from a set of defining equations or, equivalently, from a set of generator vectors. In this section we give sets of generator vectors that can be used to construct small regular fractional factorial designs with all factors with 2 levels.

For two vectors $\mathbf{a}$ and $\mathbf{b}$ we will define $\mathbf{a} + \mathbf{b}$ using component-wise addition modulo 2 (that is, $0 + 0 = 1 + 1 = 0$ and $0 + 1 = 1 + 0 = 1$). Thus for the two vectors

$$\mathbf{a} = (0,0,0,0,1,1,1,1) \text{ and } \mathbf{b} = (0,0,1,1,0,0,1,1)$$

we see that

$$\mathbf{a} + \mathbf{b} = (0,0,1,1,1,1,0,0).$$

We will write a set of *generator vectors* in a standard order. If the design has 2^{k-p} treatments then we need to define $k - p$ generators, $\mathbf{b}_i$ say. We let $\mathbf{b}_1$ have its first 2^{k-p-1} entries as 0 and the remaining 2^{k-p-1} entries as 1. For $\mathbf{b}_2$ we let the first 2^{k-p-2} entries be 0, the next 2^{k-p-2} entries be 1, the next 2^{k-p-2} be 0 and the final 2^{k-p-2} be 1. For $\mathbf{b}_3$ we let the first 2^{k-p-3} entries be 0, the next 2^{k-p-3} entries be 1, the next 2^{k-p-3} be 0 and so on. We continue defining generators in this way until we get to $\mathbf{b}_{k-p}$ which alternates 0s and 1s.

For instance when $k = 4$ and $p = 1$ we get the three generator vectors

$$\mathbf{b}_1 = (0,0,0,0,1,1,1,1), \mathbf{b}_2 = (0,0,1,1,0,0,1,1) \text{ and } \mathbf{b}_3 = (0,1,0,1,0,1,0,1).$$

We can now define the levels of all the other factors in the design in terms of the $\mathbf{b}_i$. Table 2.9 contains designs with up to 10 factors of resolution 3 and Table 2.10 contains

Table 2.9 Smallest Known 2-Level Designs with Resolution at Least 3

k	N	Other Factors
4	8	$\mathbf{b}_4 = \mathbf{b}_1 + \mathbf{b}_2 + \mathbf{b}_3$
5	8	$\mathbf{b}_4 = \mathbf{b}_1 + \mathbf{b}_2, \mathbf{b}_5 = \mathbf{b}_1 + \mathbf{b}_3$
6	8	$\mathbf{b}_4 = \mathbf{b}_1 + \mathbf{b}_2, \mathbf{b}_5 = \mathbf{b}_1 + \mathbf{b}_3, \mathbf{b}_6 = \mathbf{b}_2 + \mathbf{b}_3$
7	8	$\mathbf{b}_4 = \mathbf{b}_1 + \mathbf{b}_2, \mathbf{b}_5 = \mathbf{b}_1 + \mathbf{b}_3, \mathbf{b}_6 = \mathbf{b}_2 + \mathbf{b}_3,$ $\mathbf{b}_7 = \mathbf{b}_1 + \mathbf{b}_2 + \mathbf{b}_3$
8	16	$\mathbf{b}_5 = \mathbf{b}_2 + \mathbf{b}_3 + \mathbf{b}_4, \mathbf{b}_6 = \mathbf{b}_1 + \mathbf{b}_3 + \mathbf{b}_4,$ $\mathbf{b}_7 = \mathbf{b}_1 + \mathbf{b}_2 + \mathbf{b}_3, \mathbf{b}_8 = \mathbf{b}_1 + \mathbf{b}_2 + \mathbf{b}_4$
9	16	$\mathbf{b}_5 = \mathbf{b}_2 + \mathbf{b}_3 + \mathbf{b}_4, \mathbf{b}_6 = \mathbf{b}_1 + \mathbf{b}_3 + \mathbf{b}_4,$ $\mathbf{b}_7 = \mathbf{b}_1 + \mathbf{b}_2 + \mathbf{b}_3, \mathbf{b}_8 = \mathbf{b}_1 + \mathbf{b}_2 + \mathbf{b}_4,$ $\mathbf{b}_9 = \mathbf{b}_1 + \mathbf{b}_2 + \mathbf{b}_3 + \mathbf{b}_4$
10	16	$\mathbf{b}_5 = \mathbf{b}_2 + \mathbf{b}_3 + \mathbf{b}_4, \mathbf{b}_6 = \mathbf{b}_1 + \mathbf{b}_3 + \mathbf{b}_4,$ $\mathbf{b}_7 = \mathbf{b}_1 + \mathbf{b}_2 + \mathbf{b}_3, \mathbf{b}_8 = \mathbf{b}_1 + \mathbf{b}_2 + \mathbf{b}_4,$ $\mathbf{b}_9 = \mathbf{b}_1 + \mathbf{b}_2 + \mathbf{b}_3 + \mathbf{b}_4, \mathbf{b}_{10} = \mathbf{b}_1 + \mathbf{b}_2$

Table 2.10 Smallest Known 2-Level Designs with Resolution at Least 5

k	N	Other Factors
5	16	$\mathbf{b}_5 = \mathbf{b}_1 + \mathbf{b}_2 + \mathbf{b}_3 + \mathbf{b}_4$
6	32	$\mathbf{b}_6 = \mathbf{b}_1 + \mathbf{b}_2 + \mathbf{b}_3 + \mathbf{b}_4 + \mathbf{b}_5$
7	64	$\mathbf{b}_7 = \mathbf{b}_1 + \mathbf{b}_2 + \mathbf{b}_3 + \mathbf{b}_4 + \mathbf{b}_5 + \mathbf{b}_6$
8	64	$\mathbf{b}_7 = \mathbf{b}_1 + \mathbf{b}_2 + \mathbf{b}_3 + \mathbf{b}_4,$ $\mathbf{b}_8 = \mathbf{b}_1 + \mathbf{b}_2 + \mathbf{b}_5 + \mathbf{b}_6$
9	128	$\mathbf{b}_8 = \mathbf{b}_1 + \mathbf{b}_3 + \mathbf{b}_4 + \mathbf{b}_6 + \mathbf{b}_7,$ $\mathbf{b}_9 = \mathbf{b}_2 + \mathbf{b}_3 + \mathbf{b}_5 + \mathbf{b}_6 + \mathbf{b}_7$
10	128	$\mathbf{b}_8 = \mathbf{b}_1 + \mathbf{b}_2 + \mathbf{b}_3 + \mathbf{b}_7,$ $\mathbf{b}_9 = \mathbf{b}_2 + \mathbf{b}_3 + \mathbf{b}_4 + \mathbf{b}_5,$ $\mathbf{b}_{10} = \mathbf{b}_1 + \mathbf{b}_3 + \mathbf{b}_4 + \mathbf{b}_6$

designs with up to 10 factors of resolution 5. We let N denote the number of treatment combinations so $N = 2^{k-p}$.

■ **EXAMPLE 2.2.7.**
We use the results in Table 2.9 to construct a binary design with $k = 6$ factors of resolution 3. The 8 runs of the design are given in Table 2.11. □

Table 2.11 A Design with 6 Binary Factors of Resolution 3

b_1	b_2	b_3	b_4	b_5	b_6
0	0	0	0	0	0
0	0	1	0	1	1
0	1	0	1	0	1
0	1	1	1	1	0
1	0	0	1	1	0
1	0	1	1	0	1
1	1	0	0	1	1
1	1	1	0	0	0

2.2.2 Three-Level Fractions

A regular 3^{k-p} fraction is one that is defined by the solutions to a set of p independent ternary equations; that is, a set of equations where all the arithmetic is done modulo 3 (so $0 + 2 = 1 + 1 = 2 + 0 = 2$, $1 + 2 = 2 + 1 = 0$ and $0 + 1 = 1 + 0 = 2 + 2 = 1$). These equations are the *defining equations* or the *defining contrasts* of the regular fractional factorial design.

■ **EXAMPLE 2.2.8.**
A regular 3^{4-2} fraction has 9 treatment combinations which are the solutions to two independent ternary equations. The solutions to $x_1 + x_2 + x_3 = 0$ and $x_1 + 2x_2 + x_4 = 0$ are given in Table 2.12. (If we had used $x_1 + x_2 + x_3 = 0$ and $x_1 + x_2 + x_4 = 0$, then we would have had $x_3 = x_4$, and so we would not have been able to estimate the effects of the third and fourth factors independently.) □

A regular 3^{k-p} fraction satisfies p independent ternary equations. When checking for independence, we must now add the original equations in pairs, triples, and so on (as you do for binary equations), but we must also check each equation plus twice every other equation, and so on. More formally, if we let $E_1, E_2, \ldots, E_p$ be the defining contrasts, then we must calculate

$$\sum_q \alpha_q E_q; \; \alpha_q = 0, 1, 2, \; q = 1, 2, \ldots, p,$$

and show that none of these 3^p equations are identically 0 except when

$$\alpha_1 = \alpha_2 = \ldots = \alpha_p = 0.$$

Table 2.12 A 3^{4-2} Fractional Factorial Design

0	0	0	0
0	1	2	1
0	2	1	2
1	0	2	2
1	1	1	0
1	2	0	1
2	0	1	1
2	1	0	2
2	2	2	0

■ **EXAMPLE 2.2.9.**
Let $k = 6$. Consider the defining contrasts

$$x_1 + x_2 + x_3 + 2x_4 = 0, \; x_1 + x_2 + 2x_3 + 2x_5 = 0 \quad \text{and} \quad x_1 + 2x_2 + x_3 + 2x_6 = 0.$$

These are all independent since each of x_4, x_5 and x_6 is involved in only one equation. □

■ **EXAMPLE 2.2.10.**
Let $k = 5$. Consider the defining contrasts

$$x_1 + x_2 + x_3 = 0 \quad \text{and} \quad x_3 + x_4 + x_5 = 0.$$

Can we find a third independent equation with at least three non-zero coefficients? The two equations that are linear combinations of the defining contrasts are

$$x_1 + x_2 + x_3 + x_3 + x_4 + x_5 = x_1 + x_2 + 2x_3 + x_4 + x_5 = 0$$

and

$$x_1 + x_2 + x_3 + 2x_3 + 2x_4 + 2x_5 = x_1 + x_2 + 2x_4 + 2x_5 = 0.$$

(We do not need to double the first equation and add the second equation — this is just double the sum of the first equation and twice the second equation; similarly, if we double both equations and add, we just get double the sum of the two equations.) Any equation other than the original two equations and the two equations we have found by addition will be independent of the two equations.

What can we say about the possible equations? Given we are using the equations to construct fractional factorial designs, we do not want an equation with only one x_q in it since the corresponding factor would not vary in the experiment. Similarly, if we use $x_{q_1} + \alpha x_{q_2}$, then there is a constant relationship between the corresponding factors and the associated main effects would be aliased.

If we think about equations with three variables, then we cannot have more than one of x_1, x_2 and x_3 (since if we used, say, $x_1 + x_2 + x_4 = 0$, then

$$x_1 + x_2 + x_3 + 2(x_1 + x_2 + x_4) = x_3 + 2x_4 = 0$$

and so $x_3 = x_4$) and we can not use more than one of x_3, x_4 and x_5. So an equation involving three of the x_q forces two of the x_q to be equal.

A similar argument shows that you can not have equations with four or five variables either. □

Since any interaction sum of squares is independent of the particular contrasts that are used to define the interaction effects, in this section we will define the interactions using contrasts between sets of the form

$$\sum_q \alpha_q x_q = \theta, \quad \theta = 0, 1, 2.$$

The only attributes whose levels determine the entries in the sets are those with non-zero coefficients. There are three sets associated with each equation:

$$\{(x_1, x_2, \ldots, x_k) | \sum_q \alpha_q x_q = 0\}, \ \{(x_1, x_2, \ldots, x_k) | \sum_q \alpha_q x_q = 1\}$$

and

$$\{(x_1, x_2, \ldots, x_k) | \sum_q \alpha_q x_q = 2\}.$$

We will let

$$P(\sum_q \alpha_q x_q) = P(\alpha_1, \alpha_2, \ldots, \alpha_k)$$

be the set of three sets associated with the equation $\sum_q \alpha_q x_q$, and we will talk about the *pencil* associated with $\sum_q \alpha_q x_q$.

For any interaction involving t attributes, there are 2^{t-1} associated pencils. For example, if we want to calculate the interaction between the first three factors we would calculate contrasts between the sets in $P(1110\ldots0)$, $P(1120\ldots0)$, $P(1210\ldots0)$ and $P(1220\ldots0)$. Each pencil gives rise to two independent contrasts. Contrasts from different pencils are orthogonal since any two sets from different pencils intersect in 3^{k-2} treatment combinations.

■ EXAMPLE 2.2.11.

Let $k = 2$. The sets in each pencil are indicated in Table 2.13. The corresponding contrasts are given in Table 2.14. □

Table 2.13 Pencils for a 3^2 Factorial Design

Pencil	$\theta = 0$	$\theta = 1$	$\theta = 2$
$P(10)$ or $x_1 = \theta$	00, 01, 02	10, 11, 12	20, 21, 22
$P(01)$ or $x_2 = \theta$	00, 10, 20	01, 11, 21	02, 12, 22
$P(11)$ or $x_1 + x_2 = \theta$	00, 12, 21	01, 10, 22	02, 11, 20
$P(12)$ or $x_1 + 2x_2 = \theta$	00, 11, 22	02, 10, 21	01, 12, 20

■ EXAMPLE 2.2.12.

Consider the design constructed in Example 2.2.8 and given in Table 2.12. The contrasts for the main effects for the design are given in Table 2.15. □

Table 2.14 Contrasts for a 3^2 Factorial Design

00	01	02	10	11	12	20	21	22	
−1	−1	−1	0	0	0	1	1	1	A_L
1	1	1	−2	−2	−2	1	1	1	A_Q
−1	0	1	−1	0	1	−1	0	1	B_L
1	−2	1	1	−2	1	1	−2	1	B_Q
−1	0	1	0	1	−1	1	−1	0	$P(11)_L$
1	−2	1	−2	1	1	1	1	−2	$P(11)_Q$
−1	1	0	0	−1	1	1	0	−1	$P(12)_L$
1	1	−2	−2	1	1	1	−1	1	$P(12)_Q$

Table 2.15 Contrasts for a 3^{4-2} Factorial Design

0000	0121	0212	1022	1110	1201	2011	2102	2220	
−1	−1	−1	0	0	0	1	1	1	A_L
1	1	1	−2	−2	−2	1	1	1	A_Q
−1	0	1	−1	0	1	−1	0	1	B_L
1	−2	1	1	−2	1	1	−2	1	B_Q
−1	1	0	1	0	−1	0	−1	1	C_L
1	1	−2	1	−2	1	−2	1	1	C_Q
−1	0	1	1	−1	0	0	1	−1	D_L
1	−2	1	1	1	−2	−2	1	1	D_Q

2.2.2.1 ***Regular Designs with All Factors at 3 Levels*** For two vectors **a** and **b**, we will define **a** + **b** using component-wise addition modulo 3. Thus

$$0+0=1+2=2+1=0,\ \ 0+1=1+0=2+2=1 \text{ and } 0+2=1+1=2+0=2.$$

Hence, for the two vectors $\mathbf{a}=(0,0,0,1,1,1,2,2,2)$ and $\mathbf{b}=(0,1,2,0,1,2,0,1,2)$, we see that $\mathbf{a}+\mathbf{b}=(0,1,2,1,2,0,2,0,1)$.

We will write a set of *generators* in a standard order. If the design has 3^{k-p} treatments, then we need to define $k-p$ generator vectors $\mathbf{b}_i$. For $\mathbf{b}_1$, we let the first 3^{k-p-1} entries be 0, the next 3^{k-p-1} entries be 1 and the remaining 3^{k-p-1} entries be 2. For $\mathbf{b}_2$, we let the first 3^{k-p-2} entries be 0, the next 3^{k-p-2} entries be 1, the next 3^{k-p-2} be 2, the next 3^{k-p-2} entries be 0, and so on. For $\mathbf{b}_3$, we let the first 3^{k-p-3} entries be 0, the next 3^{k-p-3} entries be 1, the next 3^{k-p-3} be 2, and so on. We continue defining generators in this way until we get to $\mathbf{b}_{k-p}$ which has 0, 1 and 2 in turn.

For instance, when $k=4$ and $p=1$, we get the three generator vectors

$$\mathbf{b}_1=(0,0,0,0,0,0,0,0,0,1,1,1,1,1,1,1,1,1,2,2,2,2,2,2,2,2,2),$$

$$\mathbf{b}_2=(0,0,0,1,1,1,2,2,2,0,0,0,1,1,1,2,2,2,0,0,0,1,1,1,2,2,2),$$

and

$$\mathbf{b}_3=(0,1,2,0,1,2,0,1,2,0,1,2,0,1,2,0,1,2,0,1,2,0,1,2,0,1,2).$$

If we write $2\mathbf{a}$, then that means we multiply each component of **a** by 2 modulo 3.

We can now define the levels of all the other factors in the design in terms of the $\mathbf{b}_i$. Table 2.16 contains generators for designs of resolution 3 with up to 10 factors and Table 2.17 contains generators for designs of resolution 5 with up to 10 factors . We let N denote the number of combinations of factor levels so $N=3^{k-p}$.

Table 2.16 Smallest Known Regular 3-Level Designs with Resolution at Least 3

k	N	Other Factors
4	9	$\mathbf{b}_3=\mathbf{b}_1+\mathbf{b}_2, \mathbf{b}_4=\mathbf{b}_1+2\mathbf{b}_2$
5	27	$\mathbf{b}_4=\mathbf{b}_1+\mathbf{b}_2+\mathbf{b}_3, \mathbf{b}_5=\mathbf{b}_1+\mathbf{b}_2+2\mathbf{b}_3$
6	27	$\mathbf{b}_4=\mathbf{b}_1+\mathbf{b}_2+\mathbf{b}_3, \mathbf{b}_5=\mathbf{b}_1+\mathbf{b}_2+2\mathbf{b}_3, \mathbf{b}_6=\mathbf{b}_1+2\mathbf{b}_2+\mathbf{b}_3$
7	27	$\mathbf{b}_4=\mathbf{b}_1+\mathbf{b}_2+\mathbf{b}_3, \mathbf{b}_5=\mathbf{b}_1+\mathbf{b}_2+2\mathbf{b}_3,$ $\mathbf{b}_6=\mathbf{b}_1+2\mathbf{b}_2+\mathbf{b}_3, \mathbf{b}_7=\mathbf{b}_1+2\mathbf{b}_2+2\mathbf{b}_3$
8	27	$\mathbf{b}_4=\mathbf{b}_1+\mathbf{b}_2+\mathbf{b}_3, \mathbf{b}_5=\mathbf{b}_1+\mathbf{b}_2+2\mathbf{b}_3, \mathbf{b}_6=\mathbf{b}_1+2\mathbf{b}_2+\mathbf{b}_3,$ $\mathbf{b}_7=\mathbf{b}_1+2\mathbf{b}_2+2\mathbf{b}_3, \mathbf{b}_8=\mathbf{b}_1+\mathbf{b}_2$
9	27	$\mathbf{b}_4=\mathbf{b}_1+\mathbf{b}_2+\mathbf{b}_3, \mathbf{b}_5=\mathbf{b}_1+\mathbf{b}_2+2\mathbf{b}_3, \mathbf{b}_6=\mathbf{b}_1+2\mathbf{b}_2+\mathbf{b}_3,$ $\mathbf{b}_7=\mathbf{b}_1+2\mathbf{b}_2+2\mathbf{b}_3, \mathbf{b}_8=\mathbf{b}_1+\mathbf{b}_2, \mathbf{b}_9=\mathbf{b}_1+\mathbf{b}_3$
10	27	$\mathbf{b}_4=\mathbf{b}_1+\mathbf{b}_2+\mathbf{b}_3, \mathbf{b}_5=\mathbf{b}_1+\mathbf{b}_2+2\mathbf{b}_3, \mathbf{b}_6=\mathbf{b}_1+2\mathbf{b}_2+\mathbf{b}_3,$ $\mathbf{b}_7=\mathbf{b}_1+2\mathbf{b}_2+2\mathbf{b}_3, \mathbf{b}_8=\mathbf{b}_1+\mathbf{b}_2, \mathbf{b}_9=\mathbf{b}_1+\mathbf{b}_3, \mathbf{b}_{10}=\mathbf{b}_2+\mathbf{b}_3$

2.2.3 A Brief Introduction to Finite Fields

The ideas developed in the previous two sections make it easy to construct fractions that confound certain effects. The same ideas will not work for all possible numbers of levels

Table 2.17 Smallest Known 3-Level Designs with Resolution at Least 5

k	N	Other Factors
5	81	$\mathbf{b}_5 = \mathbf{b}_1 + \mathbf{b}_2 + \mathbf{b}_3 + \mathbf{b}_4$
6	243	$\mathbf{b}_6 = \mathbf{b}_1 + \mathbf{b}_2 + \mathbf{b}_3 + \mathbf{b}_4 + \mathbf{b}_5$
7	243	$\mathbf{b}_6 = 2\mathbf{b}_1 + 2\mathbf{b}_2 + 2\mathbf{b}_3 + 2\mathbf{b}_4, \mathbf{b}_7 = 2\mathbf{b}_1 + 2\mathbf{b}_2 + \mathbf{b}_4 + 2\mathbf{b}_5$
8	243	$\mathbf{b}_6 = \mathbf{b}_1 + \mathbf{b}_2 + 2\mathbf{b}_3 + \mathbf{b}_5, \mathbf{b}_7 = \mathbf{b}_1 + 2\mathbf{b}_2 + \mathbf{b}_4 + \mathbf{b}_5,$ $\mathbf{b}_8 = \mathbf{b}_1 + \mathbf{b}_3 + \mathbf{b}_4 + 2\mathbf{b}_5$
9	243	$\mathbf{b}_6 = \mathbf{b}_1 + \mathbf{b}_2 + 2\mathbf{b}_3 + \mathbf{b}_5, \mathbf{b}_7 = \mathbf{b}_1 + 2\mathbf{b}_2 + \mathbf{b}_4 + \mathbf{b}_5,$ $\mathbf{b}_8 = \mathbf{b}_1 + \mathbf{b}_3 + \mathbf{b}_4 + 2\mathbf{b}_5, \mathbf{b}_9 = \mathbf{b}_1 + \mathbf{b}_2 + \mathbf{b}_3 + 2\mathbf{b}_4$
10	243	$\mathbf{b}_6 = \mathbf{b}_1 + \mathbf{b}_2 + 2\mathbf{b}_3 + \mathbf{b}_5, \mathbf{b}_7 = \mathbf{b}_1 + 2\mathbf{b}_2 + \mathbf{b}_4 + \mathbf{b}_5,$ $\mathbf{b}_8 = \mathbf{b}_1 + \mathbf{b}_3 + \mathbf{b}_4 + 2\mathbf{b}_5, \mathbf{b}_9 = \mathbf{b}_1 + \mathbf{b}_2 + \mathbf{b}_3 + 2\mathbf{b}_4,$ $\mathbf{b}_{10} = 2\mathbf{b}_1 + \mathbf{b}_2 + \mathbf{b}_3 + \mathbf{b}_4 + \mathbf{b}_5$

nor for situations in which not all factors have the same number of levels. In this section we develop the idea of a *finite field*. This is an algebraic structure that we need in order to be able to extend the techniques of the previous two sections.

We begin by looking at four levels. Suppose that there are $k = 2$ factors. Suppose that we work modulo 4 so that $1 + 1 = 2, 1 + 2 = 3, 1 + 3 = 0, 2 + 2 = 0, 2 + 3 = 1$, and so on. Think about the pencil $P(12)$. It contains the four sets

$$\{(x_1, x_2)|x_1 + 2x_2 = 0\} = \{00, 02, 21, 23\},$$

$$\{(x_1, x_2)|x_1 + 2x_2 = 1\} = \{10, 12, 31, 33\},$$
$$\{(x_1, x_2)|x_1 + 2x_2 = 2\} = \{01, 03, 20, 22\},$$
$$\{(x_1, x_2)|x_1 + 2x_2 = 3\} = \{11, 13, 30, 32\}.$$

Note that each of these sets contains either two entries or zero entries from the sets of the pencil $P(10)$ (from which we calculate the A main effect). Since these intersections are not of a constant size, the arguments that we have used in the previous sections, to establish the orthogonality of the contrasts used to estimate the effects, do not work. How can we overcome this problem?

The reason that all the sets from different pencils intersected in a constant number of points when working modulo 2 and 3 is that every non-zero element could be multiplied by some other element to get 1; that is, all non-zero elements have multiplicative inverses. For instance, $1 \times 1 = 1$ modulo both 2 and 3 and $2 \times 2 = 1$ modulo 3. When we work modulo 4 we see that $2 \times 1 = 2, 2 \times 2 = 0$ and $2 \times 3 = 2$. Thus 2 does not have a multiplicative inverse. So we want to get a set of 4 elements in which multiplicative inverses exist.

We do this by using an appropriate polynomial which we evaluate over the integers modulo 2, Z_2. To get such a polynomial, consider the quadratics over Z_2, namely, x^2, $x^2 + x = x(x + 1)$, $x^2 + 1 = (x + 1)^2$ and $x^2 + x + 1$. The first three quadratic equations factor over Z_2, but $1^2 + 1 + 1 = 1$ and $0^2 + 0 + 1 = 1$ in Z_2, and hence the quadratic polynomial $x^2 + x + 1$ does not factor over Z_2. It is said to be *irreducible* over Z_2. We now try to embed Z_2 in a larger field in which $x^2 + x + 1$ will factor.

Suppose we let α be a solution of $x^2 + x + 1 = 0$. So $\alpha^2 + \alpha + 1 = 0$ and hence $\alpha^2 = \alpha + 1$. Since we are working modulo 2, we have

$$(\alpha + 1)^2 + (\alpha + 1) + 1 = (\alpha^2 + 1) + (\alpha + 1) + 1 = \alpha^2 + \alpha + 1 = 0.$$

Consequently, $\alpha + 1$ is the other solution of our equation. Thus we get the addition and multiplication tables shown in Table 2.18.

Table 2.18 The Finite Field with 4 Elements

$+$	0	1	α	$\alpha+1$
0	0	1	α	$\alpha+1$
1	1	0	$\alpha+1$	α
α	α	$\alpha+1$	0	1
$\alpha+1$	$\alpha+1$	α	1	0

$\times$	0	1	α	$\alpha+1$
0	0	0	0	0
1	0	1	α	$\alpha+1$
α	0	α	$\alpha+1$	1
$\alpha+1$	0	$\alpha+1$	1	α

We are, in fact, taking the ring of polynomials over Z_2 and working with them modulo $x^2 + x + 1$ to give the field of order 4. We will write $GF[4]$ for the field of order 4 (where GF stands for "Galois field" after the French mathematician Evariste Galois (1811–1832)). The integers modulo n form a field if and only if n is prime. In the same way, if we start from Z_p for some prime p, and consider the ring of polynomials $Z_p[x]$ over Z_p, modulo a polynomial $f(x)$, this forms a field if and only if $f(x)$ is irreducible over Z_p.

We will write the entries in $GF[4]$ as ordered pairs: $(11) = \alpha + 1$ and $(01) = 1$, for instance. The only other thing that we need to know about finite fields is that the multiplicative group of the finite field is cyclic. Thus, if we take all the elements in the field other than 0, then each element can be expressed as a power of a *primitive element* of the field. In the case of $GF[4]$ the primitive element can be taken as α since $\alpha^2 = \alpha + 1$ and $\alpha^3 = 1$. For $GF[5]$, we can use 2 as the primitive element since $2^2 = 4$, $2^3 = 3$ and $2^4 = 1$, or we can use 3 as the primitive element ($3^2 = 4$, $3^3 = 2$ and $3^4 = 1$) but we can not use 4 (since $4^2 = 1$).

Of course if we represent the entries in $GF[4]$ as ordered pairs then we have the same representation for the levels of a four-level factor as we would have from having two factors each with two levels. This can be a useful way of thinking about a four-level factor; we say that the two new factors are the *pseudo-factors* corresponding to the original four-level factor. Three orthogonal contrasts for the four-level factor can be represented by linear, quadratic, and cubic orthogonal polynomials, or they can be represented by the main effects and the two-factor interaction of the pseudo-factors. We will see later that both representations are useful, depending on the circumstances.

Armed with these finite fields, we can now give constructions of regular fractions for any symmetric design with a prime or prime-power number of levels.

2.2.4 Fractions for Prime-Power Levels

In this section we will use ℓ to represent either a prime or a prime power. We will write $GF[\ell]$ as $\alpha_0 = 0, \alpha_1 = 1, \alpha_2 = x, \ldots, \alpha_{\ell-1} = x^{\ell-2}$ where x is a primitive element of $GF[\ell]$.

2.2.4.1 Regular Resolution 3 Fractions

■ **CONSTRUCTION 2.2.1.**
There is a resolution 3 design with $\ell + 1$ factors on ℓ symbols.

Proof. We will let $\mathbf{b}_1$ be a vector with ℓ 0's, then ℓ 1's, then ℓ α_2's and so on. We will let $\mathbf{b}_2$ be a vector with the elements of $GF[\ell]$ repeated in order ℓ times. Thus the pairs from the corresponding positions of $\mathbf{b}_1$ and $\mathbf{b}_2$ give one copy of each of the possible ordered pairs with entries from $GF[\ell]$. We can construct a resolution 3 design with up to $\ell + 1$ factors and with ℓ^2 treatment combinations by using $\mathbf{b}_1$ for the levels of the first factor, $\mathbf{b}_2$ for the levels of the second factor and $\mathbf{b}_1 + \alpha_q \mathbf{b}_2$ for the levels of the $(q + 2)$th factor, $q = 1, 2, \ldots, \ell - 1$. □

■ EXAMPLE 2.2.13.
Let $\ell = 4$. Let $x = \alpha$. Then $\alpha_1 = 1, \alpha_2 = \alpha$ and $\alpha_3 = \alpha^2 = \alpha + 1$. Thus we get

$$\mathbf{b}_1 = (0, 0, 0, 0, 1, 1, 1, 1, \alpha, \alpha, \alpha, \alpha, \alpha + 1, \alpha + 1, \alpha + 1, \alpha + 1)$$

and

$$\mathbf{b}_2 = (0, 1, \alpha, \alpha + 1, 0, 1, \alpha, \alpha + 1, 0, 1, \alpha, \alpha + 1, 0, 1, \alpha, \alpha + 1).$$

The third factor has levels given by $\mathbf{b}_1 + \mathbf{b}_2$, the fourth factor has levels given by $\mathbf{b}_1 + \alpha \mathbf{b}_2$ and the fifth factor has levels given by $\mathbf{b}_1 + (\alpha + 1)\mathbf{b}_2$. The final design is given in Table 2.19. □

Table 2.19 A Resolution 3 4^{5-3} Fractional Factorial Design

Factors				
1	2	3	4	5
0	0	0	0	0
0	1	1	α	$\alpha+1$
0	α	α	$\alpha+1$	1
0	$\alpha+1$	$\alpha+1$	1	α
1	0	1	1	1
1	1	0	$\alpha+1$	α
1	α	$\alpha+1$	α	0
1	$\alpha+1$	α	0	$\alpha+1$
α	0	α	α	α
α	1	$\alpha+1$	0	1
α	α	0	1	$\alpha+1$
α	$\alpha+1$	1	$\alpha+1$	0
$\alpha+1$	0	$\alpha+1$	$\alpha+1$	$\alpha+1$
$\alpha+1$	1	α	1	0
$\alpha+1$	α	1	0	α
$\alpha+1$	$\alpha+1$	0	α	1

This construction gives us resolution 3 designs for up to five 4-level factors, up to six 5-level factors, up to eight 7-level factors, up to 9 8-level factors and so on. If we need more factors then we will need other constructions; see Section 2.3.

2.2.4.2 Regular Resolution 5 Fractions Here is an easy construction for regular resolution 5 designs that works only when the number of levels is a prime or a prime power. It is a special case of a more general construction given by Bush (1952); see also Hedayat et al. (1999).

■ **CONSTRUCTION 2.2.2.**
Let ℓ be a prime or a prime power with $\ell \leq 4$. Then there is a resolution 5 fractional factorial design with 5 factors and ℓ^4 treatment combinations.

Proof. Let the elements of the Galois field be $\alpha_0 = 0, \alpha_1 = 1, \ldots, \alpha_{\ell-1}$. Construct the generators $\mathbf{b}_1$, $\mathbf{b}_2$, $\mathbf{b}_3$, and $\mathbf{b}_4$ in the usual way. Then construct one further factor, where $\mathbf{b}_5 = \mathbf{b}_1 + \mathbf{b}_2 + \mathbf{b}_3 + \mathbf{b}_4$. Verifying that this construction works is left as an exercise. □

The three designs from this construction have five 2-level factors or five 3-level factors or five 4-level factors.

There are other constructions for such designs available when the restriction about regularity is removed; see Section 2.3.

2.2.5 Exercises

1. Construct a regular design of resolution 3 with five factors each with two levels.

2. Construct a regular design of resolution 3 with five factors each with two levels and with $N = 16$. How many inequivalent designs can you get for this situation? (Two designs are said to be *inequivalent* if you cannot get from one to the other by permuting factors or levels within factors.)

3. Let $k = 6$ and suppose that all factors have two levels. Consider the fractional factorial design given by $x_1 + x_2 + x_4 = 0, x_1 + x_3 + x_5 = 0$ and $x_2 + x_3 + x_6 = 0$. Find the sum of the three pairs of equations and the sum of all three equations. Hence give the resolution of the design.

4. Give the 9 treatment combinations in the regular 3-level design with $k = 4$ factors.

5. List all the quadratic polynomials over $GF[3]$. Hence construct the Galois field $GF[9]$.

6. Use Construction 2.2.1 to construct a resolution 3 design for 6 factors each with 5 levels.

7. Proof that the construction given in Construction 2.2.2 works.

2.3 IRREGULAR FRACTIONS

When they exist, the defining equations of a regular fraction provide a convenient way of summarizing the treatment combinations in a fraction. For asymmetric factorials, it is not possible to define fractions in such a neat way; instead, we must list explicitly the treatment combinations in the fraction. Even for a symmetric factorial, it is sometimes more convenient to list the treatment combinations rather than just the defining contrasts.

We begin by considering the properties that the treatment combinations in regular fractions have and consider how to apply these to asymmetric factorials.

We first consider fractions of resolution 3. Because the treatment combinations in a regular fraction are the solutions to a set of independent linear equations, each with at least three non-zero coefficients, each level of each factor appears equally often. So this is one feature that we would like to be true for any irregular fraction as well.

In fact, since each of the independent linear equations has at least three non-zero coefficients in it, the levels of any two factors can be specified and there are the same number of solutions to the equations with the specified levels of these two factors. This means that, in the regular fraction for any two factors, each pair of levels appears in the same number of treatment combinations. This is the second feature that we would like for an irregular fraction.

If an irregular fraction has these two features then the main effects contrasts from any two factors will be orthogonal, and so the corresponding effects will be independently estimated. Thus irregular fractions in which each level of each factor appears equally often and any pair of levels from any two factors appears equally often are the natural generalization of regular fractions of resolution 3. In such fractions, the main effects can be estimated independently.

In a similar way, we can generalize the features of a regular fraction of resolution 5. In addition to the two features discussed above, we know that the defining equations for a design of resolution 5 all have at least 5 non-zero coefficients. Thus we can independently specify levels for 4 of the factors, and there will be the same number of treatment combinations for any such combination of levels for each set of four factors. This becomes the third feature that we need to get a fraction that is a generalization of a regular fraction of resolution 5. Because for each pair of factors each possible level combination appears with each possible level combination for any other pair of factors, the two-factor interactions can be estimated independently of each other.

For symmetric designs we can formalize these observations in the following definition which is a natural generalization of the requirements for fractions of resolution 3 and 5.

An *orthogonal array* OA$[N, k, \ell, t]$ is a $N \times k$ array with elements from a set of ℓ symbols such that any $N \times t$ subarray has each t-tuple appearing as a row N/ℓ^t times. Often N/ℓ^t is called the *index* of the array, t the *strength* of the array, k is the *number of constraints* and ℓ is the *number of levels*. The fractional factorial design in Table 2.5 is an example of an OA with $k = 4$ and $n = 8$. We see that in each column there are 4 0s and 4 1s. In any pair of columns, there are 2 copies of each of the pairs (0,0), (0,1), (1,0) and (1,1); thus the array has strength 2. We know that this array has resolution 3 since the defining equation has three non-zero coefficients. This illustrates a general result: An orthogonal array of strength t is a fractional factorial design of resolution $t+1$. To establish that a design is of resolution 4, it may be easier to establish that any set of three columns has each of the possible ordered triples appearing as rows equally often. Similarly, a design has resolution 5 if any set of four columns has each of the possible ordered quadruples appearing as rows equally often.

This gives us a definition that we can easily generalize to asymmetric factorials. The estimability properties of these asymmetric orthogonal arrays are the same as those of symmetric orthogonal arrays of the same strength; see Hedayat et al. (1999) for a formal proof.

An *asymmetric orthogonal array* OA$[N; \ell_1, \ell_2, \dots, \ell_k; t]$ is a $N \times k$ array with elements from a set of ℓ_q symbols in column q such that any $N \times t$ subarray has each t-tuple appearing as a row an equal number of times. Such an array is said to have *strength* t.

We will usually use "orthogonal array" for either an asymmetric or a symmetric array.

Orthogonal arrays of strength two are a subset of the class of orthogonal main effect plans. We let n_{xq} be the number of times that level x appears in column q of the array. A k factor, N run, ℓ_q-level, $1 \leq q \leq k$, *orthogonal main effects plan* (OMEP) is an $N \times k$ array with symbols $0, 1, \dots, \ell_q - 1$ in column q such that, for any pair of columns q and p, the number of times that the ordered pair (x, y) appears in the columns is $n_{xq}n_{yp}/N$.

It can be shown that the main effects can be estimated orthogonally from the results of an OMEP; see Dey (1985).

If we represent the two levels of each binary factor in an OMEP by -1 and 1 then the inner product of any two binary columns of the OMEP is 0.

Sometimes several factors will have the same number of levels and this is often indicated by powers. So an OA[32;2,2,2,4,4;4] is written as OA[32;2^3,4^2;4]. Another common notation for an OA or an OMEP is to use $\ell_1 \times \ell_2 \times \ldots \times \ell_k // N$ for an OA[$N; \ell_1, \ell_2, \ldots, \ell_k; t$], most often when $t = 2$ or the fact that $t > 2$ is not relevant.

2.3.1 Two Constructions for Symmetric OAs

Various constructions for orthogonal arrays have been found; a nice summary is given in Hedayat et al. (1999). We give two of the most useful constructions in this section.

The next construction gives a resolution 3 design with $2\ell^2$ treatment combinations for up to $2\ell + 1$ factors when ℓ is odd.

■ **CONSTRUCTION 2.3.1.**
If ℓ is an odd prime or prime power then there is an $OA[2\ell^2, 2\ell + 1, \ell, 2]$.

Proof. Let the elements in $GF[\ell] = \{\alpha_0 = 0, \alpha_1 = 1, \alpha_2, \ldots, \alpha_{\ell-1}\}$. The design is constructed in two parts. As in Sections 2.2.1 and 2.2.2 we let $\mathbf{b}_1$ be a vector of length ℓ^2 with the first ℓ entries equal to α_0, the next ℓ entries equal to α_1 and so on until the final ℓ entries are equal to $\alpha_{\ell-1}$. We let $\mathbf{b}_2$ have first entry α_0, second entry α_1 and so on until entry ℓ is $\alpha_{\ell-1}$ and these entries are repeated in order ℓ times. For the first ℓ^2 treatment combinations we use $\mathbf{b}_1$ for the levels of the first factor, $\mathbf{b}_2$ for the levels of the second factor, and $\mathbf{b}_1 + \alpha_q \mathbf{b}_2$ for the levels of the $(q + 2)$th factor, $q = 1, 2, \ldots, \ell - 1$. To get the levels for the remaining ℓ factors, we use $\mathbf{b}_1^2 + \alpha_q \mathbf{b}_1 + \mathbf{b}_2$, $q = 0, \ldots, \ell - 1$. For the remaining ℓ^2 treatment combinations, we use $\mathbf{b}_1$ for the levels of the first factor, $\mathbf{b}_2$ for the levels of the second factor, $\mathbf{b}_1 + \alpha_q \mathbf{b}_2 + \nu_q$, $q = 1, \ldots, \ell - 1$, for the levels of the next $\ell - 1$ factors and $\theta \mathbf{b}_1^2 + \mathbf{b}_2$, $\theta \mathbf{b}_1^2 + \theta_q \mathbf{b}_1 + \mathbf{b}_2 + \rho_q$, $q = 1, \ldots, \ell - 1$, for the levels of the final ℓ factors, where we have to determine $\nu_1, \nu_2, \ldots, \nu_{\ell-1}, \rho_1, \rho_2, \ldots, \rho_{\ell-1}, \theta, \theta_1, \theta_2, \ldots, \theta_{\ell-1}$.

We let θ be any non-square element of $GF[\ell]$. We let $\nu_q = (\theta - 1)(4\theta\alpha_q)^{-1}$, $\theta_q = \theta\alpha_q$ and $\rho_q = \alpha_q^2(\theta - 1)4^{-1}$, $q = 1, 2, \ldots, \ell - 1$. □

The result sometimes holds for ℓ even; see Hedayat et al. (1999) (p. 47) for a discussion and another way of constructing OAs with 2^n symbols. The smallest design for ℓ a power of 2 that would be given by this construction is an OA with 9 factors each with 4 levels with $N = 32$.

■ **EXAMPLE 2.3.1.**
Let $\ell = 3$. Since 2 is not a square in $GF[3]$, we let $\theta = 2$. So we have $\nu_1 = (4 \times 2 \times 1)^{-1} = 2$, $\nu_2 = (4 \times 2 \times 2)^{-1} = 1$, $\theta_1 = 2$, $\theta_2 = 1$, $\rho_1 = 4 - 1 = 1$ and $\rho_2 = 4^{-1} = 1$. Thus the treatment combinations in the second half are given by $\mathbf{b}_1$, $\mathbf{b}_2$, $\mathbf{b}_1 + \mathbf{b}_2 + 2$, $\mathbf{b}_1 + 2\mathbf{b}_2 + 1$, $2\mathbf{b}_1^2 + \mathbf{b}_2$, $2\mathbf{b}_1^2 + 2\mathbf{b}_1 + \mathbf{b}_2 + 1$ and $2\mathbf{b}_1^2 + \mathbf{b}_1 + \mathbf{b}_2 + 1$. The final 18 treatment combinations are given in Table 2.20. □

The next result is a special case of a result in Bush (1952) and gives an orthogonal array of strength 4, equivalently a resolution 5 fractional factorial design.

■ **CONSTRUCTION 2.3.2.**
If ℓ is a prime or a prime power and $\ell \geq 3$, then there is an $OA[\ell^4, \ell + 1, \ell, 4]$.

Table 2.20 A Resolution 3 Fractional Factorial Design for Seven 3-Level Factors

0	0	0	0	0	0	0
0	1	1	2	1	1	1
0	2	2	1	2	2	2
1	0	1	1	1	2	0
1	1	2	0	2	0	1
1	2	0	2	0	1	2
2	0	2	2	1	0	2
2	1	0	1	2	1	0
2	2	1	0	0	2	1
0	0	2	1	0	1	1
0	1	0	0	1	2	2
0	2	1	2	2	0	0
1	0	0	2	2	2	1
1	1	1	1	0	0	2
1	2	2	0	1	1	0
2	0	1	0	2	1	2
2	1	2	2	0	2	0
2	2	0	1	1	0	1

Proof. Use the elements of $GF[\ell]$ to label the columns of the array and the polynomials of degree at most 3 to label the rows of the array. Suppose that ψ_i is the polynomial associated with row i and that field element α_j is associated with column j. Then the (i,j) entry is $\psi_i(\alpha_j)$. The final column contains the coefficient of x^3 in ψ_i. The verification that this array has the desired properties is left as an exercise. □

■ **EXAMPLE 2.3.2.**
Let $\ell = 3$. Then we can use Construction 2.3.2 to construct an OA[81,4,3,4]. There are three polynomials of order 0: 0, 1 and 2. There are 6 polynomials of order 1: x, $x+1$, $x+2$, $2x$, $2x+1$ and $2x+2$. There are 18 polynomials of order 2: the 9 with leading coefficient 1 are x^2, x^2+1, x^2+2, x^2+x, x^2+x+1, x^2+x+2, x^2+2x, x^2+2x+1 and x^2+2x+2. There are a further 9 quadratics with leading coefficient equal to 2. Finally each of these 27 polynomials can have x^3 or $2x^3$ added to it to give the 81 polynomials required in the construction. □

2.3.2 Constructing OA[2^k; 2^{k_1}, 4^{k_2}; 4]

Addelman (1972) gave some useful constructions for these designs. He described each 4-level factor by three 2-level generators and he gave a set of generators for the 2-level factors. This set had to be such that no linear combination of two or three of the generators in the set was also in the set.

We give the six designs that he gave in Table 2.21.

■ **EXAMPLE 2.3.3.**
To construct an OA[64;2^3,4^2;4] we need the generators of length 64. So $\mathbf{b}_1$ has 32 0s and then 32 1s, $\mathbf{b}_2$ has 16 0s, then 16 1s then 16 0s then 16 1s and so on until $\mathbf{b}_6$ which alternates 0s and 1s. Then we use $\mathbf{b}_5$, $\mathbf{b}_6$ and $\mathbf{b}_1+\mathbf{b}_3+\mathbf{b}_5+\mathbf{b}_6$ to determine the 2-level factors and

Table 2.21 Generators for $OA[2^k; 2^{k_1}, 4^{k_2}; 4]$

OA	Two-Level Factors	Four-Level Factors
$OA[32;2^4,4;4]$	$\mathbf{b}_3, \mathbf{b}_4, \mathbf{b}_5$, $\mathbf{b}_1 + \mathbf{b}_3 + \mathbf{b}_4 + \mathbf{b}_5$	$(\mathbf{b}_1, \mathbf{b}_2, \mathbf{b}_1 + \mathbf{b}_2)$
$OA[64;2^6,4;4]$	$\mathbf{b}_3, \mathbf{b}_4, \mathbf{b}_5, \mathbf{b}_6$ $\mathbf{b}_1 + \mathbf{b}_3 + \mathbf{b}_4 + \mathbf{b}_5$, $\mathbf{b}_2 + \mathbf{b}_3 + \mathbf{b}_4 + \mathbf{b}_6$	$(\mathbf{b}_1, \mathbf{b}_2, \mathbf{b}_1 + \mathbf{b}_2)$
$OA[64;2^3,4^2;4]$	$\mathbf{b}_5, \mathbf{b}_6$, $\mathbf{b}_1 + \mathbf{b}_3 + \mathbf{b}_5 + \mathbf{b}_6$	$(\mathbf{b}_1, \mathbf{b}_2, \mathbf{b}_1 + \mathbf{b}_2)$ $(\mathbf{b}_3, \mathbf{b}_4, \mathbf{b}_3 + \mathbf{b}_4)$
$OA[128;2^9,4;4]$	$\mathbf{b}_3, \mathbf{b}_4, \mathbf{b}_5, \mathbf{b}_6, \mathbf{b}_7$, $\mathbf{b}_1 + \mathbf{b}_3 + \mathbf{b}_4 + \mathbf{b}_5$, $\mathbf{b}_1 + \mathbf{b}_3 + \mathbf{b}_6 + \mathbf{b}_7$, $\mathbf{b}_2 + \mathbf{b}_3 + \mathbf{b}_4 + \mathbf{b}_6$, $\mathbf{b}_2 + \mathbf{b}_4 + \mathbf{b}_5 + \mathbf{b}_7$	$(\mathbf{b}_1, \mathbf{b}_2, \mathbf{b}_1 + \mathbf{b}_2)$
$OA[128;2^6,4^2;4]$	$\mathbf{b}_5, \mathbf{b}_6, \mathbf{b}_7$, $\mathbf{b}_1 + \mathbf{b}_3 + \mathbf{b}_5 + \mathbf{b}_6$, $\mathbf{b}_1 + \mathbf{b}_4 + \mathbf{b}_5 + \mathbf{b}_7$, $\mathbf{b}_2 + \mathbf{b}_3 + \mathbf{b}_6 + \mathbf{b}_7$	$(\mathbf{b}_1, \mathbf{b}_2, \mathbf{b}_1 + \mathbf{b}_2)$ $(\mathbf{b}_3, \mathbf{b}_4, \mathbf{b}_3 + \mathbf{b}_4)$
$OA[128;2^3,4^3;4]$	$\mathbf{b}_7$, $\mathbf{b}_1 + \mathbf{b}_3 + \mathbf{b}_5 + \mathbf{b}_7$, $\mathbf{b}_2 + \mathbf{b}_4 + \mathbf{b}_6 + \mathbf{b}_7$	$(\mathbf{b}_1, \mathbf{b}_2, \mathbf{b}_1 + \mathbf{b}_2)$ $(\mathbf{b}_3, \mathbf{b}_4, \mathbf{b}_3 + \mathbf{b}_4)$ $(\mathbf{b}_5, \mathbf{b}_6, \mathbf{b}_5 + \mathbf{b}_6)$

use $\mathbf{b}_1$, $\mathbf{b}_2$, $\mathbf{b}_1 + \mathbf{b}_2$ and $\mathbf{b}_3$, $\mathbf{b}_4$, $\mathbf{b}_3 + \mathbf{b}_4$ for the two 4-level factors. The resulting design is given Table 2.22. □

Table 2.22 The OA[64;2^3,4^2;4]

0 0 0 0 0	0 1 1 0 0	1 0 1 0 0	1 1 0 0 0
0 0 0 0 1	0 1 1 0 1	1 0 1 0 1	1 1 0 0 1
0 0 1 0 2	0 1 0 0 2	1 0 0 0 2	1 1 1 0 2
0 0 1 0 3	0 1 0 0 3	1 0 0 0 3	1 1 1 0 3
0 0 0 1 0	0 1 1 1 0	1 0 1 1 0	1 1 0 1 0
0 0 0 1 1	0 1 1 1 1	1 0 1 1 1	1 1 0 1 1
0 0 1 1 2	0 1 0 1 2	1 0 0 1 2	1 1 1 1 2
0 0 1 1 3	0 1 0 1 3	1 0 0 1 3	1 1 1 1 3
0 0 1 2 0	0 1 0 2 0	1 0 0 2 0	1 1 1 2 0
0 0 1 2 1	0 1 0 2 1	1 0 0 2 1	1 1 1 2 1
0 0 0 2 2	0 1 1 2 2	1 0 1 2 2	1 1 0 2 2
0 0 0 2 3	0 1 1 2 3	1 0 1 2 3	1 1 0 2 3
0 0 1 3 0	0 1 0 3 0	1 0 0 3 0	1 1 1 3 0
0 0 1 3 1	0 1 0 3 1	1 0 0 3 1	1 1 1 3 1
0 0 0 3 2	0 1 1 3 2	1 0 1 3 2	1 1 0 3 2
0 0 0 3 3	0 1 1 3 3	1 0 1 3 3	1 1 0 3 3

2.3.3 Obtaining New Arrays from Old

Given an orthogonal array, symmetric or asymmetric, it is possible to get other arrays from it by deleting one, or more, factors and by equating some of the levels for one, or more, of the factors. Sometimes it is possible to replace the entries in one factor by several factors, each with fewer levels, and sometimes it is possible to combine several factors to get one factor with more levels. In this section we will discuss each of these situations in turn.

■ **CONSTRUCTION 2.3.3. Collapsing Levels**
Consider an orthogonal array with a factor with ℓ_q levels. Suppose that we want to make this into a factor with $\ell_s < \ell_q$ levels. We can do this by changing level ℓ_s to 0, $\ell_s + 1$ to 1, $\ell_s + 2$ to 2, and so on, until all the levels in the original factor have been changed to ones for the new factor. □

This way of making the changes guarantees that each level of the new factor appears about the same number of times (exactly the same number of times if $\ell_s | \ell_q$) and the properties of the original array guarantee that the final array is an OMEP. But any collapsing that results in levels appearing about the same number of times is just as good.

This procedure is almost impossible to reverse.

■ **EXAMPLE 2.3.4.**
Consider the array in Table 2.20. Suppose that we want an array with 7 factors, 3 with two levels and 4 with three levels. Then we could collapse the levels in the first three factors to get the OMEP shown in Table 2.23. Note that the first three factors have level 0 repeated 12 times and level 1 repeated 6 times. □

The next construction only applies to orthogonal arrays of strength 2.

Table 2.23 A Resolution 3 Fractional Factorial Design for Three 2-Level Factors and Four 3-Level Factors

0	0	0	0	0	0	0
0	1	1	2	1	1	1
0	0	0	1	2	2	2
1	0	1	1	1	2	0
1	1	0	0	2	0	1
1	0	0	2	0	1	2
0	0	0	2	1	0	2
0	1	0	1	2	1	0
0	0	1	0	0	2	1
0	0	0	1	0	1	1
0	1	0	0	1	2	2
0	0	1	2	2	0	0
1	0	0	2	2	2	1
1	1	1	1	0	0	2
1	0	0	0	1	1	0
0	0	1	0	2	1	2
0	1	0	2	0	2	0
0	0	0	1	1	0	1

■ **CONSTRUCTION 2.3.4. Expansive Replacement**
Suppose that we have an orthogonal array of strength 2 in which there is one factor with ℓ_1 levels. Suppose that there is an orthogonal array of strength 2 which has ℓ_1 runs. Label these runs from 0 to $\ell_1 - 1$. Then we can replace each level of the factor with ℓ_1 levels in the first array by the corresponding run of the second array. □

The name arises because the new array has more factors than the original array.

■ **EXAMPLE 2.3.5.**
Consider the array in Table 2.24. Suppose that we want to replace one of the factors with the runs from the OA[4,3,2,2]. The runs of this array are given in Table 2.25. The resulting OA[16;2,2,2,4,4,4,4;2] is given in Table 2.26. If we make this replacement for all five of the 4-level factors, then we get the array in Table 2.27. □

The most common examples of expansive replacement are replacement of a factor with 4 levels by three factors each with 2 levels and replacement of a factor with 8 levels by seven factors each with 2 levels.

It is sometimes possible to replace several factors with one factor with more levels. Before we describe this *contractive replacement* we need to define the idea of a a *tight* or *saturated* orthogonal array. In a *tight* or *saturated* orthogonal array with k factors, $N = 1 + \sum_{q=1}^{k}(\ell_q - 1)$. Thus we see that there is no room for any more factors.

■ **CONSTRUCTION 2.3.5. Contractive Replacement**
Let A be an OA[$N; \ell_1, \ell_2, \ldots, \ell_k; 2$] such that the first s columns of A form N/N_1 copies of an OA[$N_1; \ell_1, \ell_2, \ldots, \ell_s; 2$], B say, that is tight. Use $0, 1, \ldots, N_1 - 1$ to label the rows of B, and then use these labels to replace the first s columns of A. The resulting design is an OA[$N; N_1, \ell_{s+1}, \ldots, \ell_k; 2$]. □

Table 2.24 An OA[16,5,4,2]

0	0	0	0	0
0	1	1	2	3
0	2	2	3	1
0	3	3	1	2
1	0	1	1	1
1	1	0	3	2
1	2	3	2	0
1	3	2	0	3
2	0	2	2	2
2	1	3	0	1
2	2	0	1	3
2	3	1	3	0
3	0	3	3	3
3	1	2	1	0
3	2	1	0	2
3	3	0	2	1

Table 2.25 An OA[4,3,2,2]

0	0	0
0	1	1
1	0	1
1	1	0

Table 2.26 An OA[16;2,2,2,4,4,4,4;2]

0	0	0	0	0	0	0
0	0	0	1	1	2	3
0	0	0	2	2	3	1
0	0	0	3	3	1	2
0	1	1	0	1	1	1
0	1	1	1	0	3	2
0	1	1	2	3	2	0
0	1	1	3	2	0	3
1	0	1	0	2	2	2
1	0	1	1	3	0	1
1	0	1	2	0	1	3
1	0	1	3	1	3	0
1	1	0	0	3	3	3
1	1	0	1	2	1	0
1	1	0	2	1	0	2
1	1	0	3	0	2	1

Table 2.27 An OA[16,15,2,2]

0	0	0	0	0	0	0	0	0	0	0	0	0	0	0
0	0	0	0	1	1	0	1	1	1	0	1	1	1	0
0	0	0	1	0	1	1	0	1	1	1	0	0	1	1
0	0	0	1	1	0	1	1	0	0	1	1	1	0	1
0	1	1	0	0	0	0	1	1	0	1	1	0	1	1
0	1	1	0	1	1	0	0	0	1	1	0	1	0	1
0	1	1	1	0	1	1	1	0	1	0	1	0	0	0
0	1	1	1	1	0	1	0	1	0	0	0	1	1	0
1	0	1	0	0	0	1	0	1	1	0	1	1	0	1
1	0	1	0	1	1	1	1	0	0	0	0	0	1	1
1	0	1	1	0	1	0	0	0	0	1	1	1	1	0
1	0	1	1	1	0	0	1	1	1	1	0	0	0	0
1	1	0	0	0	0	1	1	0	1	1	0	1	1	0
1	1	0	0	1	1	1	0	1	0	1	1	0	0	0
1	1	0	1	0	1	0	1	1	0	0	0	1	0	1
1	1	0	1	1	0	0	0	0	1	0	1	0	1	1

■ EXAMPLE 2.3.6.

Consider the design in Table 2.26. The first three columns form four copies of a tight OA[4, 3, 2, 2] and so replacing 000 with 0, 011 with 1, 101 with 2 and 110 with 3 gives the OA[16,5,4,2] in Table 2.24. □

The same construction will work if the array A could have columns adjoined so that B would be tight. The construction does not work if the original array is not tight, or cannot be made tight, or if only some columns of the tight subarray are replaced, as the next example shows.

■ EXAMPLE 2.3.7.

Consider the first two columns of the orthogonal array in Table 2.26. Then these two columns form 16/4=4 copies of an OA[4,2,2,2], although it is not tight since

$$1 + (\ell_1 - 1) + (\ell_2 - 1) = 1 + 1 + 1 = 3 \neq 4.$$

If we replace 00 by 0, 01 by 1, 10 by 2 and 11 by 3 we see that the 0s in the third column of the original array only appear with 0 and 2 in the proposed new 4-level column. This is because the third column is the binary sum of the first two columns and so its levels are not independently determined. The first three columns of the array do form a tight OA[4,3,2,2]. □

By juxtaposing several copies of an array it is possible to get a larger array with $k+1$ factors.

■ CONSTRUCTION 2.3.6. Adding One More Factor

Take any OA[$N;\ell_1,\ell_2,\ldots,\ell_k;t$] and write down ℓ_{k+1} copies of the array, one above the other, and then adjoin a $(k+1)$st factor with N copies of 0, then N copies of 1, and so on, until there are N copies of $\ell_{k+1}-1$. The result is an OA$[N\times\ell_{k+1};\ell_1,\ell_2,\ldots,\ell_k,\ell_{k+1};t]$. □

The levels of the last factor appear equally often with every pair of levels for any other two factors and so the array has strength 3 with respect to the last factor; this property is useful when constructing an OA to estimate two-factor interactions all of which involve one factor.

■ EXAMPLE 2.3.8.
Consider the OA[4,3,2,2] given in Table 2.25. By writing down three copies of this array and adjoining one further column for the three-level factor, we obtain the OA[12;2,2,2,3;2] in Table 2.28. □

Table 2.28 An OA[12;2,2,2,3;2]

0	0	0	0
0	1	1	0
1	0	1	0
1	1	0	0
0	0	0	1
0	1	1	1
1	0	1	1
1	1	0	1
0	0	0	2
0	1	1	2
1	0	1	2
1	1	0	2

The next construction extends this idea by adding several factors simultaneously.

■ CONSTRUCTION 2.3.7. Juxtaposing Two OAs
Take an OA[$N_1; \ell_{11}, \ell_{12}, \ldots, \ell_{1k}; t$] and an OA[$N_2; \ell_{21}, \ell_{22}, \ldots, \ell_{2k}; t$] and write down N_2 copies of the first array, one above the other. Adjoin the first row of the second array to the first copy of the first array, adjoin the second row of the second array to the second copy of the first array and so on. The result is an OA[$N_1N_2; \ell_{11}, \ell_{12}, \ldots, \ell_{1k}, \ell_{21}, \ell_{22}, \ldots, \ell_{2k}; t$]. □

In an OA obtained from Construction 2.3.7 there are several sets of three factors in which every triple of levels appears equally often. These include every set of two factors from the original OA and one factor from the adjoined OA and every set with one factor from the original OA and two factors from the adjoined OA.

■ EXAMPLE 2.3.9.
Consider the OA[4;3,2;2] given in Table 2.25. Use this as both the original OA and as the adjoined OA. Then we get the OA[16;6,2;2] in Table 2.29. Notice that the only two sets of three factors which do not contain all the triples of levels are factors 1, 2, and 3 and factors 4, 5, and 6. □

■ CONSTRUCTION 2.3.8. Obtaining One Factor with Many Levels
Take any orthogonal array OA[$N; \ell_1, \ell_2, \ldots, \ell_k; t$] and write down ℓ copies of the array, one above the other, and then leave the levels of the kth factor unaltered in the first copy of the array, use ℓ_k different levels in the second copy of the array, use a further ℓ_k levels in the third copy of the array and so on. The resulting array is an OA[$N\ell; \ell_1, \ell_2, \ldots, \ell_{k-1}, \ell\ell_k; t$]. □

Table 2.29 An OA[16;6,2;2]

0	0	0	0	0	0
0	1	1	0	0	0
1	0	1	0	0	0
1	1	0	0	0	0
0	0	0	0	1	1
0	1	1	0	1	1
1	0	1	0	1	1
1	1	0	0	1	1
0	0	0	1	0	1
0	1	1	1	0	1
1	0	1	1	0	1
1	1	0	1	0	1
0	0	0	1	1	0
0	1	1	1	1	0
1	0	1	1	1	0
1	1	0	1	1	0

■ **EXAMPLE 2.3.10.**
The OA[12;2,2,6;2] in Table 2.30 is obtained by writing down three copies of the OA[4,3,2,2] given in Table 2.25 and replacing the final column with three distinct sets of two symbols for the final factor. □

Table 2.30 An OA[12;2,2,6;2]

0	0	0
0	1	1
1	0	1
1	1	0
0	0	2
0	1	3
1	0	3
1	1	2
0	0	4
0	1	5
1	0	5
1	1	4

The next construction shows how to remove an unrealistic (or unwanted) treatment combination from a fractional factorial.

■ **CONSTRUCTION 2.3.9. Avoiding an Unrealistic Treatment Combination**
Given an orthogonal array OA[$N;\ell_1,\ell_2,\ldots,\ell_{k-1},\ell_k;t$] which contains an unrealistic treatment combination, the treatment combination can be removed by adding another treatment combination to every treatment combination in the array. This addition is done component-wise mod ℓ_q in position q. □

The treatment combination to add needs to be chosen so all the resulting treatment combinations are acceptable; often such a treatment combination can only be found by trial and error.

■ **EXAMPLE 2.3.11.**
Suppose that we want an OA[12;2,2,6;2] without the treatment combination 000. Then we need to add a treatment combination to those in the array in Table 2.30 so that 000 is avoided. This means that we need to add some treatment combination that is not the negative of any of the treatment combinations that are already there. So we try using 113. This gives the array in Table 2.31. □

Table 2.31 An OA[12;2,2,6;2] without 000

1	1	3
1	0	4
0	1	4
0	0	3
1	1	5
1	0	0
0	1	0
0	0	5
1	1	1
1	0	2
0	1	2
0	0	1

2.3.4 Exercises

1. Use Construction 2.3.2 to construct an OA[81,4,3,4].

2. Prove that the array constructed in Construction 2.3.2 has the desired properties.

3. Consider the design in Table 2.32.

   (a) Verify that the design is of resolution 3.

   (b) Is it resolution 4?

   (c) Show that the final three columns can be replaced by one column with 4 levels.

   (d) Are there any other sets of three columns for which you could make this replacement?

   (e) In particular, can you write the design as a resolution 3 array with two factors each with 4 levels?

4. Construct an OA[32;2^4,4;4].

Table 2.32 An OA[8,6,2,2]

0	0	0	0	0	0
0	1	1	0	1	1
1	0	1	1	0	1
1	1	0	1	1	0
0	0	1	1	1	0
0	1	0	1	0	1
1	0	0	0	1	1
1	1	1	0	0	0

2.4 OTHER USEFUL DESIGNS

There are various other combinatorial designs that have been used in the construction of stated preference choice experiments. We collect the definitions here for convenience.

Consider a set of v items. From this set of v items construct b subsets, or *blocks*, each with u distinct items. Suppose that any two items appear together in exactly λ of the b blocks. Then the set of b blocks form a *balanced incomplete block design* (BIBD). Let r_i be the number of times that item T_i appears in the BIBD. If we count the pairs that item T_i appears in the BIBD we have that T_i appears in r_i blocks and there are $u - 1$ other items in the blocks. So there are $r_i(u - 1)$ pairs involving item T_i. On the other hand there are $v - 1$ other items in the BIBD and item T_i appears with each of these λ times in the BIBD. So there are $\lambda(v - 1)$ pairs involving item T_i. Equating we get that $r_i(u - 1) = \lambda(v - 1)$ and so we see that all of the items appear equally often in the BIBD. We call this replication number r and talk about a (v, b, r, u, λ) BIBD. If $v = b$ then $r = u$ and the design is said to be a *symmetric* BIBD (SBIBD), written (v, r, λ). Tables of balanced incomplete block designs may be found in Mathon and Rosa (2006) and Abel and Greig (2006).

■ **EXAMPLE 2.4.1.**

Let $v = 7$ and consider the blocks in Table 2.33. These blocks form a BIBD on 7 items with 7 blocks each of size 3 and with each item appearing in 3 blocks. There is a unique block which contains every pair of items so $\lambda = 1$. □

Table 2.33 The Blocks a (7,3,1) BIBD

1	2	3
1	4	5
1	6	7
2	4	6
2	5	7
3	4	7
3	5	6

One easy way to construct BIBDs is to use *difference sets*. Let $X = \{x_1, x_2, \ldots, x_u\}$ be a set of integers modulo v. Suppose that every non-zero value mod v can be represented as a difference $x_i - x_j$, $x_i, x_j \in X$, in exactly λ ways. Then X is said to be a (v, u, λ) *difference set*.

■ EXAMPLE 2.4.2.

Let $v = 7$ and $u = 3$ and let $X = \{0, 1, 3\}$. Then X is a (7,3,1)-difference set since each of the non-zero integers mod 7 can be represented as a difference in exactly one way using the elements from X. In fact we have $1 = 1 - 0$, $2 = 3 - 1$, $3 = 3 - 0$, $4 = 0 - 3$, $5 = 1 - 3$ and $6 = 0 - 1$. □

We give some small difference sets in Table 2.34. More extensive tables of difference sets are given by Jungnickel et al. (2006).

Table 2.34 Some Small Difference Sets

v	u	λ	Set
7	3	1	0,1,3
7	4	2	0,1,2,4
13	4	1	0,1,3,9
21	5	1	3,6,7,12,14
11	5	2	1,3,4,5,9
11	6	3	0,2,6,7,8,10
15	7	3	0,1,2,4,5,8,10

Of course, if $X = \{x_1, x_2, \ldots, x_u\}$ is a (v, u, λ) difference set then so is

$$\{x_1 + 1, x_2 + 1, \ldots, x_u + 1\}.$$

It is also true that $\bar{X} = \{0, 1, \ldots, v - 1\} \backslash X$ is an $(v, v - u, v - 2u + \lambda)$ difference set. We can see that this is true by counting the number of times that each non-zero value mod v appears as a difference. The total number of differences in $\bar{X}$ is $(v - u)(v - u - 1)$. The total number of differences in X is $u(u - 1)$ and each of the $v - 1$ non-zero values mod v appear as a difference λ times so $u(u - 1) = \lambda(v - 1)$. Thus we see that

$$\begin{aligned}
(v-u)(v-u-1) &= v(v-1-u) - u(v-1-u) \\
&= v(v-1) - vu - uv - +u(u+1) \\
&= v(v-1) + \lambda(v-1) + 2u - 2uv \\
&= v(v-1) + \lambda(v-1) - 2u(v-1) \\
&= (v-1)(v+\lambda-2u).
\end{aligned}$$

Since $\{0, 1, \ldots, v - 1\}$ is a (v, v, v) difference set the result follows.

It is not essential for a difference set to be defined over the integers mod v. Any group can be used although the only extension that we will make here is to give one difference set defined on ordered pairs where differences on each element of the pair are evaluated mod 4. Thus there are 16 levels in total and each is represented by an ordered pair. If $u = 6$ then the ordered pairs $\{(0,0), (0,1), (1,0), (1,2), (2,0), (2,3)\}$ form a (16,6,2) difference set. For example we see that the element (1,3) arises as a difference twice: $(2,3) - (1,0)$ and $(1,0) - (0,1)$.

A further extension is to have several sets and look at the differences across all the sets. Formally suppose that $X_i = \{x_{i1}, x_{i2}, \ldots, x_{iu}\}$ is a set of integers modulo v. Then

$X_1, X_2, \ldots, X_f$ form a (v, u, λ) *difference family* if every non-zero value mod v can be represented as a difference $x_{ia} - x_{ib}$, $x_{ia}, x_{ib} \in X_i$, $1 \leq i \leq f$, in exactly λ ways.

EXAMPLE 2.4.3.
The sets (0,1,4) and (0,2,7) form a (13,3,1) difference family. □

Table 2.35 contains some small difference families.

Table 2.35 Some Small Difference Families

v	u	λ	Sets				
13	3	1	0,1,4	0,2,7			
19	3	1	0,1,4	0,2,9	0,5,11		
16	3	2	0,1,2	0,2,8	0,3,7	0,4,7	0,5,10
11	4	6	0,1,8,9	0,2,5,7	0,1,4,5	0,2,3,5	0,4,5,9
15	4	6	0,1,2,3	0,2,4,6	0,4,8,12	0,1,8,9	
			0,3,6,9	0,1,5,10	0,2,5,10		
19	4	2	0,1,3,12	0,1,5,13	0,4,6,9		
13	5	5	0,1,2,4,8	0,1,3,6,12	0,2,5,6,10		
17	5	5	0,1,4,13,16	0,3,5,12,14	0,2,8,9,15	0,6,7,10,11	

A *Hadamard matrix* of order h is an $h \times h$ matrix with entries 1 and -1 satisfying $HH' = I_h$. A table of Hadamard matrices is maintained by Sloane (2006a).

EXAMPLE 2.4.4.
Here is a Hadamard matrix of order 4.

$$\begin{bmatrix} 1 & 1 & 1 & 1 \\ 1 & 1 & -1 & -1 \\ 1 & -1 & 1 & -1 \\ 1 & -1 & -1 & 1 \end{bmatrix}$$

□

2.5 TABLES OF FRACTIONAL FACTORIAL DESIGNS AND ORTHOGONAL ARRAYS

There are two extensive tables of orthogonal arrays available on the web. One is the table of orthogonal arrays maintained by Sloane (2006b). This website uses "oa.N.k.s.t.name" to denote an orthogonal array with N runs, k factors, s levels (for each of the factors), and strength t (equivalently resolution $t + 1$). This site sometimes gives mathematically inequivalent designs with the same parameters and it makes no claim that it lists all possible parameters, even for small designs. It has many designs of resolution 3 and some designs for larger resolutions.

The other tables are maintained by Kuhfeld (2006) and list all parent designs of resolution 3 with up to 143 treatment combinations as well as many designs with more. A *parent* design is an orthogonal array in which no further contractive replacement is possible. As well as listing the parent arrays, Kuhfeld has a list of the number of parent designs for a given value of N and the number of designs that exist for that value of N. For example, there are 4 parent designs and 4 designs with $N = 12$ (OA[12,11,2,2],

OA[12;2,2,2,2,3;2], OA[12;2,2,6;2] and OA[12;3,4;2]), while there are 2 parent designs with $N = 16$ (OA[16;2,2,2,2,2,2,2,2,8;2] and OA[16,5,4,2]) but 7 designs obtainable from them by expansive replacement. This site has no designs of resolution other than 3.

When trying to find a design that you need from these or any other tables, remember that you can omit factors without affecting the resolution of the design. Frequently, designs that are not tight are not listed and such designs are often what is required. Thus you need to think in terms of collapsing levels and omitting factors from the tabulated designs to get the ones that you want.

2.5.1 Exercises

1. Go to the website http://www.research.att.com/~njas/oadir/ and find a design with $k = 5$ factors, four with 2 levels and one with 3 levels. How many treatment combinations does it have? How would you define the main effects for each of these factors?

2. Consider the column $(1, 1, 1, 0, 1, 0, 0)'$. Obtain 6 further columns by rotating one position for each new column. Adjoin a row of 0's. Verify that the resulting array has 7 binary factors and is of resolution 3.

3. Use one of the websites mentioned above to find an OA of strength 2 for eight 4-level factors and an 8-level factor in 32 runs. Now make an OA with 288 runs with one more factor with nine levels. Show how to use this design to get an OA with two 2-level factors, four 4-level factors, an 8-level factor and one factor with 36 levels still in 288 runs.

2.6 REFERENCES AND COMMENTS

Although used in some agricultural experiments from the mid-1800s, the first formal exposition of factorial experiments was given by Yates (1935) and they appeared in Fisher (1935).

There are various books that develop factorial designs for use in comparative experiments including Kuehl (1999), Mason et al. (2003) and Montgomery (2001). These books also discuss the derivation of orthogonal polynomials, for equally spaced levels, whose use in model fitting was first pointed out by Fisher. The links between fractional factorial designs and finite geometries were pointed out by Bose and Kishen (1940) and exploited by Bose (1947). Much of the early work in the area is summarized in Raghavarao (1971) and Raktoe et al. (1981).

There is a close relationship between fractional factorial designs and orthogonal arrays, and an extensive treatment of results pertaining to orthogonal arrays can be found in Hedayat et al. (1999). They give a number of constructions that rely on the structural properties of the original OA to allow further factors to be added. The simplest of these ideas was presented in Section 2.3.3. The initial theoretical development of regular fractional factorial designs appeared in Fisher (1945) and Finney (1945). Dey (1985) provides a number of constructions for fractional factorial designs and OMEPs.

There has been some debate about the appropriate contrasts to use in a fractional design and some of the issues are addressed in Beder (2004) and John and Dean (1975).

CHAPTER 3

THE MNL MODEL AND COMPARING DESIGNS

As we explained in Chapter 1, a stated preference choice experiment consists of a finite set of choice sets and each choice set consists of a finite number of options. The options within each choice set must be distinct and they must be exhaustive, either because there is an "other" option, a "none of these" option or because the subjects are asked to assume that they have narrowed down the possible options to those given in the choice set. This final assumption is often described by saying that the subjects have participated in a *forced choice* stated preference experiment.

We assume that each subject chooses, from each choice set, the option that is "best" for them. The researcher knows which options have been compared in each choice set and which option has been selected but has no idea how the subject has decided the relative value of each option. However, the researcher assumes that these relative values are a function of the levels of the attributes of the options under consideration, some of which have been deliberately varied by the researcher.

In this chapter we begin by defining utility and showing how a choice process that bases choices on the principle of random utility maximization can result in the multinomial logit (MNL) model for a specific assumption about the errors. We then derive the choice probabilities for the MNL model. Next we consider the Bradley–Terry model which arises when all choice sets have only two options, before looking at the MNL model for any choice set size. We assume that the MNL model is the discrete choice model that is the decision rule subjects are employing. The results on optimality that we present will depend on this assumption. Should a different model be used, then different designs may well prove to be optimal. We also discuss how to decide objectively which of a set of proposed choice experiments is the best one to use, from a statistical perspective, for a given situation.

At times in this chapter, and indeed throughout the rest of the book, we will talk about options that are described by the levels of k attributes. These attributes are precisely the same as the factors of the previous chapter.

3.1 UTILITY AND CHOICE PROBABILITIES

In this section we define utility and show how a choice process that bases choices on the principle of random utility maximization can result in the MNL model for a specific assumption about the errors.

3.1.1 Utility

Train (2003) has defined *utility* as "the net benefit derived from taking some action"; in a choice experiment, we assume that each subject chooses the option that has maximum utility from the ones available in each choice set. Thus each subject assigns some utility to each option in a choice set and then the subject chooses the option with the maximum utility.

If we let $U_{j\alpha}$ be the utility assigned by subject α to option $j, j = 1, \ldots, m$, in a choice set with m options, then option i is chosen if and only if $U_{i\alpha} > U_{j\alpha} \forall j \neq i$. The researcher does not see the utilities but only the options offered and the choice made (from each of the choice sets). These options are usually described by levels of several attributes. The systematic component of the utility that the researcher captures will be denoted by $V_{j\alpha}$, and we assume that $U_{j\alpha} = V_{j\alpha} + \epsilon_{j\alpha}$, where $\epsilon_{j\alpha}$ includes all the things that affect the utility that have not been included in $V_{j\alpha}$. Thus the $\epsilon_{j\alpha}$ are random terms, and we get different choice models depending on the assumptions that we make about the distribution of the $\epsilon_{j\alpha}$.

We know that

$$\begin{aligned} Pr(\text{option } i \text{ is chosen by subject } \alpha) &= Pr(U_{i\alpha} > U_{j\alpha} \forall j \neq i) \\ &= Pr(V_{i\alpha} + \epsilon_{i\alpha} > V_{j\alpha} + \epsilon_{j\alpha} \forall j \neq i) \\ &= Pr(V_{i\alpha} - V_{j\alpha} > \epsilon_{j\alpha} - \epsilon_{i\alpha} \forall j \neq i). \end{aligned}$$

If the $\epsilon_{j\alpha}$ are independently identically distributed extreme value type 1 random variables then the resulting model is the multinomial logit (MNL) model. This assumption on the $\epsilon_{j\alpha}$ is equivalent to assuming that the unobserved attributes have the same variance for all options in each choice set and that these attributes are uncorrelated over all the options in each choice set (Train (2003)). Train (2003) gives an example for choosing travel options where this independence assumption is not reasonable: If a subject does not want to travel by bus because of the presence of strangers on the bus, then that subject is also unlikely to choose to travel by train; and discusses other models that have been proposed which avoid this independence assumption.

These models are the generalized extreme value (GEV) models, the probit model and the mixed logit (ML) model. In the generalized extreme value models, the unobserved portions of the utility are assumed to be jointly distributed as generalized extreme value. Consequently, correlation in the unobserved attributes is allowed, but this model collapses to an extreme value model if the correlation is 0 for all attributes. In the probit model, the unobserved attributes are assumed to be normally distributed. This means that unobserved attributes can be assumed to be jointly normal and be modeled with any appropriate correlation structure over both options and time (for example, with panel data where the same individuals are asked to respond on several successive occasions). Of course it is rarely appropriate for a price attribute to come from a normal distribution since few people want to pay more for the same or worse features (but consider those who like to emphasize their wealth through conspicuous consumption). This imposes some limitations on the application of the probit model. In the mixed logit model, the unobserved attributes can be decomposed into two parts: one with the correlation, which may have any distribution, and one which is extreme value. Variations on these models, such as nested logit, are also possible. While there are many different choice models, we will only discuss designing choice experiments for the multinomial logit model in this book.

3.1.2 Choice Probabilities

We assume that each unobserved component of the utility $\epsilon_{i\alpha}$ is independently identically distributed Type I extreme value, sometimes called the Gumbel distribution. It has density

$$f(\epsilon) = e^{-\epsilon}e^{-e^{-\epsilon}}, -\infty < \epsilon < \infty$$

and cumulative distribution function

$$F(\epsilon) = e^{-e^{-\epsilon}}.$$

We now derive the choice probabilities using the argument given in Train (2003, pp. 40–44, 78–79). From the results above, we have that

$$P_{i\alpha} = Pr(\text{option } i \text{ is chosen by subject } \alpha) = Pr(V_{i\alpha} - V_{j\alpha} + \epsilon_{i\alpha} > \epsilon_{j\alpha}\ \forall j \neq i).$$

If we regard $\epsilon_{i\alpha}$ as given, then $P_{i\alpha}$ is just $F(\epsilon_{i\alpha} + V_{i\alpha} - V_{j\alpha}) = e^{-e^{-(\epsilon_{i\alpha}+V_{i\alpha}-V_{j\alpha})}}$, again for all $j \neq i$. Now the ϵ's are independent and hence the cumulative distribution over all $j \neq i$ is just the product of the cumulative distributions over all $j \neq i$. This gives

$$Pr(\text{option } i \text{ is chosen by subject } \alpha | \epsilon_{i\alpha}) = \prod_{j \neq i} e^{-e^{-(\epsilon_{i\alpha}+V_{i\alpha}-V_{j\alpha})}}.$$

In practice, we do not actually know $\epsilon_{i\alpha}$ and so we need to integrate out $\epsilon_{i\alpha}$, weighting the values by their density, to give

$$P_{i\alpha} = \int_{\epsilon_{i\alpha}=-\infty}^{\epsilon_{i\alpha}=\infty} \left(\prod_{j \neq i} e^{-e^{-(\epsilon_{i\alpha}+V_{i\alpha}-V_{j\alpha})}} \right) e^{-\epsilon_{i\alpha}} e^{-e^{-\epsilon_{i\alpha}}} d\epsilon_{i\alpha}.$$

To evaluate this integral, observe that $V_{i\alpha} - V_{i\alpha} = 0$; so $\epsilon_{i\alpha} + V_{i\alpha} - V_{i\alpha} = \epsilon_{i\alpha}$. Hence we can remove the restriction on the product by using the last term. This gives

$$\begin{aligned}
P_{i\alpha} &= \int_{\epsilon_{i\alpha}=-\infty}^{\epsilon_{i\alpha}=\infty} \left(\prod_j e^{-e^{-(\epsilon_{i\alpha}+V_{i\alpha}-V_{j\alpha})}} \right) e^{-\epsilon_{i\alpha}} d\epsilon_{i\alpha} \\
&= \int_{\epsilon_{i\alpha}=-\infty}^{\epsilon_{i\alpha}=\infty} \exp\left(-\sum_j e^{-(\epsilon_{i\alpha}+V_{i\alpha}-V_{j\alpha})} \right) e^{-\epsilon_{i\alpha}} d\epsilon_{i\alpha} \\
&= \int_{\epsilon_{i\alpha}=-\infty}^{\epsilon_{i\alpha}=\infty} \exp\left(-e^{-\epsilon_{i\alpha}} \sum_j e^{-(V_{i\alpha}-V_{j\alpha})} \right) e^{-\epsilon_{i\alpha}} d\epsilon_{i\alpha}.
\end{aligned}$$

Now we let $z = e^{-\epsilon_{i\alpha}}$. Then $dz = -e^{-\epsilon_{i\alpha}} d\epsilon_{i\alpha}$. As $\epsilon_{i\alpha}$ tends to ∞, z tends to 0, and as $\epsilon_{i\alpha}$ tends to $-\infty$, z approaches ∞. So we get

$$\begin{aligned}
P_{i\alpha} &= \int_{z=\infty}^{z=0} \exp\left(-z \sum_j e^{-(V_{i\alpha}-V_{j\alpha})} \right) (-dz) \\
&= \int_{z=0}^{z=\infty} \exp\left(-z \sum_j e^{-(V_{i\alpha}-V_{j\alpha})} \right) dz \\
&= \left. \frac{\exp(-z \sum_j e^{-(V_{i\alpha}-V_{j\alpha})})}{-\sum_j e^{-(V_{i\alpha}-V_{j\alpha})}} \right|_0^\infty \\
&= \frac{\exp(0)}{\sum_j e^{-(V_{i\alpha}-V_{j\alpha})}} \\
&= \frac{1}{\sum_j e^{(V_{j\alpha}-V_{i\alpha})}} \\
&= \frac{e^{V_{i\alpha}}}{\sum_j e^{V_{j\alpha}}}.
\end{aligned}$$

For the time being we let $\pi_i = e^{V_i}$, constant for all subjects, and we write $\ln(\pi_i) = \gamma_i$.

3.2 THE BRADLEY–TERRY MODEL

A *paired comparison* experiment is a choice experiment in which subjects are shown two options at a time and are asked to say which one they prefer. Thus it is simply a choice experiment in which all choice sets are of size 2.

Such experiments were being used by psychophysicists in the mid-nineteenth century. They were interested in looking at how much two objects had to differ to be perceived as different. For example, Thurstone (1927) describes an experiment in which 19 offences, ranging (alphabetically) from abortion to vagrancy, were shown, in pairs, to subjects who were asked to decide which offence was more serious. He developed various models to analyze data of this type. Zermelo (1929) independently used one of these models to rank chess players. More details may be found in MacKay (1988).

Unaware of the earlier work by Thurstone (1927) and Zermelo (1929), Bradley and Terry (1952) proposed using the Zermelo model in a psychological setting and developed likelihood estimates and appropriate test procedures for the unknown parameters. The model that they proposed is now known as the *Bradley–Terry* model. We will consider this model, and develop appropriate estimates and distribution results, in the remainder of this section.

In the Bradley–Terry model, we assume that altogether there are t items, $T_1, T_2, \ldots, T_t$, to be compared, although any choice set will only compare two of the items. We assume that each of the items has a constant

$$\pi_i = e^{V_i}$$

associated with it so the MNL model becomes a special case of the Bradley–Terry model. As we saw in the previous section, the logarithms of these constants are the utilities of the items and these constants may be called the *merits* of the items. Each choice set consists of a pair of items which are called the *options* in the choice set.

Using the argument of the previous section, we get

$$Pr(T_i \text{ is preferred to } T_j) = \frac{\pi_i}{(\pi_i + \pi_j)}, \; i \neq j, \; i, j = 1, \ldots, t.$$

We assume that all subjects see the same pairs and that there are no repeated pairs in the experiment. We let

$$n_{ij} = \begin{cases} 1 & \text{when the pair } (T_i, T_j) \text{ is in the choice experiment,} \\ 0 & \text{when the pair } (T_i, T_j) \text{ is not in the choice experiment.} \end{cases}$$

We will assume that the order of presentation of options within a pair and of pairs within the experiment is unimportant although for some situations there is evidence that order effects exist (Timmermans et al. (2006)). In practice we would randomly permute the order of the choice sets within the experiment before presenting them to respondents.

■ **EXAMPLE 3.2.1.**

Suppose that $t = 4$. Then the 6 possible pairs are (T_1, T_2), (T_1, T_3), (T_1, T_4), (T_2, T_3), (T_2, T_4), and (T_3, T_4). If all pairs are shown to all subjects, then $n_{ij} = 1$ for each of the 6 pairs. If we only use the pairs (T_1, T_2), (T_1, T_3), and (T_1, T_4) then $n_{1j} = 1$ for $j = 2, 3, 4$ and $n_{ij} = 0$ if $i, j \neq 1$. □

We want to be able to find estimates of the π_i since these can be used to assess the relative attractiveness of the items under consideration. Since the properties of the maximum likelihood estimates (MLEs) of the π_i values have been well-established (see, for instance, David (1988)), we find such estimates for the Bradley–Terry model in what follows.

3.2.1 The Likelihood Function

We begin by evaluating the likelihood function. For subject α, we let

$$w_{ij\alpha} = \begin{cases} 1 & \text{when } T_i \text{ is preferred to } T_j, \\ 0 & \text{when } T_j \text{ is preferred to } T_i. \end{cases}$$

Note that $1 - w_{ij\alpha} = w_{ji\alpha}$. We let $\boldsymbol{\pi}=(\pi_1, \pi_2, \ldots, \pi_t)$ and we let $f_{ij\alpha}(w_{ij\alpha}, \boldsymbol{\pi})$ be the probability density function for subject α and choice set (T_i, T_j), where, for ease of presentation, we assume that $w_{ij\alpha} = w_{ji\alpha} = 0$ if $n_{ij} = 0$. Thus

$$f_{ij\alpha}(w_{ij\alpha}, \boldsymbol{\pi}) = \frac{\pi_i^{w_{ij\alpha}} \pi_j^{w_{ji\alpha}}}{(\pi_i + \pi_j)^{n_{ij}}}.$$

We assume there are s subjects in total and we let $\sum_\alpha w_{ij\alpha} = w_{ij}$. Then the likelihood function is given by

$$L(\boldsymbol{\pi}) = \prod_{i<j} \prod_{\alpha=1}^{s} f_{ij\alpha}(w_{ij\alpha}, \boldsymbol{\pi}) = \prod_{i<j} \left(\frac{\pi_i^{w_{ij}} \pi_j^{sn_{ij}-w_{ij}}}{(\pi_i + \pi_j)^{sn_{ij}}} \right).$$

We can simplify this further if we let $w_i = \sum_j w_{ij}$ be the total number of times that T_i is chosen from all choice sets containing T_i. Then

$$L(\boldsymbol{\pi}) = \frac{\pi_1^{w_1} \pi_2^{w_2} \cdots \pi_t^{w_t}}{\prod_{i<j} (\pi_i + \pi_j)^{sn_{ij}}}.$$

3.2.2 Maximum Likelihood Estimation

The usual way to estimate π is to find the maximum likelihood estimators. This requires setting the first derivative of the likelihood function, or, equivalently, the first derivative of the log-likelihood function, to zero and solving iteratively.

We have that the log-likelihood is

$$\ln(L(\boldsymbol{\pi})) = \sum_{i=1}^{t} w_i \ln(\pi_i) - \sum_{i<j} sn_{ij} \ln(\pi_i + \pi_j).$$

To get the maximum likelihood estimates, we solve

$$\frac{\partial \ln(L(\boldsymbol{\pi}))}{\partial \pi_i} = \frac{w_i}{\hat{\pi}_i} - \sum_{j \neq i} \frac{sn_{ij}}{\hat{\pi}_i + \hat{\pi}_j} = 0, \; i = 1, 2, \ldots, t,$$

together with the normalizing constraint $\prod_i \hat{\pi}_i = 1$. Rearranging, we need to solve

$$\hat{\pi}_i = \frac{w_i}{\sum_{j \neq i} \frac{sn_{ij}}{\hat{\pi}_i + \hat{\pi}_j}} = \frac{w_i}{\sum_{j \neq i} sn_{ij} (\hat{\pi}_i + \hat{\pi}_j)^{-1}} \tag{3.1}$$

iteratively. We let $\hat{\pi}_i^{(r)}$ be the estimate of π_i at the rth iteration. We continue iterations until agreement between $\hat{\pi}_i^{(r-1)}$ and $\hat{\pi}_i^{(r)}$ is close enough.

Bradley (1985) says that convergence is fast and initial estimates of 1 suffice. In practice most statistical software does not even require that initial estimates be provided.

■ **EXAMPLE 3.2.2.**

Suppose that $t = 4$ and that we show all 6 possible pairs (T_1, T_2), (T_1, T_3), (T_1, T_4), (T_2, T_3), (T_2, T_4), and (T_3, T_4) to each of $s = 5$ subjects. The results of this experiment are given in Table 3.1. We see that

$$w_{121} = 1 = w_{131} = w_{141} = w_{231} = w_{241} = w_{341}$$

and hence

$$w_{211} = 0 = w_{311} = w_{411} = w_{321} = w_{421} = w_{431}.$$

Similarly

$$w_{122} = 0 = w_{132} = w_{142} = w_{232} = w_{242} = w_{342}$$

and hence

$$w_{212} = 1 = w_{312} = w_{412} = w_{322} = w_{422} = w_{432}.$$

Using all the results, we get that $w_1 = 8$, $w_2 = 6$, $w_3 = 7$ and $w_4 = 9$.

Table 3.1 Choice Sets and Choices for Example 3.2.2, with $t = 4$, $s = 5$

Choice Set	Subject 1	2	3	4	5
(T_1, T_2)	T_1	T_2	T_1	T_1	T_2
(T_1, T_3)	T_1	T_3	T_1	T_3	T_1
(T_1, T_4)	T_1	T_4	T_4	T_1	T_4
(T_2, T_3)	T_2	T_3	T_2	T_3	T_3
(T_2, T_4)	T_2	T_4	T_2	T_4	T_4
(T_3, T_4)	T_3	T_4	T_3	T_4	T_4

Thus we have that

$$L(\boldsymbol{\pi}) = \frac{\pi_1^8 \pi_2^6 \pi_3^7 \pi_4^9}{(\pi_1 + \pi_2)^5 (\pi_1 + \pi_3)^5 (\pi_1 + \pi_4)^5 (\pi_2 + \pi_3)^5 (\pi_2 + \pi_4)^5 (\pi_3 + \pi_4)^5},$$

and the corresponding log-likelihood is given by

$$\ln(L(\boldsymbol{\pi})) = \sum_{i=1}^{4} w_i \ln(\pi_i) - 5 \sum_{i<j} \ln(\pi_i + \pi_j).$$

The maximum likelihood estimates are the solutions to the partial derivatives of the log-likelihood and we get the following equations:

$$\begin{aligned}
\frac{8}{\hat{\pi}_1} - 5 \sum_{j \neq 1} \frac{1}{\hat{\pi}_1 + \hat{\pi}_j} &= 0; \\
\frac{6}{\hat{\pi}_2} - 5 \sum_{j \neq 2} \frac{1}{\hat{\pi}_2 + \hat{\pi}_j} &= 0; \\
\frac{7}{\hat{\pi}_3} - 5 \sum_{j \neq 3} \frac{1}{\hat{\pi}_3 + \hat{\pi}_j} &= 0; \\
\frac{9}{\hat{\pi}_4} - 5 \sum_{j \neq 4} \frac{1}{\hat{\pi}_4 + \hat{\pi}_j} &= 0.
\end{aligned}$$

We start the recursion by assuming that all the items are equally attractive; thus we are assuming that $\hat{\pi}_i^{(0)} = 1, i = 1, 2, 3, 4$.

Using Equation (3.1), we get

$$\begin{aligned}
\hat{\pi}_1^{(1)} &= \frac{w_1}{5}\left[\frac{1}{\hat{\pi}_1^{(0)}+\hat{\pi}_2^{(0)}}+\frac{1}{\hat{\pi}_1^{(0)}+\hat{\pi}_3^{(0)}}+\frac{1}{\hat{\pi}_1^{(0)}+\hat{\pi}_4^{(0)}}\right]^{-1} \\
&= \frac{8}{5}\left[\frac{1}{1+1}+\frac{1}{1+1}+\frac{1}{1+1}\right]^{-1} \\
&= \frac{16}{15}.
\end{aligned}$$

We now use this estimate of π_1 to help estimate π_2. We get

$$\begin{aligned}
\hat{\pi}_2^{(1)} &= \frac{w_2}{5}\left[\frac{1}{\hat{\pi}_1^{(1)}+\hat{\pi}_2^{(0)}}+\frac{1}{\hat{\pi}_2^{(0)}+\hat{\pi}_3^{(0)}}+\frac{1}{\hat{\pi}_2^{(0)}+\hat{\pi}_4^{(0)}}\right]^{-1} \\
&= \frac{6}{5}\left[\frac{1}{\frac{16}{15}+1}+\frac{1}{1+1}+\frac{1}{1+1}\right]^{-1} \\
&= \frac{93}{115}.
\end{aligned}$$

To get $\hat{\pi}_3^{(1)}$, we use $\hat{\pi}_1^{(1)}$, $\hat{\pi}_2^{(1)}$, $\hat{\pi}_3^{(0)}$ and $\hat{\pi}_4^{(0)}$. The results of the first 8 iterations, before normalization, are given in Table 3.2.

Table 3.2 Estimates of π_i from All Six Choice Sets

Iteration	$\hat{\pi}_1$	$\hat{\pi}_2$	$\hat{\pi}_3$	$\hat{\pi}_4$
0	1.000	1.000	1.000	1.000
1	1.067	0.8087	0.9110	1.154
2	1.074	0.7398	0.8784	1.225
3	1.068	0.7143	0.8658	1.259
4	1.061	0.7047	0.8608	1.275
5	1.056	0.7010	0.8588	1.282
6	1.053	0.6996	0.8580	1.286
7	1.052	0.6990	0.8577	1.287
8	1.051	0.6988	0.8575	1.288
8(normalized)	1.107	0.7363	0.9036	1.357

To normalize these estimates we divide each estimate by

$$\sqrt[4]{\prod_{i=1}^{4}\pi_i} = 0.94905.$$

We get $\hat{\pi}_1 = 1.107$, $\hat{\pi}_2 = 0.7363$, $\hat{\pi}_3 = 0.9036$, and $\hat{\pi}_4 = 1.357$. So, based on these results, we would rank the items (from best to worst) as 4, 1, 3, and 2. (In general we would divide by the tth root of the product.)

If we suppose that only the pairs involving T_1 were shown to the subjects, we would then have

$$L(\boldsymbol{\pi}) = \frac{\pi_1^8\pi_2^2\pi_3^2\pi_4^3}{(\pi_1+\pi_2)^5(\pi_1+\pi_3)^5(\pi_1+\pi_4)^5},$$

and the corresponding log-likelihood would be given by

$$\ln(L(\boldsymbol{\pi})) = \sum_{i=1}^{4} w_i \ln(\pi_i) - 5\sum_{1<j} \ln(\pi_1 + \pi_j).$$

The maximum likelihood estimates are the solutions to the partial derivatives of the log-likelihood, and we get the following equations:

$$\begin{aligned}
\frac{8}{\hat{\pi}_1} - 5\sum_{j\neq 1}\frac{1}{\hat{\pi}_1 + \hat{\pi}_j} &= 0;\\
\frac{2}{\hat{\pi}_2} - 5\frac{1}{\hat{\pi}_2 + \hat{\pi}_1} &= 0;\\
\frac{2}{\hat{\pi}_3} - 5\frac{1}{\hat{\pi}_3 + \hat{\pi}_1} &= 0;\\
\frac{3}{\hat{\pi}_4} - 5\frac{1}{\hat{\pi}_4 + \hat{\pi}_1} &= 0.
\end{aligned}$$

Again we start the recursion by assuming that all the items are equally attractive; so again we have $\hat{\pi}_i^{(0)} = 1, i = 1, 2, 3, 4$. We get

$$\begin{aligned}
\hat{\pi}_1^{(1)} &= \frac{w_1}{5}\left[\frac{1}{\hat{\pi}_1^{(0)} + \hat{\pi}_2^{(0)}} + \frac{1}{\hat{\pi}_1^{(0)} + \hat{\pi}_3^{(0)}} + \frac{1}{\hat{\pi}_1^{(0)} + \hat{\pi}_4^{(0)}}\right]^{-1}\\
&= \frac{8}{5}\left[\frac{1}{1+1} + \frac{1}{1+1} + \frac{1}{1+1}\right]^{-1}\\
&= \frac{16}{15}.
\end{aligned}$$

We now use this estimate of π_1 to help estimate π_2. We get

$$\begin{aligned}
\hat{\pi}_2^{(1)} &= \frac{w_2}{5}\left[\frac{1}{\hat{\pi}_1^{(1)} + \hat{\pi}_2^{(0)}}\right]^{-1}\\
&= \frac{2}{5}\left[\frac{1}{\frac{16}{15} + 1}\right]^{-1}\\
&= \frac{62}{75}.
\end{aligned}$$

To get $\hat{\pi}_3^{(1)}$, we use $\hat{\pi}_1^{(1)}$ and $\hat{\pi}_3^{(0)}$; to get $\hat{\pi}_4^{(1)}$, we use $\hat{\pi}_1^{(1)}$ and $\hat{\pi}_4^{(0)}$. The results of three of the first 10 iterations, before normalization, are given in Table 3.3.

Normalizing, we get $\hat{\pi}_1 = 1.107$, $\hat{\pi}_2 = 0.738 = \hat{\pi}_3$, and $\hat{\pi}_4 = 1.658$. Based on these results we would rank the items (from best to worst) as 4, 1 and then 2 and 3 equal. □

Two questions still need to be addressed. How do we know that the iterations will converge? What do we know about the properties of the maximum likelihood estimates?

3.2.3 Convergence

Convergence is assured if the n_{ij} are all equal to 1, although it may be slow (p 62, David (1988) quoting Dykstra (1956)). If the n_{ij} are not equal then Zermelo (1929) and Ford Jr.

Table 3.3 Estimates of π_i from the First Three Choice Sets Only

Iteration	$\hat{\pi}_1$	$\hat{\pi}_2$	$\hat{\pi}_3$	$\hat{\pi}_4$
0	1.000	1.000	1.000	1.000
5	1.061	0.7115	0.7115	1.550
9	1.056	0.7045	0.7045	1.581
10	1.056	0.7042	0.7042	1.582
10 (normalized)	1.107	0.738	0.738	1.658

(1957), quoted in David (1988), have established that the iterative process will converge provided that: "In every possible partition of the objects into two non-empty subsets, some object in the first set has been preferred at least once to some object in the second set." If this condition is not satisfied, then either there are two sets of items which have never been compared (so how could they be ranked relative to each other), or every comparison favours one set, P_1 say, and so the π_i values of all the items in the other set, P_2 say, must be 0. This follows by noting that, if the π_i associated with the items in P_2 were not 0, then the likelihood function could be increased by multiplying all these π_i by a constant less than 1 and dividing the π_i values of the items in P_1 by a constant greater than 1 so that the constraint $\prod_i \pi_i = 1$ is preserved (from David (1988), pp. 63–64)).

We choose the pairs that we present to ensure that there are no sets of items that are not compared. This does not require that all pairs of items be compared directly, as we saw in Example 3.2.2, but it does require that for any two items, T_i and T_j, it is possible to form a list of items $T_i, T_{i_1}, T_{i_2}, \ldots, T_j$, such that adjacent items in the list correspond to pairs in the design. Such a design is said to be *connected*.

■ **EXAMPLE 3.2.3.**
If we have four items of interest and we use the pairs (T_1, T_2), (T_1, T_3) and (T_1, T_4), then any item is directly compared to item T_1. Consider the pairs of items not involving T_1. Then the list for the pair T_2, T_3 is T_2, T_1, T_3, and for the pair T_2, T_4 is T_2, T_1, T_4, and for the pair T_3, T_4 is T_3, T_1, T_4. Thus the original set of three pairs forms a connected design. □

Even with a connected design, it can still be possible for every comparison to favour one set.

■ **EXAMPLE 3.2.4.**
In Example 3.2.2, for instance, there is at least one subject who has chosen each of the items. So, if we divide the items into two sets, say $P_1 = \{T_1, T_2\}$ and $P_2 = \{T_3, T_4\}$, then there is at least one time when an object in P_1 has been preferred, and at least one time that an object in P_2 has been preferred. On the other hand, if we let $P_1 = \{T_1, T_2, T_3\}$ and $P_2 = \{T_4\}$ and consider only subject 1, then the option from P_2 is never chosen. The likelihood for the first subject is given by

$$L(\boldsymbol{\pi}) = \frac{\pi_1^3 \pi_2^2 \pi_3^1 \pi_4^0}{(\pi_1 + \pi_2)(\pi_1 + \pi_3)(\pi_1 + \pi_4)(\pi_2 + \pi_3)(\pi_2 + \pi_4)(\pi_3 + \pi_4)}.$$

Observe that the value of the likelihood function is largest when $\pi_4 = 0$ and that the likelihood function gets smaller for larger values of π_4. So, if we start by assuming the null hypothesis that all the $\pi_i = 1$, then we get $L(1, 1, 1, 1) = 0.015625$. If we let $\pi_4 = 0$, then we have $L(1,1,1,0) = 0.125$. If we try to have a small positive value for π_4, say $\pi_4 = 1/1000$, then $L(10,10,10,1/1000) = 0.124963$ which is somewhat smaller than the likelihood when $\pi_4 = 0$. If instead we give π_4 a value which suggests that item T_4 was very popular, say $\pi_4 = 8$, then $L(1/2,1/2,1/2,8) = 0.0000254427$ which is very much smaller than the value with all the π_is equal to 1 and certainly smaller than the largest value of the likelihood possible. This illustrates the discussion about maximizing the likelihood when the items chosen have resulted in a disconnected design. □

3.2.4 Properties of the MLEs

Assuming that we have used a connected design and that each item has been chosen at least once, we know that the maximum likelihood estimates will exist. In this section, we outline the results that lead to the asymptotic distribution of the maximum likelihood estimates, $\hat{\pi}_i$. We briefly recall the results on the distribution of the MLEs for a random sample from one population and then we outline how the results need to be modified to apply to the Bradley–Terry model.

Let $\boldsymbol{x_i}$ be a random vector of length t. Let $\boldsymbol{x_1}, \boldsymbol{x_2}, \ldots, \boldsymbol{x_s}$ be a random sample of size s from a common distribution $f(\boldsymbol{x}, \boldsymbol{\theta})$ with unknown parameter vector $\boldsymbol{\theta}$. Suppose that the distribution satisfies some mild smoothness assumptions that are stated explicitly in Cramer (1946) or Scholz (1985). Then, for large values of s, the distribution of $\sqrt{s}\hat{\boldsymbol{\theta}}_{\boldsymbol{s}}$ is approximately t-variate normal with mean vector $\boldsymbol{\theta}$ and variance-covariance matrix $[I(\boldsymbol{\theta})]^{-1}$, where $I(\boldsymbol{\theta})$ is the *Fisher information matrix*. The Fisher information matrix has entries given by

$$(I(\boldsymbol{\theta}))_{ij} = \mathcal{E}\left(\left(\frac{\partial \ln f(\boldsymbol{x},\boldsymbol{\theta})}{\partial\theta_i}\right)\left(\frac{\partial \ln f(\boldsymbol{x},\boldsymbol{\theta})}{\partial\theta_j}\right)\right) = -\mathcal{E}\left(\frac{\partial^2 \ln f(\boldsymbol{x},\boldsymbol{\theta})}{\partial\theta_i\partial\theta_j}\right).$$

When we consider finding parameter estimates from a choice experiment, choice sets that contain treatment T_i also share the parameter π_i in the corresponding distribution function. Thus these distributions are not independent but are said to be *associated*. In this situation, the result quoted above needs to be modified; see Bradley and Gart (1962) for details. We now outline the relevant result for the Bradley–Terry model when all subjects see the same choice sets.

As in El-Helbawy and Bradley (1978), we view each choice set as one of the associated populations. We have sn_{ij} observations from the pair (T_i, T_j) and $sN = s\sum_{i<j} n_{ij}$ observations in total. Thus the proportion of observations that come from any choice set is $\lambda_{ij} = n_{ij}/N$. Then the elements of the information matrix for $\sqrt{sN}\hat{\boldsymbol{\pi}}$ are given by

$$(I(\boldsymbol{\pi}))_{ij} = \sum_{a<b} \lambda_{ab}\mathcal{E}_{\pi}\left(\left(\frac{\partial \ln f_{ab\alpha}(w_{ab\alpha},\boldsymbol{\pi})}{\partial\pi_i}\right)\left(\frac{\partial \ln f_{ab\alpha}(w_{ab\alpha},\boldsymbol{\pi})}{\partial\pi_j}\right)\right).$$

If $i \neq j$ the only pair which has both of these partial derivatives non-zero is the pair in which T_i and T_j are compared. In that case we have

$$\ln\left(f_{ij\alpha}(w_{ij\alpha},\boldsymbol{\pi})\right) = \left(w_{ij\alpha}\ln(\pi_i) + w_{ji\alpha}\ln(\pi_j) - n_{ij}\ln(\pi_i+\pi_j)\right)$$

from which we get

$$\frac{\partial \ln[f(w_{ij\alpha},\boldsymbol{\pi})]}{\partial\pi_i} = \frac{w_{ij\alpha}}{\pi_i} - \frac{1}{\pi_i+\pi_j}.$$

Substituting, we have

$$(I(\boldsymbol{\pi}))_{ij} = \lambda_{ij}\mathcal{E}_\pi \left[\left(\frac{w_{ij\alpha}}{\pi_i} - \frac{1}{\pi_i+\pi_j}\right)\left(\frac{w_{ji\alpha}}{\pi_j} - \frac{1}{\pi_i+\pi_j}\right)\right].$$

Now $w_{ij\alpha}$ is a Bernoulli random variable which takes the value 1 with probability

$$\pi_i/(\pi_i+\pi_j).$$

So

$$\mathcal{E}(w_{ij\alpha}) = \pi_i/(\pi_i+\pi_j)$$

and

$$\mathrm{Var}(w_{ij\alpha}) = \pi_i/(\pi_i+\pi_j)(1-\pi_i/(\pi_i+\pi_j)) = \pi_i\pi_j/(\pi_i+\pi_j)^2.$$

If $n_{ij}=0$, then $w_{ij\alpha}=0$ and is not a random variable. For fixed but arbitrary i, any two $w_{ih\alpha}$ random variables are independent by assumption (that is, choices made in one choice set are independent of choices made in any other choice set). Hence we have

$$\begin{aligned}
(I(\boldsymbol{\pi}))_{ij} &= \lambda_{ij}\mathcal{E}_\pi \left[\left(\frac{w_{ij\alpha}}{\pi_i} - \frac{1}{\pi_i+\pi_j}\right)\left(\frac{w_{ji\alpha}}{\pi_j} - \frac{1}{\pi_i+\pi_j}\right)\right] \\
&= \lambda_{ij}\left[\mathcal{E}_\pi\left(\frac{w_{ij\alpha}w_{ji\alpha}}{\pi_i\pi_j}\right) - \frac{\mathcal{E}_\pi(w_{ij\alpha})}{(\pi_i+\pi_j)\pi_i} - \frac{\mathcal{E}_\pi(w_{ji\alpha})}{(\pi_i+\pi_j)\pi_j} + \frac{1}{(\pi_i+\pi_j)^2}\right] \\
&= \lambda_{ij}\left[0 - \frac{\pi_i}{(\pi_i+\pi_j)^2\pi_i} - \frac{\pi_j}{(\pi_i+\pi_j)^2\pi_j} + \frac{1}{(\pi_i+\pi_j)^2}\right] \\
&= -\lambda_{ij}\frac{1}{(\pi_i+\pi_j)^2},
\end{aligned}$$

when $i \neq j$.

If $i=j$, then

$$\begin{aligned}
(I(\boldsymbol{\pi}))_{ii} &= \sum_{a<b}\lambda_{ab}\mathcal{E}_\pi\left(\left(\frac{\partial \ln f_{ab\alpha}(w_{ab\alpha},\boldsymbol{\pi})}{\partial \pi_i}\right)\left(\frac{\partial \ln f_{ab\alpha}(w_{ab\alpha},\boldsymbol{\pi})}{\partial \pi_i}\right)\right) \\
&= \sum_{a\neq i}\lambda_{ia}\mathcal{E}_\pi\left(\left(\frac{\partial \ln f_{ia\alpha}(w_{ia\alpha},\boldsymbol{\pi})}{\partial \pi_i}\right)^2\right) \\
&= \sum_{a\neq i}\lambda_{ia}\left[\mathcal{E}_\pi\left(\left(\frac{w_{ia\alpha}}{\pi_i}\right)^2\right) - \frac{2}{(\pi_i+\pi_a)}\frac{1}{\pi_i}\frac{\pi_i}{(\pi_a+\pi_i)} + \frac{1}{(\pi_i+\pi_a)^2}\right] \\
&= \sum_{a\neq i}\lambda_{ia}\left[\frac{1}{\pi_i(\pi_i+\pi_a)} - \frac{1}{(\pi_i+\pi_a)^2}\right] \\
&= \sum_{a\neq i}\lambda_{ia}\frac{\pi_a}{(\pi_a+\pi_i)^2\pi_i}
\end{aligned}$$

since

$$\begin{aligned}
\mathcal{E}(w_{ia\alpha}^2) &= \mathrm{Var}(w_{ia\alpha}) + (\mathcal{E}(w_{ia\alpha}))^2 \\
&= \frac{\pi_a\pi_i}{(\pi_i+\pi_a)^2} + \frac{\pi_i^2}{(\pi_i+\pi_a)^2} \\
&= \frac{\pi_i(\pi_i+\pi_a)}{(\pi_i+\pi_a)^2}.
\end{aligned}$$

We also want to calculate the information matrix for the γ_i values. Since $\gamma_i = ln(\pi_i)$, $\partial\gamma_i/\partial\pi_i = 1/\pi_i$; so $\partial\pi_i/\partial\gamma_i = \pi_i$. Hence $I(\gamma) = PI(\boldsymbol{\pi})P$, where P is a diagonal matrix with ith entry equal to π_i. Following El-Helbawy et al. (1994), we let $\Lambda(\boldsymbol{\pi}) = I(\gamma)$ be the information matrix for $\sqrt{sN}\hat{\gamma}$.

Putting this altogether we have that

$$\Lambda_{ii}(\boldsymbol{\pi}) = \pi_i \sum_{j,j\neq i} \lambda_{ij} \frac{\pi_j}{(\pi_i + \pi_j)^2}, \quad i = 1, \ldots, t,$$

and

$$\Lambda_{ij}(\boldsymbol{\pi}) = -\lambda_{ij} \frac{\pi_i \pi_j}{(\pi_i + \pi_j)^2}, \quad i \neq j, \quad i, j = 1, \ldots, t.$$

The $\Lambda(\pi)$ matrix for the 6 pairs in Example 3.2.2 is

$$\frac{1}{6}\begin{bmatrix} \pi_1 \sum_{j\neq 1} \frac{\pi_j}{(\pi_1+\pi_j)^2} & \frac{-\pi_1\pi_2}{(\pi_1+\pi_2)^2} & \frac{-\pi_1\pi_3}{(\pi_1+\pi_3)^2} & \frac{-\pi_1\pi_4}{(\pi_1+\pi_4)^2} \\ \frac{-\pi_1\pi_2}{(\pi_1+\pi_2)^2} & \pi_2 \sum_{j\neq 2} \frac{\pi_j}{(\pi_2+\pi_j)^2} & \frac{-\pi_2\pi_3}{(\pi_2+\pi_3)^2} & \frac{-\pi_2\pi_4}{(\pi_2+\pi_4)^2} \\ \frac{-\pi_1\pi_3}{(\pi_1+\pi_3)^2} & \frac{-\pi_2\pi_3}{(\pi_2+\pi_3)^2} & \pi_3 \sum_{j\neq 3} \frac{\pi_j}{(\pi_3+\pi_j)^2} & \frac{-\pi_3\pi_4}{(\pi_3+\pi_4)^2} \\ \frac{-\pi_1\pi_4}{(\pi_1+\pi_4)^2} & \frac{-\pi_2\pi_4}{(\pi_2+\pi_4)^2} & \frac{-\pi_3\pi_4}{(\pi_3+\pi_4)^2} & \pi_4 \sum_{j\neq 4} \frac{\pi_j}{(\pi_4+\pi_j)^2} \end{bmatrix}.$$

(Recall that $\lambda_{ij} = 1/6$ for every pair.)

The $\Lambda(\pi)$ matrix for the three pairs in Example 3.2.2 is

$$\frac{1}{3}\begin{bmatrix} \pi_1 \sum_{j\neq 1} \frac{\pi_j}{(\pi_1+\pi_j)^2} & \frac{-\pi_1\pi_2}{(\pi_1+\pi_2)^2} & \frac{-\pi_1\pi_3}{(\pi_1+\pi_3)^2} & \frac{-\pi_1\pi_4}{(\pi_1+\pi_4)^2} \\ \frac{-\pi_1\pi_2}{(\pi_1+\pi_2)^2} & \frac{\pi_1\pi_2}{(\pi_1+\pi_2)^2} & 0 & 0 \\ \frac{-\pi_1\pi_3}{(\pi_1+\pi_3)^2} & 0 & \frac{\pi_1\pi_3}{(\pi_1+\pi_3)^2} & 0 \\ \frac{-\pi_1\pi_4}{(\pi_1+\pi_4)^2} & 0 & 0 & \frac{\pi_1\pi_4}{(\pi_1+\pi_4)^2} \end{bmatrix}.$$

(Recall that $\lambda_{ij} = 1/3$ or 0 for every pair.)

The estimated Λ matrix would be obtained by substituting the estimates for the π_i values. (This illustrates one of the main differences between the Bradley–Terry model (indeed any non-linear model) and the familiar least squares models; here the estimate of the covariance matrix depends on the unknown parameters, not just on the design that has been chosen.) For example, the $\Lambda(\pi)$ matrix under the null hypothesis of equally attractive options, that is, assuming $H_0 : \pi_1 = \pi_2 = \ldots = \pi_t = 1$, is given by

$$\Lambda_{ii}(\mathbf{j}_t) = \frac{1}{4} \sum_{j,j\neq i} \lambda_{ij}, \quad i = 1, \ldots, t,$$

and

$$\Lambda_{ij}(\mathbf{j}_t) = -\frac{1}{4}\lambda_{ij}, \quad i \neq j, \quad i, j = 1, \ldots, t,$$

where $\mathbf{j}_t$ is a $t \times 1$ vector of 1s.

The $\Lambda(\mathbf{j}_4)$ matrix for the 6 pairs in Example 3.2.2, when we assume that all the options are equally attractive ($\pi_1 = \pi_2 = \pi_3 = \pi_4 = 1$), is given by

$$\Lambda = \frac{1}{6\times 4}\begin{bmatrix} 3 & -1 & -1 & -1 \\ -1 & 3 & -1 & -1 \\ -1 & -1 & 3 & -1 \\ -1 & -1 & -1 & 3 \end{bmatrix}.$$

The $\Lambda(\mathbf{j}_4)$ matrix for the three pairs in Example 3.2.2, when we assume that all the options are equally attractive, is given by

$$\Lambda = \frac{1}{6\times 4}\begin{bmatrix} 3 & -1 & -1 & -1 \\ -1 & 1 & 0 & 0 \\ -1 & 0 & 1 & 0 \\ -1 & 0 & 0 & 1 \end{bmatrix}.$$

Further discussion about the use of prior information when designing choice experiments may be found in Section 7.4.

3.2.5 Representing Options Using k Attributes

Usually choice experiments are conducted using items that are described by the levels of each of k attributes. With such a representation it then becomes natural to ask whether the main effects of the attributes are significant, whether the two-factor interaction effects of pairs of attributes are significant and so on. In this section, we derive the parameter estimates for the factorial effects of interest and establish the asymptotic distributional properties of these estimates.

Consider the k attributes. We assume that attribute q has ℓ_q levels. Thus there are $L = \prod_q \ell_q$ items all together. We will assume that the level combinations are ordered lexicographically, just as we did in Chapter 2, so T_1 corresponds to $(0, 0, \ldots, 0)$, T_2 corresponds to $(0, 0, \ldots, 0, 1)$, and so on until T_L corresponds to $(\ell_1 - 1, \ell_2 - 1, \ldots, \ell_k - 1)$. Sometimes we will replace the subscript "i" on T_i by the corresponding vector of attribute levels.

We want this representation to be reflected in the way we represent the γ_i; so we write

$$\gamma_{(f_1, f_2, \ldots, f_k)} = \sum_q \beta_{q, f_q} + \sum_{q_1 < q_2} \beta_{q_1 q_2, f_{q_1} f_{q_2}} + \cdots + \beta_{12 \ldots k, f_1 f_2 \ldots f_k},$$

where β_{q,f_q} is the effect of attribute q when at level f_q, $\beta_{q_1 q_2, f_{q_1} f_{q_2}}$ is the joint effect of attributes q_1 and q_2 at levels f_{q_1} and f_{q_2} and so on. This representation has more parameters than it requires. Specifically it has

$$\sum_q \ell_q + \sum_{q_1 < q_2} \ell_{q_1} \ell_{q_2} + \cdots + L = \prod_q (\ell_q + 1) - 1$$

parameters; so we must impose $\prod_q (\ell_q + 1) - \prod_q \ell_q$ independent constraints. We see that this is exactly analogous to the situation for the usual representation of a full factorial model in an ordinary least squares setting and we impose the familiar constraints

$$\sum_{f_q} \beta_{q, f_q} = 0, \quad \sum_{f_{q_1}} \beta_{q_1 q_2, f_{q_1} f_{q_2}} = \sum_{f_{q_2}} \beta_{q_1 q_2, f_{q_1} f_{q_2}} = 0$$

and so on.

Many authors represent $\gamma_{(f_1, f_2, \ldots, f_k)}$ by $V_i = \boldsymbol{\beta}' \mathbf{x}_i$, where $(f_1, f_2, \ldots, f_k)$ represents the ith treatment combination in lexicographic order, $\boldsymbol{\beta}$ contains all the possible β values and $\mathbf{x}_i$ is a (0,1) indicator vector.

Another common representation incorporates the constraints into the vector $\boldsymbol{\beta}$. Thus instead of having ℓ_q elements in $\boldsymbol{\beta}$ for attribute q, there are only $\ell_q - 1$ entries and there are $(\ell_{q_1} - 1)(\ell_{q_2} - 1)$ parameters for the joint effect of attributes q_1 and q_2 and so on. This means that $\mathbf{x}_i$ uses $\ell_q - 1$ entries for each attribute, with all entries 0 except for the position corresponding to f_q, which is 1, for the first $\ell_q - 1$ levels and level $f_{\ell_q - 1}$ corresponding to $\ell_q - 1$ entries of -1. In either case, the null hypothesis $\boldsymbol{\pi} = \mathbf{j}_L$ is equivalent to having all entries in $\boldsymbol{\beta}$ equal to 0.

We illustrate these ideas in the next example.

■ **EXAMPLE 3.2.5.**

Let $k = 2$ and let $\ell_1 = 2$ and $\ell_2 = 3$. Then, for the first representation, we have

$$\boldsymbol{\beta}' = (\beta_{1,0}, \beta_{1,1}, \beta_{2,0}, \beta_{2,1}, \beta_{2,2}, \beta_{12,00}, \beta_{12,01}, \beta_{12,02}, \beta_{12,10}, \beta_{12,11}, \beta_{12,12}).$$

The treatment combinations and the corresponding $\mathbf{x}_i$ and γ_i are given in Table 3.4. In this representation the constraints are

$$\beta_{1,0} + \beta_{1,1} = 0, \;\; \beta_{2,0} + \beta_{2,1} + \beta_{2,2} = 0,$$

$$\beta_{12,00} + \beta_{12,10} = 0, \;\; \beta_{12,01} + \beta_{12,11} = 0, \;\; \beta_{12,02} + \beta_{12,12} = 0,$$

and

$$\beta_{12,00} + \beta_{12,01} + \beta_{12,02} = 0, \;\; \beta_{12,10} + \beta_{12,11} + \beta_{12,12} = 0.$$

To get the second representation we need to incorporate these constraints into $\boldsymbol{\beta}$. We see that

$$\begin{aligned} \beta_{1,1} &= -\beta_{1,0}, \\ \beta_{2,2} &= -\beta_{2,0} - \beta_{2,1}, \\ \beta_{12,10} &= -\beta_{12,00}, \\ \beta_{12,11} &= -\beta_{12,01}, \\ \beta_{12,02} &= -\beta_{12,00} - \beta_{12,01}, \end{aligned}$$

and

$$\beta_{12,12} = -\beta_{12,02} = \beta_{12,00} + \beta_{12,01}.$$

Thus we have

$$\boldsymbol{\beta}' = (\beta_{1,0}, \beta_{2,0}, \beta_{2,1}, \beta_{12,00}, \beta_{12,01}).$$

The treatment combinations, and the corresponding $\mathbf{x}_i$ and γ_i, are given in Table 3.5. □

Table 3.4 Unconstrained Representation of Treatment Combinations

Treatment Combination	$\mathbf{x}_i$	γ_i
0,0	1,0,1,0,0,1,0,0,0,0,0	$\beta_{1.0} + \beta_{2.0} + \beta_{12.00}$
0,1	1,0,0,1,0,0,1,0,0,0,0	$\beta_{1.0} + \beta_{2.1} + \beta_{12.01}$
0,2	1,0,0,0,1,0,0,1,0,0,0	$\beta_{1.0} + \beta_{2.2} + \beta_{12.02}$
1,0	0,1,1,0,0,0,0,0,1,0,0	$\beta_{1.1} + \beta_{2.0} + \beta_{12.10}$
1,1	0,1,0,1,0,0,0,0,0,1,0	$\beta_{1.1} + \beta_{2.1} + \beta_{12.11}$
1,2	0,1,0,0,1,0,0,0,0,0,1	$\beta_{1.1} + \beta_{2.2} + \beta_{12.12}$

We will be using the first representation in this book.

Table 3.5 Constrained Representation of Treatment Combinations

Treatment Combination	$\mathbf{x}_i$	γ_i
0,0	1,1,0,1,0	$\beta_{1.0}+\beta_{2.0}+\beta_{12.00}$
0,1	1,0,1,0,1	$\beta_{1.0}+\beta_{2.1}+\beta_{12.01}$
0,2	1,−1,−1,−1,−1	$\beta_{1.0}-\beta_{2.0}-\beta_{2.1}-\beta_{12.00}-\beta_{12.01}$
1,0	−1,1,0,−1,0	$-\beta_{1.0}+\beta_{2.0}-\beta_{12.00}$
1,1	−1,0,1,0,−1	$-\beta_{1.0}+\beta_{2.1}-\beta_{12.01}$
1,2	−1,−1,−1,1,1	$-\beta_{1.0}-\beta_{2.0}-\beta_{2.1}+\beta_{12.00}+\beta_{12.01}$

Questions about the main effects of attribute q are answered by considering contrasts of the β_{q,f_q}, questions about the two-factor interaction of attributes q_1 and q_2 are answered by considering contrasts of the $\beta_{q_1q_2,f_{q_1}f_{q_2}}$ and so on. If we want to develop the best design to test if some particular effects are zero, we do this by specifying which effects we want to test are zero, and which effects (if any) we are going to assume are zero, and then we find the parameter estimates and determine the asymptotic properties of these estimates.

We write the p contrasts that we want to test as the rows of the matrix B_h. We choose the coefficients in the contrast matrix B_h so that $B_hB_h'=I_p$ where I_p is the identity matrix of order p. That is, we say that the matrix B_h is *orthonormal*. If any row of B_h is not already of unit length then we divide the entries in that row by the square root of the sum of the squares of the entries in that row. All contrasts that we assume are 0 appear as rows of the orthonormal matrix B_a which we will assume has a rows. Finally, B_r contains any remaining contrasts from the complete set of $L-1$ independent contrasts. The next example illustrates these ideas for three binary attributes.

■ **EXAMPLE 3.2.6.**
Let $k=3$ and let $\ell_1=\ell_2=\ell_3=2$. So the eight items are represented by 000, 001, 010, 011, 100, 101, 110 and 111, in lexicographic order. Then

$$\gamma_{011}=\beta_{1,0}+\beta_{2,1}+\beta_{3,1}+\beta_{12,01}+\beta_{13,01}+\beta_{23,11}+\beta_{123,011}.$$

The constraints that we impose are

$$\beta_{q,0}+\beta_{q,1}=0,\quad q=1,2,3,$$

$$\beta_{q_1q_2,00}+\beta_{q_1q_2,01}=\beta_{q_1q_2,10}+\beta_{q_1q_2,11}=0,\ q_1,q_2=1,2;1,3;2,1;2,3;3,1;3,2,$$

$$\beta_{123,000}+\beta_{123,001}=\beta_{123,010}+\beta_{123,011}=\beta_{123,100}+\beta_{123,101}=\beta_{123,110}+\beta_{123,111}=0,$$

and similarly for each of the other two attributes for the components of the three-factor interaction.

Suppose that we want to test if the main effects are 0. Then

$$B_h=\frac{1}{2\sqrt{2}}\begin{bmatrix}-1&-1&-1&-1&1&1&1&1\\-1&-1&1&1&-1&-1&1&1\\-1&1&-1&1&-1&1&-1&1\end{bmatrix}.$$

Observe that $B_hB_h'=I_3$.

Suppose that we do this assuming that the three-factor interaction is 0. Then

$$B_a = \frac{1}{2\sqrt{2}} \begin{bmatrix} -1 & 1 & 1 & -1 & 1 & -1 & -1 & 1 \end{bmatrix}.$$

That means that we have made no assumptions about the two-factor interaction contrasts; so

$$B_r = \frac{1}{2\sqrt{2}} \begin{bmatrix} 1 & 1 & -1 & -1 & -1 & -1 & 1 & 1 \\ 1 & -1 & 1 & -1 & -1 & 1 & -1 & 1 \\ 1 & -1 & -1 & 1 & 1 & -1 & -1 & 1 \end{bmatrix}.$$

The likelihood function itself is independent of any constraints that we impose on the parameters but the solution will have to reflect the constraints. We summarize the results in Bradley and El-Helbawy (1976) below.

We want to maximize the likelihood equation

$$L(\boldsymbol{\pi}) = \frac{\pi_1^{w_1} \pi_2^{w_2} \cdots \pi_t^{w_t}}{\prod_{i<j} (\pi_i + \pi_j)^{sn_{ij}}}$$

subject to the normalizing constraint $\sum_i \ln(\pi_i) = \sum_i \gamma_i = 0$ and the assumed constraints $B_a \gamma = \mathbf{0}_a$ where $\mathbf{0}_a$ is a column of 0s of length a. Working with the log-likelihood and using Lagrange multipliers we maximize

$$\sum_{i=1}^{L} w_i \gamma_i - \sum_{i<j} sn_{ij} \ln(\pi_i + \pi_j) + \kappa_1 \sum_{i=1}^{L} \gamma_i + [\kappa_2, \ldots, \kappa_{a+1}] B_a \gamma$$

subject to $\sum_{i=1}^{L} \gamma_i = 0$ and $B_a \gamma = \mathbf{0}_a$. As before we assume that every item has been chosen at least once.

To maximize the function we differentiate with respect to each π_i in turn. For π_i this gives

$$\frac{w_i}{\pi_i} - \sum_{j \neq i} \frac{sn_{ij}}{\pi_i + \pi_j} + \frac{\kappa_1}{\pi_i} + \sum_{j=1}^{a} \kappa_{j+1} (B_a)_{ji} \frac{1}{\pi_i} = 0.$$

Let $w_i - \sum_{j \neq i} \frac{sn_{ij}\pi_i}{\pi_i + \pi_j} = z_i$, $\mathbf{z} = (z_1, z_2, \ldots, z_L)'$ and $\boldsymbol{\kappa} = (\kappa_2, \ldots, \kappa_{a+1})'$. Then, writing the derivative in matrix notation, we have

$$\mathbf{z} + \kappa_1 \mathbf{j}_L + B_a' \boldsymbol{\kappa} = \mathbf{0}_L.$$

Pre-multiplying by $\mathbf{j}_L'$ we get

$$\mathbf{j}_L' \mathbf{z} + \kappa_1 \mathbf{j}_L' \mathbf{j}_L + \mathbf{j}_L' B_a' \boldsymbol{\kappa} = \mathbf{0}_L.$$

Since $\mathbf{j}_L' \mathbf{z} = sN - sN$ and $\mathbf{j}_L' B_a = 0$, we have that $L\kappa_1 = 0$ and since $B_a B_a' = I_a$ we have

$$B_a \mathbf{z} + \boldsymbol{\kappa} = \mathbf{0}_L.$$

Hence $\boldsymbol{\kappa} = -B_a \mathbf{z}$ and

$$\mathbf{z} - B_a' B_a \mathbf{z} = (I - B_a' B_a)\mathbf{z} = \mathbf{0}_L.$$

As before these equations are solved iteratively. Their convergence is guaranteed, provided that there are no items that are never chosen; see El-Helbawy and Bradley (1977) for details.

What about the distribution of these constrained maximum likelihood parameter estimates? To find this we use the ideas of El-Helbawy and Bradley (1978) who apply the results of Bradley and Gart (1962) to the parameters in $\begin{bmatrix} B_h \\ B_r \end{bmatrix} \gamma = B_{hr}\gamma = \boldsymbol{\theta}$. Then the information matrix C of $\sqrt{sN}\hat{\boldsymbol{\theta}}$ has entries C_{uq} given by

$$\sum_{a<b} \lambda_{ab} \mathcal{E}_\theta \left(\left(\frac{\partial \ln f_{ab\alpha}(w_{ab\alpha}, \boldsymbol{\pi})}{\partial \theta_u} \right) \left(\frac{\partial \ln f_{ab\alpha}(w_{ab\alpha}, \boldsymbol{\pi})}{\partial \theta_q} \right) \right)$$

$$= \sum_i \sum_j \left[\sum_{a<b} \lambda_{ab} \mathcal{E}_\pi \left(\left(\frac{\partial \ln f_{ab\alpha}(w_{ab\alpha}, \boldsymbol{\pi})}{\partial \gamma_i} \right) \left(\frac{\partial \ln f_{ab\alpha}(w_{ab\alpha}, \boldsymbol{\pi})}{\partial \gamma_j} \right) \right) \right] (B_{hr})_{ui} (B_{hr})_{qj}.$$

Thus we have

$$C = B_{hr} \Lambda(\boldsymbol{\pi}) B'_{hr}.$$

Unless we need to emphasize the value of $\boldsymbol{\pi}$ we are considering, we will write Λ for $\Lambda(\boldsymbol{\pi})$ from now on.

We are really interested in $B_h\hat{\gamma}$ of course, and these are the first p entries of $\hat{\boldsymbol{\theta}}$. The variance-covariance matrix of $\sqrt{sN}\hat{\boldsymbol{\theta}} = \sqrt{sN} \begin{bmatrix} B_h\hat{\gamma} \\ B_r\hat{\gamma} \end{bmatrix}$ is C^{-1}. If we write

$$C = \begin{bmatrix} C_{hh} & C_{hr} \\ C_{rh} & C_{rr} \end{bmatrix}$$

then the variance-covariance matrix of $\sqrt{sN}B_h\hat{\gamma}$ is given by C_{hh}^{-1} provided, of course, $C_{hr} = \mathbf{0}_{p,L-1-p-a}$. If $C_{hr} \neq \mathbf{0}_{p,L-1-p-a}$ then the variance-covariance matrix of $\sqrt{sN}B_h\hat{\gamma}$ is obtained from the principal minor of order p of C^{-1}; that is, the submatrix of order p starting in position (1,1).

■ EXAMPLE 3.2.7.

Consider the situation of Example 3.2.6. Suppose that we use the pairs (000, 111), (001, 110), (010, 101), (011, 100), (000, 011), (000, 101), (000, 110), (111, 001), (111, 010), and (111, 100) and calculate Λ assuming that $\boldsymbol{\pi} = \mathbf{j}_8$. We get

$$\Lambda = \frac{1}{40} \begin{bmatrix} 4 & 0 & 0 & -1 & 0 & -1 & -1 & -1 \\ 0 & 2 & 0 & 0 & 0 & 0 & -1 & -1 \\ 0 & 0 & 2 & 0 & 0 & -1 & 0 & -1 \\ -1 & 0 & 0 & 2 & -1 & 0 & 0 & 0 \\ 0 & 0 & 0 & -1 & 2 & 0 & 0 & -1 \\ -1 & 0 & -1 & 0 & 0 & 2 & 0 & 0 \\ -1 & -1 & 0 & 0 & 0 & 0 & 2 & 0 \\ -1 & -1 & -1 & 0 & -1 & 0 & 0 & 4 \end{bmatrix}$$

and

$$C = B_{hr} \Lambda B'_{hr} = \frac{1}{40} \begin{bmatrix} 4 & 1 & 1 & \vdots & 0 & 0 & 0 \\ 1 & 4 & 1 & \vdots & 0 & 0 & 0 \\ 1 & 1 & 4 & \vdots & 0 & 0 & 0 \\ \cdots & \cdots & \cdots & \cdots & \cdots & \cdots & \cdots \\ 0 & 0 & 0 & \vdots & 2 & 1 & 1 \\ 0 & 0 & 0 & \vdots & 1 & 2 & 1 \\ 0 & 0 & 0 & \vdots & 1 & 1 & 2 \end{bmatrix}.$$

The variance-covariance matrix for $\sqrt{sN}B_{hr}\hat{\gamma}$ is $\frac{1}{sN}C^{-1}$, where

$$C^{-1} = \frac{1}{9}\left[\begin{array}{ccc:ccc} 100 & -20 & -20 & 0 & 0 & 0 \\ -20 & 100 & -20 & 0 & 0 & 0 \\ -20 & -20 & 100 & 0 & 0 & 0 \\ \hdashline 0 & 0 & 0 & 270 & -90 & -90 \\ 0 & 0 & 0 & -90 & 270 & -90 \\ 0 & 0 & 0 & -90 & -90 & 270 \end{array}\right].$$

Thus we see that the elements of $B_h\hat{\gamma}$ and $B_r\hat{\gamma}$ are independently estimated (since the covariance of any element of $B_h\hat{\gamma}$ and $B_r\hat{\gamma}$ is 0) and so the variance-covariance matrix of $\sqrt{sN}B_h\hat{\gamma}$ is the principal minor of order 3 of C^{-1}. □

However it is not always the case that the elements of $B_h\hat{\gamma}$ and $B_r\hat{\gamma}$ are independently estimated, as the next example shows.

■ EXAMPLE 3.2.8.
Let $k = \ell_1 = \ell_2 = 2$. Suppose that we want to test hypotheses about the main effects so

$$B_h = \frac{1}{2}\begin{bmatrix} -1 & -1 & 1 & 1 \\ -1 & 1 & -1 & 1 \end{bmatrix}$$

and we do not want to make any assumptions about the two-factor interaction effect. Then

$$B_r = \frac{1}{2}\begin{bmatrix} 1 & -1 & -1 & 1 \end{bmatrix}.$$

Suppose that we use the pairs (00, 11), (01, 10), and (00, 01). Then

$$\Lambda = \frac{1}{12}\begin{bmatrix} 2 & -1 & 0 & -1 \\ -1 & 2 & -1 & 0 \\ 0 & -1 & 1 & 0 \\ -1 & 0 & 0 & 1 \end{bmatrix},$$

$$C = \frac{1}{12}\left[\begin{array}{cc:c} 2 & 0 & 0 \\ 0 & 3 & -1 \\ \hdashline 0 & -1 & 1 \end{array}\right] \text{ and } C^{-1} = \left[\begin{array}{cc:c} 6 & 0 & 0 \\ 0 & 6 & 6 \\ \hdashline 0 & 6 & 18 \end{array}\right].$$

If we assume instead that

$$B_r\hat{\gamma} = 0,$$

then

$$C = B_h\Lambda B_h' = \frac{1}{12}\begin{bmatrix} 2 & 0 \\ 0 & 3 \end{bmatrix} \text{ with } C^{-1} = \begin{bmatrix} 6 & 0 \\ 0 & 4 \end{bmatrix}.$$

Hence we see that with these choice sets the variance estimate of the estimate of the main effect of the second factor, $(\gamma_{11} + \gamma_{01} - (\gamma_{10} + \gamma_{00}))/2$, depends on the assumption, if any, made about the two-factor interaction. □

The off-diagonal elements of the variance-covariance matrix C^{-1} give the covariance of pairs of contrasts in $\sqrt{sN}B_h\hat{\gamma}$. If one of these entries is 0 then the corresponding contrasts are independently estimated. So if the variance-covariance matrix is diagonal, then all pairs of contrasts are independently estimated. Of course the inverse of a diagonal matrix is diagonal; so we prefer designs in which the C matrix is diagonal.

If the attributes that are used to describe the options in a choice experiment have more than two levels then there are often several different ways in which the contrasts in B_h can be represented. We can move between these different representations of the contrasts by multiplying appropriate matrices together. Thus we are not concerned if the C matrix has non-zero off-diagonal entries that correspond to contrasts for the same effect, main or interaction.

If we think of C as being made up of block matrices of order $(\ell_{q_1} - 1) \times (\ell_{q_2} - 1)$, for the main effects, block matrices of order $(\ell_{q_1} - 1)(\ell_{q_2} - 1) \times (\ell_{q_3} - 1)(\ell_{q_4} - 1)$ for the two-factor interaction effects, $q_1, q_2, q_3, q_4 = 1, 2, \ldots, k$, and so on for higher-order interaction effects, then we want the off-diagonal blocks to be a zero matrix of the appropriate size, but the diagonal blocks are not restricted in their form. In the case when all the off-diagonal block matrices are zero, we say that C is *block diagonal* and we note that since the inverse of a block diagonal matrix is block diagonal, the covariance of any two contrasts from different attributes is 0.

If we want the components of the contrasts for a particular effect to be independent, then we would have to choose a suitable set of contrasts to achieve that. While this may be possible, it depends on the choice sets used in the experiment and it is sometimes true that the contrasts that result do not have an easily interpretable meaning.

Can we say under what circumstances the estimates of the variance-covariance matrix of $B_h\hat{\gamma}$ is independent of the form of B_r? The short answer is yes, when $C_{hr} = 0$, and we discuss this further in Chapters 4, 5, and 6.

It is not essential that the elements in B_h be independent contrasts, or even that they be contrasts. For example, if we want to work with the second representation of $\boldsymbol{\beta}$ in a main effects only setting, then we merely need to use as B_h a matrix which calculates $\beta_{q,f_q} - \beta_{q,\ell_q-1}$ for the first $\ell_q - 1$ levels of attribute q. The corresponding information matrix is still $B_h \Lambda B_h'$, given that all other interaction effects are assumed to be 0. We illustrate this in the next example.

■ EXAMPLE 3.2.9.

Consider the situation of Example 3.2.5. Suppose that we assume that the two-factor interaction is 0. Then for the first representation

$$\boldsymbol{\beta}' = (\beta_{1,0}, \beta_{1,1}, \beta_{2,0}, \beta_{2,1}, \beta_{2,2})$$

and hence

$$\gamma' = (\beta_{1,0} + \beta_{2,0}, \beta_{1,0} + \beta_{2,1}, \beta_{1,0} + \beta_{2,2}, \beta_{1,1} + \beta_{2,0}, \beta_{1,1} + \beta_{2,1}, \beta_{1,1} + \beta_{2,2}).$$

If we let

$$B_h = \begin{bmatrix} \frac{-1}{\sqrt{6}} & \frac{-1}{\sqrt{6}} & \frac{-1}{\sqrt{6}} & \frac{1}{\sqrt{6}} & \frac{1}{\sqrt{6}} & \frac{1}{\sqrt{6}} \\ \frac{-1}{2} & 0 & \frac{1}{2} & \frac{-1}{2} & 0 & \frac{1}{2} \\ \frac{1}{2\sqrt{3}} & \frac{-2}{2\sqrt{3}} & \frac{1}{2\sqrt{3}} & \frac{1}{2\sqrt{3}} & \frac{-2}{2\sqrt{3}} & \frac{1}{2\sqrt{3}} \end{bmatrix},$$

we can test hypotheses about the main effects using $B_h\gamma$.

For the second representation the parameters of interest are

$$\beta_{1,0} - \beta_{1,1}, \; \beta_{2,0} - \beta_{2,2}, \; \beta_{2,1} - \beta_{2,2}.$$

If

$$B_\Omega = \begin{bmatrix} 1 & 1 & 1 & -1 & -1 & -1 \\ 1 & 0 & -1 & 1 & 0 & -1 \\ 0 & 1 & -1 & 0 & 1 & -1 \end{bmatrix},$$

then

$$B_\Omega \gamma = \begin{bmatrix} 3(\beta_{1,0} - \beta_{1,1}) \\ 2(\beta_{2,0} - \beta_{2,2}) \\ 2(\beta_{2,1} - \beta_{2,2}) \end{bmatrix}.$$

Suppose that we use the pairs (00, 11), 01, 12), (02, 10), (10, 01), (11, 02), and (12, 00). Then, under the null hypothesis, we have

$$\Lambda = \frac{1}{4 \times 6} \begin{bmatrix} 2 & 0 & 0 & 0 & -1 & -1 \\ 0 & 2 & 0 & -1 & 0 & -1 \\ 0 & 0 & 2 & -1 & -1 & 0 \\ 0 & -1 & -1 & 2 & 0 & 0 \\ -1 & 0 & -1 & 0 & 2 & 0 \\ -1 & -1 & 0 & 0 & 0 & 2 \end{bmatrix}$$

and the information matrix for $B_\Omega \gamma$ is $B_\Omega \Lambda B'_\Omega = C_\Omega$.

The information matrix for $B_\Omega \gamma$ is often expressed in terms of the treatment combinations in each choice set (see Huber and Zwerina (1996), for example). We show the relationship between these two approaches here. For the second representation the $\mathbf{x}_i$ are represented by triples; see Example 3.2.5. These triples are the columns of B_Ω and so we will write

$$B_\Omega = \begin{bmatrix} \mathbf{x}_1, & \mathbf{x}_2, & \ldots, & \mathbf{x}_6 \end{bmatrix}.$$

Then

$$\begin{aligned} B_\Omega \Lambda B'_\Omega &= \frac{1}{4 \times 6} \left[\textstyle\sum_j n_{1j}(\mathbf{x}_1 - \mathbf{x}_j), \sum_j n_{2j}(\mathbf{x}_2 - \mathbf{x}_j), \ldots, \sum_j n_{6j}(\mathbf{x}_6 - \mathbf{x}_j) \right] B'_\Omega \\ &= \frac{1}{4 \times 6} \sum_i \sum_j n_{ij}(\mathbf{x}_i - \mathbf{x}_j)\mathbf{x}'_i \\ &= \frac{1}{4 \times 6} \sum_{i<j} n_{ij}(\mathbf{x}_i - \mathbf{x}_j)(\mathbf{x}_i - \mathbf{x}_j)'. \end{aligned}$$

Huber and Zwerina (1996) write the information matrix for $B_\Omega \gamma$, under the null hypothesis, as

$$\Omega_0 = \frac{1}{m} \sum_{n=1}^{N} \sum_{j=1}^{m} \mathbf{z}_{jn} \mathbf{z}'_{jn},$$

where there are m options in each of N choice sets and

$$\mathbf{z}_{jn} = \mathbf{x}_{jn} - \bar{\mathbf{x}}_n \text{ and } \bar{\mathbf{x}}_n = \frac{1}{m} \sum_{a=1}^{m} \mathbf{x}_{an}.$$

In this case $m = 2$, and hence we have

$$\mathbf{z}_{1n} = \mathbf{x}_{1n} - (\mathbf{x}_{1n} + \mathbf{x}_{2n})/2 = \frac{1}{2}(\mathbf{x}_{1n} - \mathbf{x}_{2n})$$

and

$$\mathbf{z}_{2n} = \mathbf{x}_{2n} - (\mathbf{x}_{1n} + \mathbf{x}_{2n})/2 = \frac{1}{2}(\mathbf{x}_{2n} - \mathbf{x}_{1n}).$$

From this we see that

$$\begin{aligned}\Omega_0 &= \frac{1}{2}\sum_{n=1}^{6}\left(\frac{1}{4}(\mathbf{x}_{1n} - \mathbf{x}_{2n})(\mathbf{x}_{1n} - \mathbf{x}_{2n})' + \frac{1}{4}(\mathbf{x}_{2n} - \mathbf{x}_{1n})(\mathbf{x}_{2n} - \mathbf{x}_{1n})'\right) \\ &= \frac{1}{4}\sum_{n=1}^{6}(\mathbf{x}_{1n} - \mathbf{x}_{2n})(\mathbf{x}_{1n} - \mathbf{x}_{2n})'.\end{aligned}$$

Since $n_{ij} = 1$ when $\mathbf{x}_i$ and $\mathbf{x}_j$ are in the same choice set and is 0 otherwise we find

$$\Omega_0 = \frac{1}{4}\sum_{i<j} n_{ij}(\mathbf{x}_i - \mathbf{x}_j)(\mathbf{x}_i - \mathbf{x}_j)' = 6B_\Omega \Lambda B_\Omega' = NC_\Omega.$$

The other question is that of connectedness. Since we are no longer interested in the t treatment combinations specifically but in the main effects (and perhaps the two-factor interactions), it is no longer necessary to use the definition of connectedness that we gave previously. Now it is only necessary for the effects that we want to estimate to be "connected". This is harder to define combinatorially, but one practical way to see if a set of choice sets is connected for the effects of interest is simply to calculate $C = B_{hr}\Lambda B_{hr}'$ and check that C is invertible. Certainly quite small designs can be useful, particularly if only the main effects are of interest, as the next example shows.

■ **EXAMPLE 3.2.10.**

If there are $k = 3$ binary attributes, then the pairs in Table 3.6 might be used as a choice experiment. (We say that an option has been *folded over* if all the 0s in an option have been replaced by 1s and vice versa to get the other option in the choice set.) The corresponding Λ matrix, under the null hypothesis that all items are equally attractive, is given by

$$\Lambda = \frac{1}{(1+1)^2} \times \frac{1}{4}\begin{bmatrix} 1 & 0 & 0 & 0 & 0 & 0 & 0 & -1 \\ 0 & 1 & 0 & 0 & 0 & 0 & -1 & 0 \\ 0 & 0 & 1 & 0 & 0 & -1 & 0 & 0 \\ 0 & 0 & 0 & 1 & -1 & 0 & 0 & 0 \\ 0 & 0 & 0 & -1 & 1 & 0 & 0 & 0 \\ 0 & 0 & -1 & 0 & 0 & 1 & 0 & 0 \\ 0 & -1 & 0 & 0 & 0 & 0 & 1 & 0 \\ -1 & 0 & 0 & 0 & 0 & 0 & 0 & 1 \end{bmatrix}.$$

Suppose that we are only interested in estimating main effects and that we can assume that the remaining contrasts of the γ_i are equal to 0. Then we have the three main effects contrasts in B_h and B_r is empty. We find that $C = \frac{1}{8}I_3$, showing that we only need these four choice sets to estimate the main effects for three binary attributes if we can assume that the interaction effects are 0. □

In this section, we have parameterized the γ_i in terms of the factorial effects, and have found the corresponding maximum likelihood estimates and variance-covariance matrices. We have seen that if $C_{hr} = \mathbf{0}$, then $B_h\hat{\gamma}$ is independent of the form of B_r and that the definition of connectedness changes to reflect the effects that we want to estimate.

Table 3.6 The Fold-over Pairs with $k = 3$

Option A	Option B
000	111
110	001
101	010
011	100

In the next section, we derive the Λ matrix for larger choice sets and briefly discuss using k attributes to describe the options in this situation. In Chapter 4 we discuss the structure of the optimal designs for binary attributes for the estimation of main effects, main effects plus two-factor interactions and main plus some two-factor interactions when all choice sets are of size 2. In Chapter 5 we allow the choice sets to have more than 2 options and in Chapter 6 we allow the attributes to have more than 2 levels each and the choice sets to be of any size.

3.2.6 Exercises

1. In Example 3.2.2 use the initial estimates $\hat{\pi}_i^{(0)} = w_i/30$ and perform 5 iterations. How do these estimates compare with the ones given in the text?
2. Suppose that $k = 4$ and that the four attributes are binary. Suppose that the four pairs (0000, 1111), (0011, 1100), (0101, 1010), and (0011, 1100) are to be used to estimate the main effects of the attributes.
   (a) Determine the matrix B_h for estimating main effects.
   (b) Determine the matrix Λ associated with these pairs.
   (c) Hence determine the information matrix C when B_a contains all of the effects except the main effects and when B_a is empty. Comment.
   (d) Find the Ω_0 matrix for these choice sets.

3.3 THE MNL MODEL FOR CHOICE SETS OF ANY SIZE

In this section we discuss the MNL model for choice sets of any size. Train (2003) gives an extensive discussion of other models and of associated estimation questions.

3.3.1 Choice Sets of Any Size

We begin by deriving the information matrix for the π_i when the choice sets may have any number of options in them. Since all subjects see the same choice sets, we only need to consider results from one subject in the derivation; so the subscript α has been suppressed in what follows.

Consider an experiment in which there are N choice sets of m options, of which $n_{i_1,i_2,\ldots,i_m}$ compare the specific options $T_{i_1}, T_{i_2}, \ldots, T_{i_m}$, where

$$n_{i_1,i_2,\ldots,i_m} = \begin{cases} 1 & \text{if } (T_{i_1}, T_{i_2}, \ldots, T_{i_m}) \text{ is a choice set,} \\ 0 & \text{otherwise.} \end{cases}$$

Then

$$N = \sum_{i_1<i_2<\cdots<i_m} n_{i_1,i_2,\ldots,i_m}.$$

We define parameters $\pi=(\pi_1, \pi_2, \ldots, \pi_t)$ associated with t items $T_1, T_2, \ldots, T_t$. Given a choice set which contains m options $T_{i_1}, T_{i_2}, \ldots, T_{i_m}$ the probability that option T_{i_1} is preferred to the other $m-1$ options in the choice set is

$$P(T_{i_1} > T_{i_2}, \ldots, T_{i_m}) = \frac{\pi_{i_1}}{\sum_{j=1}^{m} \pi_{i_j}}$$

for $i_j = 1, 2, \ldots, t$ (assuming that all options in each choice set are distinct). We also assume that choices made in one choice set do not affect choices made in any other choice set. If $m = 2$, this is just the model of the previous section.

We use the method in Bradley (1955) and Pendergrass and Bradley (1960) to derive the form of the entries in Λ for any value of m.

Let $w_{i_1,i_2,\ldots,i_m}$ be an indicator variable where

$$w_{i_1,i_2,\ldots,i_m} = \begin{cases} 1 & \text{if } T_{i_1} > T_{i_2}, T_{i_3} \ldots, T_{i_m}, \\ 0 & \text{otherwise.} \end{cases}$$

Then

$$\mathcal{E}(w_{i_1,i_2,\ldots,i_m}) = \frac{\pi_{i_1}}{\sum_{j=1}^{m} \pi_{i_j}} \quad \text{and} \quad \text{Var}(w_{i_1,i_2,\ldots,i_m}) = \frac{\pi_{i_1} \sum_{j=2}^{m} \pi_{i_j}}{(\sum_{j=1}^{m} \pi_{i_j})^2}.$$

We let w_{i_1} be the number of times option T_{i_1} is preferred to the other options available in any choice set in which T_{i_1} appears. Then

$$w_{i_1} = \sum_{i_2<i_3<\cdots<i_m} w_{i_1,i_2,\ldots,i_m},$$

where the summation is over $i_2 < i_3 < \cdots < i_m$ and $i_j \neq i_1$ for $j = 2, \ldots, m$.

It follows that

$$\mathcal{E}(w_{i_1}) = \pi_{i_1} \sum_{i_2<i_3<\cdots<i_m} \frac{1}{\sum_{j=1}^{m} \pi_{i_j}},$$

$$\text{Var}(w_{i_1}) = \pi_{i_1} \sum_{i_2<i_3<\cdots<i_m} \frac{\sum_{j=2}^{m} \pi_{i_j}}{(\sum_{j=1}^{m} \pi_{i_j})^2},$$

$$\text{and} \quad \text{Cov}(w_{i_1}, w_{i_2}) = \sum_{i_3<\cdots<i_m} \frac{-\pi_{i_1}\pi_{i_2}}{(\sum_{j=1}^{m} \pi_{i_j})^2}.$$

■ **EXAMPLE 3.3.1.**

Suppose that the four choice sets (T_1, T_2, T_3), (T_1, T_2, T_4), (T_1, T_3, T_4), and (T_2, T_3, T_4) are to be used to compare $t = 4$ items. Then we know that $w_1 = w_{123} + w_{124} + w_{134}$ and $w_2 = w_{213} + w_{214} + w_{234}$. Thus we have that

$$\begin{aligned} \mathcal{E}(w_1) &= \mathcal{E}(w_{123} + w_{124} + w_{134}) \\ &= \mathcal{E}(w_{123}) + \mathcal{E}(w_{124}) + \mathcal{E}(w_{134}) \\ &= \frac{\pi_1}{(\pi_1 + \pi_2 + \pi_3)} + \frac{\pi_1}{(\pi_1 + \pi_2 + \pi_4)} + \frac{\pi_1}{(\pi_1 + \pi_3 + \pi_4)}. \end{aligned}$$

We can get a similar expression for $\mathcal{E}(w_2)$. Next we derive the covariance of w_1 and w_2.

$$\begin{aligned}
\text{Cov}(w_1, w_2) &= \text{Cov}(w_{123} + w_{124} + w_{134}, w_{213} + w_{214} + w_{234}) \\
&= \text{Cov}(w_{123}, w_{213}) + \text{Cov}(w_{123}, w_{214}) + \text{Cov}(w_{123}, w_{234}) \\
&\quad + \text{Cov}(w_{124}, w_{213}) + \text{Cov}(w_{124}, w_{214}) + \text{Cov}(w_{124}, w_{234}) \\
&\quad + \text{Cov}(w_{134}, w_{213}) + \text{Cov}(w_{134}, w_{214}) + \text{Cov}(w_{134}, w_{234}).
\end{aligned}$$

The only two non-zero covariances are those of w_{123} and w_{213} and of w_{124} and w_{214}, as each of these pairs of random variables come from the same choice set. We have

$$\begin{aligned}
\text{Cov}(w_{123}, w_{213}) &= \mathcal{E}\left[\left(w_{123} - \frac{\pi_1}{(\pi_1 + \pi_2 + \pi_3)}\right) \times \left(w_{213} - \frac{\pi_2}{(\pi_1 + \pi_2 + \pi_3)}\right)\right] \\
&= \mathcal{E}\left(w_{123} w_{213}\right) - \frac{\pi_1 \pi_2}{(\pi_1 + \pi_2 + \pi_3)^2} \\
&= \frac{-\pi_1 \pi_2}{(\pi_1 + \pi_2 + \pi_3)^2}
\end{aligned}$$

since the product $w_{123}w_{213}$ is always 0 (because only one option can be chosen from the choice set). So either 1 is chosen, and thus w_{213} is 0; or 2 is chosen, and w_{123} is 0; or 3 is chosen, in which case w_{123} and w_{213} are both 0.

The same argument shows that

$$\text{Cov}(w_{124}, w_{214}) = \frac{-\pi_1 \pi_2}{(\pi_1 + \pi_2 + \pi_4)^2},$$

giving the result that

$$\text{Cov}(w_1, w_2) = \frac{-\pi_1 \pi_2}{(\pi_1 + \pi_2 + \pi_3)^2} + \frac{-\pi_1 \pi_2}{(\pi_1 + \pi_2 + \pi_4)^2},$$

as expected. □

We define $\lambda_{i_1,i_2,\ldots,i_m} = n_{i_1,i_2,\ldots,i_m}/N$. Then the entries of Λ are given by

$$\Lambda_{i_1,i_1} = \pi_{i_1} \sum_{i_2 < i_3 < \cdots < i_m} \frac{\lambda_{i_1,i_2,\ldots,i_m} \sum_{j=2}^{m} \pi_{i_j}}{(\sum_{j=1}^{m} \pi_{i_j})^2} \tag{3.2}$$

and

$$\Lambda_{i_1,i_2} = -\pi_{i_1}\pi_{i_2} \sum_{i_3 < i_4 < \cdots < i_m} \frac{\lambda_{i_1,i_2,\ldots,i_m}}{(\sum_{j=1}^{m} \pi_{i_j})^2}. \tag{3.3}$$

If we assume that $\pi_1 = \pi_2 = \cdots = \pi_t = 1$ (that is, all items are equally attractive, the usual null hypothesis), then

$$\Lambda_{i_1,i_1} = \frac{m-1}{m^2} \sum_{i_2 < i_3 < \cdots < i_m} \lambda_{i_1,i_2,\ldots,i_m} \tag{3.4}$$

and

$$\Lambda_{i_1,i_2} = -\frac{1}{m^2} \sum_{i_3 < i_4 < \cdots < i_m} \lambda_{i_1,i_2,\ldots,i_m}. \tag{3.5}$$

■ EXAMPLE 3.3.2.

Suppose that we have $t = 4$ items and that we compare them using the sets (T_1, T_2, T_3), (T_1, T_2, T_4), and (T_2, T_3, T_4). Then $n_{1,2,3} = 1$, $n_{1,2,4} = 1$, $n_{1,3,4} = 0$, and $n_{2,3,4} = 1$. Also observe that $N = 3$. Under the null hypothesis of equally attractive items, we get the Λ matrix

$$\Lambda = \frac{1}{9 \times 3} \begin{bmatrix} 4 & -2 & -1 & -1 \\ -2 & 6 & -2 & -2 \\ -1 & -2 & 4 & -1 \\ -1 & -2 & -1 & 4 \end{bmatrix}.$$

□

3.3.2 Representing Options Using k Attributes

These ideas can be extended naturally to items that are described by k attributes. We can calculate the matrices B_a, B_h, and B_r in the same way as we have before.

■ EXAMPLE 3.3.3.

Suppose that we have three binary attributes and that we conduct a choice experiment using the following four choice sets each of size 3: (000, 111, 010), (001, 110, 011), (010, 101, 000), (011, 100, 001). We use the same B_h matrix for main effects as in Example 3.2.6, but the new Λ matrix is

$$\frac{1}{9 \times 4} \begin{bmatrix} 4 & 0 & -2 & 0 & 0 & -1 & 0 & -1 \\ 0 & 4 & 0 & -2 & -1 & 0 & -1 & 0 \\ -2 & 0 & 4 & 0 & 0 & -1 & 0 & -1 \\ 0 & -2 & 0 & 4 & -1 & 0 & -1 & 0 \\ 0 & -1 & 0 & -1 & 2 & 0 & 0 & 0 \\ -1 & 0 & -1 & 0 & 0 & 2 & 0 & 0 \\ 0 & -1 & 0 & -1 & 0 & 0 & 2 & 0 \\ -1 & 0 & -1 & 0 & 0 & 0 & 0 & 2 \end{bmatrix}.$$

The corresponding C matrix for estimating main effects is $\frac{1}{9} I_3$. (This matrix is calculated assuming B_a contains all the contrasts other than those for main effects, but as we shall show in Chapter 5 the information matrix for any set of effects to be tested is independent of the way that the remaining contrasts are divided between B_a and B_r.)

If we use these choice sets to estimate main effects plus two-factor interactions, then we get the following C matrix.

$$C = \frac{1}{18} \begin{bmatrix} 2 & 0 & 0 & : & 0 & 0 & 0 \\ 0 & 2 & 0 & : & -1 & 0 & 0 \\ 0 & 0 & 2 & : & 0 & 0 & 0 \\ \cdot & \cdot & \cdot & \cdot & \cdot & \cdot & \cdot \\ 0 & -1 & 0 & : & 2 & 0 & 0 \\ 0 & 0 & 0 & : & 0 & 0 & 0 \\ 0 & 0 & 0 & : & 0 & 0 & 2 \end{bmatrix}.$$

We see that this matrix is not of full rank and hence the choice sets are not connected for the estimation of these effects. □

3.3.3 The Assumption of Independence from Irrelevant Alternatives

One of the most controversial properties of the MNL model is that of Independence from Irrelevant Alternatives (IIA). We define this phrase and look at its consequences briefly.

Recall that the log odds function or logit function for a binomial random variable with probability π of success is defined by $\text{logit}(\pi) = \ln(\pi/(1-\pi))$. Now consider the odds of subject α choosing T_{i_1} over T_{i_2} if the MNL model is used for modelling the choices. This is

$$\ln(P_{\alpha i_1}/P_{\alpha i_2}) = \ln\left[\frac{e^{V_{i_1\alpha}}}{\sum_j e^{V_{j\alpha}}} \frac{\sum_j e^{V_{j\alpha}}}{e^{V_{i_2\alpha}}}\right] = \ln\left[\frac{e^{V_{i_1\alpha}}}{e^{V_{i_2\alpha}}}\right].$$

We see that these odds depend only on the two options being compared and not on any of the other options in the choice set. Thus this model is said to have *independence from irrelevant alternatives* (IIA), a phrase originally used by Luce (1959).

This assumption does not always makes sense. The classic counter-example is the red bus–blue bus example. Suppose that there are two ways of getting to work: driving a car or travelling by (blue) bus. Now suppose that a third option becomes available — a red bus that in all other ways is exactly the same as the blue bus. Then logic suggests that users of the red bus will all have been users of the blue bus and none of the drivers will be tempted by the new service. So the IIA assumption does not make sense here.

The IIA assumption can be viewed as a restriction or as the natural outcome of a well-specified model (that is, one in which all sources of correlation are captured). For example, the generalized extreme value (GEV) model allows for different substitution patterns and so does not have the IIA property because it allows for the correlation structure to be modelled; see Train (2003) for more details.

3.3.4 Exercises

1. Consider the following four choice sets of size 3 for four binary attributes: (0000, 1111, 0011), (0011, 1100, 0000), (0101, 1010, 0110), (0110, 1001, 0101).

   (a) Give the Λ and C matrices corresponding to these choice sets, assuming that you want to estimate the main effects only. Does it matter how you split the remaining effects between B_a and B_r? Comment.

   (b) Give the Ω_0 matrix for these choice sets assuming that all interaction terms are 0.

2. Suppose that $t = 4$ and that the 4 items to be compared are T_1, T_2, T_3, and T_4. Suppose that we use the four choice sets (T_1, T_2), (T_1, T_3), (T_1, T_4), and (T_2, T_3, T_4).

   (a) Find the likelihood function $L(\pi)$ for this choice experiment.

   (b) Hence find Λ. Comment.

   (c) Would you be able to estimate all of the π_i?

3. Show that the Ω_0 matrix is always NC for the estimation of main effects only assuming that all interaction effects are 0.

3.4 COMPARING DESIGNS

In the univariate, ordinary least squares, normal errors situation the performance of estimates is often judged by the width of the resulting confidence interval. Estimates that are minimum variance unbiased are often deemed to be the best. In the multivariate, ordinary least squares, normal errors situation the performance of estimates is often judged by some property of the asymptotic variance-covariance matrix. In the case of a discrete choice experiment we have a non-linear multivariate situation where the variance-covariance matrix of the unknown parameter estimates depends on the values of those estimates. We find the variance-covariance matrix as a function of the parameter estimates and define the optimal design to be the design with the "best" variance-covariance matrix given H_0. Possible ways of defining "best" are given below.

This means that we have a method to compare different choice experiments objectively using the variance-covariance matrix, C^{-1}. However, we could use structural properties of the either the variance-covariance matrix or of the designs.

For the variance-covariance matrix the property that is most of interest is whether or not the parameters have been independently estimated. This is so if the matrix is diagonal. When we estimate main effects, for an attribute with ℓ_q levels, there are $\ell_q - 1$ contrasts and these are not uniquely defined. So, in fact, we really only require that C^{-1} be block diagonal for the effects from different attributes to be independently estimated.

Readers can find software to calculate the information matrix and variance-covariance matrix of any set of choice sets at http://maths.science.uts.edu.au/maths/wiki/SPExpts.

For the designs, desirable structural properties typically arise from a link between the structure of the design and the properties of the resulting estimates. Structural properties of interest could include the frequency with which each level of each attribute appears in the design or the relationship between the options in each choice set.

We discuss all of these approaches below.

3.4.1 Using Variance Properties to Compare Designs

In the univariate setting, where there is only one parameter to estimate, the variance of the parameter estimate is often used as a measure of a good estimation procedure. One talks about a "minimum variance unbiased estimator" for instance, when describing an unbiased estimator that achieves the Cramér–Rao lower bound.

In the choice experiment setting, we are interested in estimating several effects, typically the main effects or the main effects plus two-factor interaction effects. So we end up with a variance-covariance matrix, C^{-1}, to describe the variability of the estimates. Thus we would like to summarize C^{-1} in a single number and several such summaries have been proposed.

The D-, A-, and E-optimality measures are appropriate to our situation and we now define these; see Atkinson and Donev (1992) for more details.

A design is *D-optimal* if it minimizes the generalized variance of the parameter estimates, that is, $\det(C^{-1})$ is as small as possible for the D-optimal designs.

A design is *A-optimal* if it minimizes the average variance of the parameter estimates, that is, $\text{tr}(C^{-1})$ is as small as possible for the A-optimal designs.

A design is *E-optimal* if it minimizes the variance of the least well-estimated parameter, that is, the largest eigenvalue of C^{-1} is as small as possible for the E-optimal designs.

It has become usual to look for designs that are D-optimal. The idea of minimizing the generalized variance seems intuitively reasonable and the D-optimal design does not depend on the scale used for any quantitative attributes. This is of course less of a consideration given that we are viewing all the attributes as qualitative, although some authors, for example Kanninen (2002), do design for quantitative attributes in the discrete choice setting. The calculations involved in evaluating the determinant are usually more easily performed than those required to determine an A- or E-optimal design.

Following El-Helbawy and Bradley (1978), we apply the optimality criteria to the information matrix $C = B_h \Lambda B_h$ and the corresponding variance-covariance matrix C^{-1}. We can make it easier to find the D-optimal designs by noting that $\det(C^{-1}) = 1/\det(C)$. Thus the design which minimizes $\det(C^{-1})$ is the same as the design which maximizes $\det(C)$. Thus we try to find designs for which $\det(C)$ is as large as possible.

From these definitions, we see that for any optimality measure it is necessary to define a class of competing designs. As we go through the book we describe the class of competing designs for each new situation that we consider.

■ EXAMPLE 3.4.1.

Suppose that we have $k = 2$ binary attributes and that we want to estimate the main effects of these attributes using 3 choice sets each with two options. We let $B_h = B_M$ be the contrast matrix for main effects. We believe that the interaction effect is 0 and hence we let B_a be the contrast for the two-factor interaction. There are $2^2 = 4$ possible items. There are $\binom{4}{2} = 6$ pairs of options and there are $\binom{6}{3} = 20$ choice experiments involving 3 sets of these pairs. These 20 sets of three pairs constitute the class of competing designs and we want to know which of these designs is (are) best. To do this, we must calculate Λ, $C = B_M \Lambda B_M'$, and hence C^{-1}, for each design and calculate the D-, A-, and E-optimality values.

For all of the competing designs

$$B_h = B_M = \frac{1}{2}\begin{bmatrix} -1 & -1 & 1 & 1 \\ -1 & 1 & -1 & 1 \end{bmatrix}.$$

We let $\nu_i, i = 1, 2$, be the two eigenvalues of C^{-1}. Then the D-optimum value is $\nu_1\nu_2$, the A-optimum value is $\nu_1 + \nu_2$ and the E-optimum value is $\max(\nu_1, \nu_2)$.

For the first design in Table 3.7, for example,

$$\Lambda = \frac{1}{12}\begin{bmatrix} 3 & -1 & -1 & -1 \\ -1 & 1 & 0 & 0 \\ -1 & 0 & 1 & 0 \\ -1 & 0 & 0 & 1 \end{bmatrix},$$

$$C = \frac{1}{12}\begin{bmatrix} 2 & 1 \\ 1 & 2 \end{bmatrix},$$

and

$$C^{-1} = \begin{bmatrix} 8 & -4 \\ -4 & 8 \end{bmatrix}.$$

Thus $\nu_1 = 12$, $\nu_2 = 4$ and we get a D-optimal value of 48, an A-optimal value of 16 and an E-optimal value of 12. This is summarized in the first line of Table 3.7.

All 20 designs and the corresponding D-, A-, and E-optimum values are given in Table 3.7.

We can see that in this case the same four designs, designs 5, 11, 17, and 18, are A-, D-, and E-optimal. □

When we know the minimum value that an optimality measure can take in the class of competing designs, then we can compare every design to this bound and we talk of the *efficiency* of a design.

For the generalized variance, we need to allow for the number of effects that are being estimated. Let p be the number of independent parameters that are being estimated in an

Table 3.7 Triples of Pairs for $k = 2$ and the Corresponding D-, A- and E- Optimum Values

Design	Triples	ν_1, ν_2	D_{opt}	A_{opt}	E_{opt}
1	00,01 00,10 00,11	4,12	48	16	12
2	00,01 00,10 01,10	4,12	48	16	12
3	00,01 00,10 01,11	6,12	72	18	12
4	00,01 00,10 10,11	6,12	72	18	12
5	00,01 00,11 01,10	4,6	24	10	6
6	00,01 00,11 01,11	4,12	48	16	12
7	00,01 00,11 10,11	$6(2-\sqrt{2}),6(2+\sqrt{2})$	72	24	$6(2+\sqrt{2})$
8	00,01 01,10 01,11	4,12	48	16	12
9	00,01 01,10 10,11	$6(2-\sqrt{2}),6(2+\sqrt{2})$	72	24	$6(2+\sqrt{2})$
10	00,01 01,11 10,11	6,12	72	18	12
11	00,10 00,11 01,10	4,6	24	10	6
12	00,10 00,11 01,11	$6(2-\sqrt{2}),6(2+\sqrt{2})$	72	24	$6(2+\sqrt{2})$
13	00,10 00,11 10,11	4,12	48	16	12
14	00,10 01,10 01,11	$6(2-\sqrt{2}),6(2+\sqrt{2})$	72	24	$6(2+\sqrt{2})$
15	00,10 01,10 10,11	4,12	48	16	12
16	00,10 01,11 10,11	6,12	72	18	12
17	00,11 01,10 01,11	4,6	24	10	6
18	00,11 01,10 10,11	4,6	24	10	6
19	00,11 01,11 10,11	4,12	48	16	12
20	01,10 01,11 10,11	4,12	48	16	12

experiment. For example, for k attributes and a design that is estimating main effects only,

$$p = \sum_q (\ell_q - 1);$$

for a design that is estimating main effects plus two-factor interactions,

$$p = \sum_q (\ell_q - 1) + \sum_{q_1} \sum_{q_2} (\ell_{q_1} - 1)(\ell_{q_2} - 1).$$

Thus the *D-efficiency* of a design is defined to be the pth root of the ratio of the determinant of the information matrix of the proposed design to that for the optimal design. So

$$\text{Eff}_D = \left(\frac{\det(C)}{\det(C_{\text{opt}})} \right)^{1/p},$$

where we have again chosen to work with C rather than with C^{-1} since this means that we do not need to invert the C matrix.

The *A-efficiency* of a design is defined to be the ratio of the trace of the covariance matrix of the proposed design to that for the optimal design. So

$$\text{Eff}_A = \frac{\text{tr}(C_{\text{opt}}^{-1})}{\text{tr}(C^{-1})}.$$

(We can see the computational appeal of D-efficiency since there is no simple way of evaluating the trace of the inverse of a matrix from the trace of the original matrix.)

The next example illustrates the importance of considering the structure of the C matrix and not just the efficiency value.

■ **EXAMPLE 3.4.2.**

Suppose that there are $k = 3$ attributes each with 4 levels. and that the choice sets are those given in Design 1 in Table 3.8. Then the corresponding C matrix is

$$C_1 = \left[\begin{array}{ccc|ccc|ccc}
\frac{5}{512} & 0 & 0 & \frac{-1}{512} & 0 & 0 & \frac{-3}{2560} & 0 & \frac{1}{640} \\
0 & \frac{3}{256} & 0 & 0 & \frac{1}{256} & 0 & 0 & 0 & 0 \\
0 & 0 & \frac{5}{512} & 0 & 0 & \frac{-1}{512} & \frac{1}{640} & 0 & \frac{3}{2560} \\
\hline
\frac{-1}{512} & 0 & 0 & \frac{5}{512} & 0 & 0 & \frac{3}{2560} & 0 & \frac{-1}{640} \\
0 & \frac{1}{256} & 0 & 0 & \frac{3}{256} & 0 & 0 & 0 & 0 \\
0 & 0 & \frac{-1}{512} & 0 & 0 & \frac{5}{512} & \frac{-1}{640} & 0 & \frac{-3}{2560} \\
\hline
\frac{-3}{2560} & 0 & \frac{1}{640} & \frac{3}{2560} & 0 & \frac{-1}{640} & \frac{5}{512} & 0 & 0 \\
0 & 0 & 0 & 0 & 0 & 0 & 0 & \frac{3}{256} & 0 \\
\frac{1}{640} & 0 & \frac{3}{2560} & \frac{-1}{640} & 0 & \frac{-3}{2560} & 0 & 0 & \frac{5}{512}
\end{array}\right]$$

and the corresponding variance-covariance matrix is

$$C_1^{-1} = \left[\begin{array}{ccc|ccc|ccc}
\frac{768}{7} & 0 & 0 & \frac{128}{7} & 0 & 0 & \frac{384}{35} & 0 & -\frac{512}{35} \\
0 & 96 & 0 & 0 & -32 & 0 & 0 & 0 & 0 \\
0 & 0 & \frac{768}{7} & 0 & 0 & \frac{128}{7} & -\frac{512}{35} & 0 & -\frac{384}{35} \\
\hline
\frac{128}{7} & 0 & 0 & \frac{768}{7} & 0 & 0 & -\frac{384}{35} & 0 & \frac{512}{35} \\
0 & -32 & 0 & 0 & 96 & 0 & 0 & 0 & 0 \\
0 & 0 & \frac{128}{7} & 0 & 0 & \frac{768}{7} & \frac{512}{35} & 0 & \frac{384}{35} \\
\hline
\frac{384}{35} & 0 & -\frac{512}{35} & -\frac{384}{35} & 0 & \frac{512}{35} & \frac{768}{7} & 0 & 0 \\
0 & 0 & 0 & 0 & 0 & 0 & 0 & \frac{256}{3} & 0 \\
-\frac{512}{35} & 0 & -\frac{384}{35} & \frac{512}{35} & 0 & \frac{384}{35} & 0 & 0 & \frac{768}{7}
\end{array}\right].$$

This design is 95.9% efficient but the only effect that is independently estimated is the quadratic component of the third attribute.

Consider instead the choice sets in Design 2 in Table 3.8. Then the corresponding C matrix is

$$C_2 = \left[\begin{array}{ccc|ccc|ccc}
\frac{3}{320} & 0 & \frac{1}{320} & 0 & 0 & 0 & 0 & 0 & 0 \\
0 & \frac{1}{128} & 0 & 0 & 0 & 0 & 0 & 0 & 0 \\
\frac{1}{320} & 0 & \frac{9}{640} & 0 & 0 & 0 & 0 & 0 & 0 \\
\hline
0 & 0 & 0 & \frac{3}{320} & 0 & \frac{1}{320} & 0 & 0 & 0 \\
0 & 0 & 0 & 0 & \frac{1}{128} & 0 & 0 & 0 & 0 \\
0 & 0 & 0 & \frac{1}{320} & 0 & \frac{9}{640} & 0 & 0 & 0 \\
\hline
0 & 0 & 0 & 0 & 0 & 0 & \frac{3}{320} & 0 & \frac{1}{320} \\
0 & 0 & 0 & 0 & 0 & 0 & 0 & \frac{1}{128} & 0 \\
0 & 0 & 0 & 0 & 0 & 0 & \frac{1}{320} & 0 & \frac{9}{640}
\end{array}\right]$$

and the corresponding variance-covariance matrix is

$$C_2^{-1} = \left[\begin{array}{ccc:ccc:ccc} \frac{576}{5} & 0 & -\frac{128}{5} & 0 & 0 & 0 & 0 & 0 & 0 \\ 0 & 128 & 0 & 0 & 0 & 0 & 0 & 0 & 0 \\ -\frac{128}{5} & 0 & \frac{384}{5} & 0 & 0 & 0 & 0 & 0 & 0 \\ \hdashline 0 & 0 & 0 & \frac{576}{5} & 0 & -\frac{128}{5} & 0 & 0 & 0 \\ 0 & 0 & 0 & 0 & 128 & 0 & 0 & 0 & 0 \\ 0 & 0 & 0 & -\frac{128}{5} & 0 & \frac{384}{5} & 0 & 0 & 0 \\ \hdashline 0 & 0 & 0 & 0 & 0 & 0 & \frac{576}{5} & 0 & -\frac{128}{5} \\ 0 & 0 & 0 & 0 & 0 & 0 & 0 & 128 & 0 \\ 0 & 0 & 0 & 0 & 0 & 0 & -\frac{128}{5} & 0 & \frac{384}{5} \end{array}\right].$$

This design is 94.5% efficient and we see that the only correlation that exists is between the linear and cubic components of the main effect of each attribute.

Thus although the first design is slightly better in terms of the Eff_D it has a number of correlations between main effects for different attributes while the other design has all main effects independently estimated. So Eff_D is a guide to the best design but the structure of the C and the C^{-1} matrices is also important. □

Table 3.8 Two Choice Experiments

Option 1	Option 2	Option 1	Option 2
310	132	000	111
201	132	011	122
302	120	022	133
222	333	033	100
103	230	101	212
120	031	110	221
012	321	123	230
111	000	132	203
230	321	202	313
023	201	213	320
031	213	220	331
012	103	231	302
333	111	303	010
000	222	312	023
213	302	321	032
310	023	330	001
Design 1		Design 2	

3.4.2 Structural Properties

There has been a tradition in the design of experiments to try and identify structural properties of designs that are linked with desirable statistical properties. Then useful designs can be found merely by considering these structural properties. Although some structural properties have been shown to be linked with desirable properties in choice experiments, to date the results have not been as clear cut as in the construction of balanced incomplete block designs for comparing treatments in a linear models setting, for example.

Huber and Zwerina (1996) describe a set of features that they believe are characteristic of optimal choice designs. These features are:

1. *level balance.* All the levels of each attribute occur with equal frequency over all options in all choice sets (often called equi-replicate in the statistical literature).

2. *Orthogonality.* - This "is satisfied when the joint occurrence of any two levels of different attributes appear in options with frequencies equal to the product of their marginal frequencies" (Huber and Zwerina (1996)). This is often described by saying that the levels of the various attributes appear "with proportional frequencies".

3. *Minimal overlap.* "The probability that an attribute level repeats itself in each choice set should be as small as possible" (Huber and Zwerina (1996)). So, if the number of items in each choice set is fewer than the number of levels for an attribute, then no attribute level is repeated within a choice set. Thus the difference between the number of times that any two levels of an attribute are replicated should be as small as possible, ideally 0, and at most 1.

4. *Utility balance.* Options within a choice set should be equally attractive to subjects.

Bunch et al. (1996) introduced the idea of *shifted designs* in which a set of initial options is chosen for each of the N choice sets in an experiment and the subsequent option(s) in each choice set are obtained by using modular arithmetic to "shift each combination of initial attribute levels by adding a constant that depends on the number of levels." In Zwerina et al. (1996) we find the following sentence and footnote: "For certain families of plans and assuming all the coefficients are zero, these shifted designs satisfy all four principles, and thus are optimal. We are not able to analytically prove this, but after examining scores of designs, we have never found more efficient designs than those that satisfy all four principles."

But satisfying these principles does not guarantee that the design is optimal, nor even that it can estimate main effects, as the following example shows.

■ **EXAMPLE 3.4.3.**

The choice sets in Table 3.9 satisfy the four Huber and Zwerina conditions when the null hypothesis is true but the main effects for the third attribute can not be estimated, and the determinant of the information matrix for main effects is 0. Studying the design, we observe that in each choice set the levels of the third attribute appear in pairs: 0 with 1; 2 with 3; and 4 with 5. So the additional requirement, observed in shifted designs, is that the levels of each attribute must be connected. □

While the previous example is in some sense pathological, the following example shows that level balance is not essential for a choice experiment to be optimal.

■ **EXAMPLE 3.4.4.**

The choice sets in Table 3.10 are an optimal set of choice sets of size 5 for the estimation of main effects. The second attribute has all three levels appearing 30 times each across the 18 choice sets, and the third attribute has all six levels appearing 15 times each across the

Table 3.9 A Choice Experiment for the Estimation of Main Effects when There Are 3 Attributes with 2, 3, and 6 Levels which Satisfies the Huber-Zwerina Conditions but for which Main Effects Cannot Be Estimated.

Option A	Option B
000	111
001	110
002	113
003	112
004	115
005	114
010	121
011	120
012	123
013	122
014	125
015	124
020	101
021	100
022	103
023	102
024	105
025	104
100	011
101	010
102	013
103	012
104	015
105	014
110	021
111	020
112	023
113	022
114	025
115	024
120	001
121	000
122	003
123	002
124	005
125	004

Table 3.10 An Optimal Choice Experiment for the Estimation of Main Effects when There Are 3 Attributes with 2, 3, and 6 Levels

Option A	Option B	Option C	Option D	Option E
000	111	123	012	024
121	002	014	103	115
112	023	005	124	100
023	104	110	005	011
004	115	121	010	022
015	120	102	021	003
025	100	112	001	013
114	025	001	120	102
010	121	103	022	004
001	112	214	013	025
103	014	020	115	121
022	103	115	004	010
013	124	100	025	001
002	113	125	014	020
024	105	111	000	012
105	010	022	111	123
011	122	104	023	005
120	001	013	102	114

18 choice sets. However, the first attribute has one level replicated 48 times and the other level replicated 42 times. □

Chapters 4, 5, and 6 give a theoretical description of the optimal designs under the null hypothesis of no option differences. These results determine the optimal designs for the estimation of main effects, and of main effects plus two-factor interactions, for any number of levels and for any choice set size. Minimal overlap is an essential feature of optimal designs for the estimation of main effects, but precludes the estimation of interaction effects, a point made in Huber and Zwerina (1996).

3.4.3 Exercises

1. Consider two binary attributes. Assume that all four treatment combinations are equally attractive; that is, assume that the π_i values are all equal.

    (a) Give the 6 possible pairs that can be used as choice sets of size 2.

    (b) Give the 15 sets of two pairs.

    (c) For each of these pairs of pairs, evaluate the corresponding C matrix for estimating main effects and give its determinant and its trace.

    (d) Which of these 15 designs is best?

2. Consider two binary attributes and suppose that $\pi_{00} = \pi_{01} = 1$, $\pi_{10} = 1/10$ and $\pi_{11} = 10$. Repeat Question 1. Compare the best design for these two situations and comment.

3. Consider four binary attributes.

   (a) Calculate the 8 pairs that come about by pairing each treatment combination with its foldover (that is, the treatment combination in which 0s and 1s are interchanged; 0101 is the foldover of 1010, for example).

   (b) Calculate the C matrix for main effects for this design. What is $\det(C)$?

   (c) Repeat using the half-replicate given by $x_1 + x_2 + x_3 + x_4 = 0$.

   (d) Comment.

4. Let $k = 4$. Find the 8 treatment combinations in the two half-replicates given by $x_1 + x_2 + x_3 + x_4 = 0$ and by $x_1 + x_2 + x_3 = 0$. Thus there are $\binom{8}{2} = 28$ pairs that can be made from the treatment combinations in each of these designs.

   (a) For each of these sets of pairs, what is the smallest design that can be used to estimate main effects?

   (b) Which is (are) the best designs for each situation, using both A- and D-optimality to compare designs.

5. Consider the designs in Question 1.

   (a) Which ones have level balance?

   (b) Which ones are orthogonal?

   (c) Which ones have minimal overlap in all choice sets?

   (d) Do all the designs have utility balance in all choice sets?

   Which design(s) would be considered best using these criteria?

6. Repeat Question 5 for the designs in Question 2. Compare with the results above and comment.

7. Comment on the designs in Question 3 with respect to level balance, orthogonality, minimum overlap and utility balance.

3.5 REFERENCES AND COMMENTS

Train (2003) discusses discrete choice methods with a focus on the use of estimation by simulation. The first part of his book gives an introduction to the various behavioral models that have been proposed of which the MNL model is only one.

There is an extensive discussion of the history of paired comparison designs in David (1988).

Atkinson and Donev (1992) provide an extensive collection of results on optimal designs for the linear model for a variety of optimality criteria. They briefly discuss some of the issues in the construction of optimal designs for models in which the covariance matrix depends on the unknown parameters (as it does for the MNL model).

Many software packages can be used to analyze discrete choice experiments, often by utilizing the link between the DCEs and Cox's proportional hazards model. Kuhfeld (2006) gives a very complete description, including many worked examples, together with appropriate macros, on how to use SAS to do the analysis. There is a description of using S-PLUS to fit a Cox's proportional hazards model in Venables and Ripley (2003).

Multinomial models, and advice about how to fit them, may be found in Agresti (2002) and Thompson (2005). Thompson (2005) gives worked examples in both S-PLUS and R. A detailed description of using GLIM to analyze paired and triple comparisons may be found in Critchlow and Fligner (1991). Long and Freese (2006) have an extensive discussion on how to use STATA to analyze MNL models and related models such as the conditional logit and multinomial probit.

CHAPTER 4

PAIRED COMPARISON DESIGNS FOR BINARY ATTRIBUTES

In this chapter we will look at various design possibilities when all attributes have two levels and all choice sets have two items in them. We start by working with sets of choice sets of size 2 obtained from the complete factorial and then show that we can get equally good designs by constructing pairs from suitably chosen fractional factorial designs.

To help set the scene, consider the following example. Severin (2000) investigated which attributes made take-out pizza outlets more attractive. In her first experiment, she used the six attributes in Table 4.1 with the levels indicated. A sample choice set for an experiment looking at these six attributes describing take-out pizza outlets is given in Table 4.2. Observe that the question has been phrased so that the respondents are asked to imagine that the two choices presented to them are the last two options that they are considering in their search for a take-out pizza outlet. This assumption means that the respondents are naturally in a setting where it does not make sense not to choose an option and so they are forced to make a selection even though the options presented are not exhaustive. In Chapter 7, we will consider the design of choice experiments when we want to allow an option not to choose.

4.1 OPTIMAL PAIRS FROM THE COMPLETE FACTORIAL

In this section we investigate the form of the optimal paired comparison design when there are k attributes, each with two levels, and the set of possible choice pairs is restricted so that each pair of treatment combinations in which there are v attributes with different levels appears equally often. This then defines the *class of competing designs*. Using this

The Construction of Optimal Stated Choice Experiments. By D. J. Street and L. Burgess

Table 4.1 Six Attributes to Be Used in an Experiment to Compare Pizza Outlets

Attributes	**Levels**
Pizza type	Traditional Gourmet
Type of Crust	Thick Thin
Ingredients	All fresh Some tinned
Size	Small only Three sizes
Price	\$17 \$13
Delivery time	30 minutes 45 minutes

Table 4.2 One Choice Set in an Experiment to Compare Pizza Outlets

	Outlet A	**Outlet B**
Pizza type	Traditional	Gourmet
Type of Crust	Thick	Thin
Ingredients	All fresh	Some tinned
Size	Small only	Small only
Price	\$17	\$13
Delivery time	30 minutes	30 minutes

Suppose that you have already narrowed down
your choice of take-out pizza outlet to
the two alternatives above.
Which of these two would you choose?
(tick one only)
Option A ◯ Option B ◯

class of competing designs means that the matrices C^{-1} and C are diagonal and hence the relevant information matrices can be determined by evaluating $B\Lambda B'$ for the relevant set of contrasts. Using this class of competing designs ensures that the estimate of the variance-covariance matrix of $B_h\hat{\gamma}$ is independent of which contrasts are in B_a and which are in B_r.

■ **EXAMPLE 4.1.1.**

Let $k = 2$. Then the pairs with $v = 2$ attributes different, specifically (00,11) and (01,10), would each appear the same number of times in the design (although that number might be 0), and the pairs with $v = 1$ attribute different, specifically (00,01), (00,10), (01,11), and (10,11), would each appear the same number of times, again possibly 0 times each, in the design. Thus the class of competing designs consists of three designs: the two pairs (00,11) and (01,10); the four pairs (00,01), (00,10), (01,11), and (10,11); and all six of these pairs. □

4.1.1 The Derivation of the Λ Matrix

We derive the Λ matrix using the method of Section 3.2.4. As we have just said, the competing designs are those in which the set of possible choice pairs is restricted so that each pair of treatment combinations in which there are v attributes with different levels appears equally often. We will let i_v be an indicator variable defined as follows.

$$i_v = \begin{cases} 1 & \text{if all the pairs with } v \text{ attributes different are to be included in the choice experiment,} \\ 0 & \text{if none of the pairs with } v \text{ attributes different is to be included in the choice experiment.} \end{cases}$$

We then define N to be the total number of choice sets in the choice experiment and let

$$a_v = i_v/N.$$

As we have done before, to make the extension to more attributes easy, we will always place the treatment combinations in *standard order*, sometimes called *Yates* standard order, or simply *lexicographic order*.

We will let $D_{k,v}$ be a (0,1) matrix of order 2^k with rows and columns labeled by the treatment combinations in a 2^k factorial design. There is a 1 in position $(\mathbf{x}, \mathbf{y})$ of $D_{k,v}$ if treatment combinations $\mathbf{x}$ and $\mathbf{y}$ have v attributes with different levels.

■ **EXAMPLE 4.1.2.**

Let $k = 2$. Then the four treatment combinations, in standard order, are 00, 01, 10, and 11. These treatment combinations, in this order, are used to label the rows and columns of all matrices associated with a design. For example, 00 and 01 have one attribute with different levels and so $D_{2,1}(00, 01) = D_{2,1}(1, 2) = 1$. Continuing in this way, we see that

$$D_{2,1} = \begin{bmatrix} 0 & 1 & 1 & 0 \\ 1 & 0 & 0 & 1 \\ 1 & 0 & 0 & 1 \\ 0 & 1 & 1 & 0 \end{bmatrix} \text{ and } D_{2,2} = \begin{bmatrix} 0 & 0 & 0 & 1 \\ 0 & 0 & 1 & 0 \\ 0 & 1 & 0 & 0 \\ 1 & 0 & 0 & 0 \end{bmatrix}.$$

Let $k = 3$. Then the eight treatment combinations in standard order are 000, 001, 010, 011, 100, 101, 110, and 111. Thus we see that

$$D_{3,1} = \begin{bmatrix} 0 & 1 & 1 & 0 & 1 & 0 & 0 & 0 \\ 1 & 0 & 0 & 1 & 0 & 1 & 0 & 0 \\ 1 & 0 & 0 & 1 & 0 & 0 & 1 & 0 \\ 0 & 1 & 1 & 0 & 0 & 0 & 0 & 1 \\ 1 & 0 & 0 & 0 & 0 & 1 & 1 & 0 \\ 0 & 1 & 0 & 0 & 1 & 0 & 0 & 1 \\ 0 & 0 & 1 & 0 & 1 & 0 & 0 & 1 \\ 0 & 0 & 0 & 1 & 0 & 1 & 1 & 0 \end{bmatrix} \text{ and } D_{3,2} = \begin{bmatrix} 0 & 0 & 0 & 1 & 0 & 1 & 1 & 0 \\ 0 & 0 & 1 & 0 & 1 & 0 & 0 & 1 \\ 0 & 1 & 0 & 0 & 1 & 0 & 0 & 1 \\ 1 & 0 & 0 & 0 & 0 & 1 & 1 & 0 \\ 0 & 1 & 1 & 0 & 0 & 0 & 0 & 1 \\ 1 & 0 & 0 & 1 & 0 & 0 & 1 & 0 \\ 1 & 0 & 0 & 1 & 0 & 1 & 0 & 0 \\ 0 & 1 & 1 & 0 & 1 & 0 & 0 & 0 \end{bmatrix}.$$

For each treatment combination, there is only one other treatment combination in which all the attributes have different levels; this is the foldover treatment. Thus $D_{3,3}$ is a matrix with 1s on the back-diagonal and all other entries equal to 0. Each treatment combination has the same levels as itself; so $D_{3,0} = I_8$. □

Consider the 2^k treatment combinations in a 2^k complete factorial. Because we have ordered the treatment combinations lexicographically, the first level in the first 2^{k-1} treatment combinations is 0, and the remaining entries form the 2^{k-1} treatment combinations in a complete 2^{k-1} factorial. The same is true of the second 2^{k-1} treatment combinations if we remove the first level, which is of course a 1. So we can define the entries in $D_{k,v}$ in terms of those in $D_{k-1,v}$ and $D_{k-1,v-1}$. Thus we get the following recursive relationship.

■ LEMMA 4.1.1.

$$D_{k,v} = \begin{bmatrix} D_{k-1,v} & D_{k-1,v-1} \\ D_{k-1,v-1} & D_{k-1,v} \end{bmatrix}, \; k-1 \geq v \geq 1, \; D_{k,0} = I_{2^k}, \; D_{k-1,k} = \mathbf{0}.$$

□

With this notation, the a_v defined above, the Λ notation of Section 3.2.4, and the assumption that the usual null hypothesis, $\pi_1 = \pi_2 = \ldots = \pi_{2^k} = 1$ is true, we get the following expression for Λ.

■ LEMMA 4.1.2.

$$\begin{aligned} \Lambda &= \frac{1}{4}\left[\left(ka_1 + \binom{k}{2}a_2 + \cdots + \binom{k}{v}a_v + \cdots + a_k\right) I_{2^k} - \sum_{v=1}^{k} a_v D_{k,v}\right] \\ &= \frac{1}{4}\left[\left(\sum_{v=1}^{k}\binom{k}{v}a_v\right) I_{2^k} - \sum_{v=1}^{k} a_v D_{k,v}\right]. \end{aligned}$$

□

■ EXAMPLE 4.1.3.

When $k = 2$ we have

$$\Lambda = \frac{1}{4}\left[(2a_1 + a_2)I_4 - a_1 D_{2,1} - a_2 D_{2,2}\right],$$

and when $k = 3$ we have

$$\Lambda = \frac{1}{4}\left[(3a_1 + 3a_2 + a_3)I_8 - a_1 D_{3,1} - a_2 D_{3,2} - a_3 D_{3,3}\right].$$

□

EXAMPLE 4.1.4.

Let $k = 2$. Then the pairs with 1 attribute different (that is, the pairs (00,01), (00,10), (01,11), and (10,11)) would each appear either $i_1 = 0$ or $i_1 = 1$ times in the design. If there are N pairs in the design in total, then $a_1 = 0/N$ or $a_1 = 1/N$. In either case, the corresponding entries in Λ would be $a_1 D_{2,1}$. The same is true for the entries corresponding to the pairs (00, 11) and (01, 10). So the possible values of N are 2, using the pairs (00, 11) and (01, 10); 4, using the pairs (00,01), (00,10), (01,11), and (10,11); or 6, using all the pairs. The corresponding values of Λ are given in Table 4.3. □

Table 4.3 The Possible Designs and Corresponding Λ Matrices for $k = 2$

Pairs	N	Λ
(00, 11), (01, 10)	2	$\frac{1}{8}I_4 - \frac{1}{8}D_{2,2}$
(00,01), (00,10), (01,11), (10,11)	4	$\frac{2}{16}I_4 - \frac{1}{16}D_{2,1}$
(00, 11), (01, 10), (00,01), (00,10), (01,11), (10,11)	6	$\frac{3}{24}I_4 - \frac{1}{24}D_{2,1} - \frac{1}{24}D_{2,2}$

EXAMPLE 4.1.5.

Let $k = 3$. The 28 pairs are listed in Table 4.4. We can see that there are 7 possible designs: the 12 pairs with 1 attribute different, the 12 pairs with 2 attributes different, the 4 pairs with 3 attributes different, the 24 pairs with 1 or 2 attributes different, the 16 pairs with 1 or 3 attributes different, the 16 pairs with 2 or 3 attributes different, and all 28 pairs. □

Table 4.4 All Possible Pairs when $k = 3$

One Attribute Different			
(000, 001)	(000, 010)	(000, 100)	(001, 011)
(001,101)	(010, 011)	(010, 110)	(011, 111)
(101, 111)	(110, 111)	(100, 101)	(100, 110)
Two Attributes Different			
(000, 011)	(000, 101)	(000,110)	(001, 010)
(001, 100)	(001, 111)	(010, 111)	(010, 100)
(011, 110)	(011, 101)	(100, 111)	(101, 110)
Three Attributes Different			
(000, 111)	(001, 110)	(010, 101)	(011, 100)

4.1.2 Calculation of the Relevant Contrast Matrices

Let B_{2^k} be the $(2^k - 1) \times 2^k$ matrix of the usual contrasts associated with a 2^k factorial design, where the column labels are the treatment combinations in standard order, and $B_{2^k} B'_{2^k} = I_{2^k - 1}$.

■ **EXAMPLE 4.1.6.**

Let $k = 2$. Then the columns of B_{2^2} are labeled by 00, 01, 10, and 11. Write the contrasts in the following order: the main effect of the first attribute, then the main effect of the second attribute, and finally the interaction effect. The result is

$$B_{2^2} = \frac{1}{2}\begin{bmatrix} -1 & -1 & 1 & 1 \\ -1 & 1 & -1 & 1 \\ 1 & -1 & -1 & 1 \end{bmatrix}.$$

□

■ **EXAMPLE 4.1.7.**

Let $k = 3$. Write the contrasts in the following order: the main effects of the first, second and third attributes; then the interaction effect of the second and third attributes; then the interaction effect of the first and second attributes; the interaction effect of the first and third attributes; and finally the three-factor interaction effect. We get

$$B_{2^3} = \frac{1}{\sqrt{8}}\begin{bmatrix} -1 & -1 & -1 & -1 & 1 & 1 & 1 & 1 \\ -1 & -1 & 1 & 1 & -1 & -1 & 1 & 1 \\ -1 & 1 & -1 & 1 & -1 & 1 & -1 & 1 \\ 1 & -1 & -1 & 1 & 1 & -1 & -1 & 1 \\ 1 & 1 & -1 & -1 & -1 & -1 & 1 & 1 \\ 1 & -1 & 1 & -1 & -1 & 1 & -1 & 1 \\ -1 & 1 & 1 & -1 & 1 & -1 & -1 & 1 \end{bmatrix}.$$

Observe that we can write

$$B_{2^3} = \frac{1}{\sqrt{8}}\begin{bmatrix} -\mathbf{j}'_{2^2} & \mathbf{j}'_{2^2} \\ 2B_{2^2} & 2B_{2^2} \\ -2B_{2^2} & 2B_{2^2} \end{bmatrix}.$$

This happens because of the lexicographic ordering imposed on the treatment combinations; the second and third attributes are repeated in the same order while the first attribute is 0, and then again while it is 1. □

The previous example illustrates the following general recursive result:

$$B_{2^k} = \frac{1}{\sqrt{2^k}}\begin{bmatrix} -\mathbf{j}'_{2^{k-1}} & \mathbf{j}'_{2^{k-1}} \\ \sqrt{2^{k-1}}B_{2^{k-1}} & \sqrt{2^{k-1}}B_{2^{k-1}} \\ -\sqrt{2^{k-1}}B_{2^{k-1}} & \sqrt{2^{k-1}}B_{2^{k-1}} \end{bmatrix}.$$

The factor of $1/\sqrt{2^k}$ is to normalize B_{2^k}. Note that, when written like this, the main effects contrasts are not the first k rows of B_{2^k} but are in rows 1, 2, 4, 8, and so on.

Sometimes we only want a contrast matrix for main effects or for main effects plus two-factor interactions. We let $B_{2^k,M}$ be a contrast matrix that contains only the contrasts for main effects and we let $B_{2^k,MT}$ be a contrast matrix that contains only the contrasts for main effects plus two-factor interactions.

4.1.3 The Model for Main Effects Only

Suppose that we want to estimate the main effects only. We know the general form of the Λ matrix and we now explicitly evaluate the $k \times k$ principal minor of $C = B_{2^k}\Lambda B_{2^k}$ associated with the main effects so that we can determine the A- and D-optimal designs. Thus we are evaluating the information matrix when $B_h = B_{2^k,M}$.

We want to get a neat expression for $B_{2^k,M}D_{k,i}$, and the following example indicates the form that such a result might take.

EXAMPLE 4.1.8.
Let $k = 2$. Then

$$B_{2^2,M} = \frac{1}{2}\begin{bmatrix} -1 & -1 & 1 & 1 \\ -1 & 1 & -1 & 1 \end{bmatrix} \text{ and } D_{2,1} = \begin{bmatrix} 0 & 1 & 1 & 0 \\ 1 & 0 & 0 & 1 \\ 1 & 0 & 0 & 1 \\ 0 & 1 & 1 & 0 \end{bmatrix}.$$

It is clear that $B_{2^2,M}D_{2,1} = 0B_{2^2,M}$. Obviously $B_{2^2,M}D_{2,0} = B_{2^2,M}I_4 = B_{2^2,M}$. By straight-forward multiplication, we can show that $B_{2^2,M}D_{2,2} = -B_{2^2,M}$. □

The following result can be proved in various ways; an alternative method of proof appears in Section 6.3.

LEMMA 4.1.3.

$$B_{2^k,M}D_{k,v} = \left[\binom{k-1}{v} - \binom{k-1}{v-1}\right] B_{2^k,M}$$

for all $k \geq 2$, for all allowable v.

Proof. This is the sort of result that lends itself to proof by induction. So we must establish the result for a small value of k and then show that assuming the result for k we can prove the result for $k + 1$. From Example 4.1.8, we know that the result holds for $k = 2$ for all allowable v.

Assume that the result is true for k and consider $k + 1$. Thus we can assume that $B_{2^k,M}D_{k,v} = \left[\binom{k-1}{v} - \binom{k-1}{v-1}\right] B_{2^k,M}$ and we must see what happens when we consider $B_{2^{k+1},M}D_{k+1,v}$. We get

$$\sqrt{2^{k+1}}B_{2^{k+1},M}D_{k+1,v} = \begin{bmatrix} -\mathbf{j}'_{2^k} & \mathbf{j}'_{2^k} \\ \sqrt{2^k}B_{2^k,M} & \sqrt{2^k}B_{2^k,M} \end{bmatrix} \begin{bmatrix} D_{k,v} & D_{k,v-1} \\ D_{k,v-1} & D_{k,v} \end{bmatrix}$$
$$= \begin{bmatrix} -\mathbf{j}'D_{k,v} + \mathbf{j}'D_{k,v-1} & \mathbf{j}'D_{k,v} - \mathbf{j}'D_{k,v-1} \\ \sqrt{2^k}B_{2^k,M}(D_{k,v} + D_{k,v-1}) & \sqrt{2^k}B_{2^k,M}(D_{k,v} + D_{k,v-1}) \end{bmatrix}.$$

Now

$$-\mathbf{j}'D_{k,v} + \mathbf{j}'D_{k,v-1} = -\mathbf{j}'\left[\binom{k}{v} - \binom{k}{v-1}\right]$$

and

$$B_{2^k,M}(D_{k,v} + D_{k,v-1}) = \left[\binom{k-1}{v} - \binom{k-1}{v-1} + \binom{k-1}{v-1} - \binom{k-1}{v-2}\right] B_{2^k,M}.$$

It can be easily shown that

$$\binom{k}{v} = \binom{k-1}{v} + \binom{k-1}{v-1}$$

and

$$\binom{k}{v-1} = \binom{k-1}{v-1} + \binom{k-1}{v-2};$$

so

$$\binom{k}{v} - \binom{k}{v-1} = \binom{k-1}{v} - \binom{k-1}{v-2},$$

and hence the result is proved for $k+1$ for all v. □

We now show that at this stage we do not need to make any assumptions about the contrasts in B_r since the class of competing designs that we have chosen ensures that $C_{hr} = \mathbf{0}$ for any choice of B_r (also see Exercise 4.1.5.5). By definition,

$$C_{hr} = \begin{bmatrix} B_h \\ B_r \end{bmatrix} \Lambda \, [B_h' \; B_r'] = \begin{bmatrix} B_h \Lambda B_h' & B_h \Lambda B_r' \\ B_r \Lambda B_h' & B_r \Lambda B_r' \end{bmatrix} = \begin{bmatrix} C_{hh} & C_{hr} \\ C_{rh} & C_{rr} \end{bmatrix}.$$

Since $B_h = B_{2^k,M}$, we can see that

$$\begin{aligned} C_{hr} &= \left[\sum_{v=1}^{k} \binom{k}{v} a_v\right] B_h I_{2^k} B_r' - \sum_{v=1}^{k} a_v B_h D_{k,v} B_r' \\ &= \left[\sum_{v=1}^{k} \binom{k}{v} a_v\right] B_{2^k,M} B_r' - \sum_{v=1}^{k} a_v \left[\binom{k-1}{v} - \binom{k-1}{v-1}\right] B_{2^k,M} B_r' \\ &= \mathbf{0}_{k,2^k}. \end{aligned}$$

This is true for any choice of B_r as long as the set of competing designs remains the same so that the form of Λ stays as a linear combination of the $D_{k,v}$.

Using the result in Lemma 4.1.3, we can get an explicit expression for the information matrix, C_M, for the k main effects.

■ **LEMMA 4.1.4.**

The information matrix for main effects under the null hypothesis is given by

$$C_M = \left[\frac{1}{2} \sum_{v=1}^{k} a_v \binom{k-1}{v-1}\right] I_k$$

and the determinant of C_M *is*

$$2^{-k} \left[\sum_{v=1}^{k} a_v \binom{k-1}{v-1}\right]^k .$$

Proof. The information matrix for main effects under the null hypothesis is

$$\begin{aligned} C_M &= B_{2^k,M} \Lambda B_{2^k,M}' \\ &= \frac{1}{4} B_{2^k,M} \left[\sum_{v=1}^{k} \binom{k}{v} a_v I_{2^k} - \sum_{v=1}^{k} a_v D_{k,v}\right] B_{2^k,M}' \\ &= \frac{1}{4} \left\{ \sum_{v=1}^{k} \left(\binom{k}{v} a_v\right) I_k - \sum_{v=1}^{k} \left(a_v \left[\binom{k-1}{v} - \binom{k-1}{v-1}\right]\right) B_{2^k,M} B_{2^k,M}' \right\} \\ &= \left\{ \frac{1}{4} \sum_{v=1}^{k} a_v \left[\binom{k}{v} - \binom{k-1}{v} + \binom{k-1}{v-1}\right] \right\} I_k \\ &= \left\{ \frac{1}{2} \sum_{v=1}^{k} a_v \binom{k-1}{v-1} \right\} I_k, \end{aligned}$$

and the determinant of C_M is

$$2^{-k}\left[\sum_{v=1}^{k} a_v \binom{k-1}{v-1}\right]^k,$$

as required. □

Recall that the a_v are the proportions of the pairs that have v attributes different. Thus we know that $2^{k-1}\sum_{v=1}^{k}\binom{k}{v}a_v = 1$.

■ **EXAMPLE 4.1.9.**
Let $k = 3$. Then

$$\det(C_M) = \frac{1}{8}\left[a_1\binom{2}{0} + a_2\binom{2}{1} + a_3\binom{2}{2}\right]^3,$$

subject to the constraint

$$4\left[\binom{3}{1}a_1 + \binom{3}{2}a_2 + \binom{3}{3}a_3\right] = 1.$$

So we want to maximize

$$\frac{1}{8}(a_1 + 2a_2 + a_3)^3$$

subject to $12a_1 + 12a_2 + 4a_3 = 1$. We can write $a_3 = (1 - 12a_1 - 12a_2)/4$ and so we want to maximize

$$\frac{1}{8}[a_1 + 2a_2 + (1 - 12a_1 - 12a_2)/4]^3 = \frac{1}{8}\left[\frac{1}{4} - 2a_1 - a_2\right]^3.$$

This is largest when $a_1 = a_2 = 0$. Thus the optimal paired comparison design for estimating main effects consists of the four pairs (000, 111), (001, 110), (010, 101), and (011, 100), each appearing once; so $a_3 = 1/4$. In Table 4.5, we give all the possible designs and the corresponding values of $\det(C_M)$. We know there are $\binom{8}{2} = 28$ pairs of distinct treatment combinations; 4 have three attributes different, 12 have two attributes different and 12 have one attribute different. Since we are including all pairs with a given number of attributes different, or none of them, there are 7 possible sets of pairs to consider. We see that the first design in Table 4.5 consists of all the pairs with only one attribute different, the second design consists of all the pairs with 2 attributes different, and so on; the final design consists of all 28 pairs. □

We now extend this idea to get the D-optimal paired comparison design for binary attributes for any value of k.

■ **THEOREM 4.1.1.**
The D-optimal paired comparison design for estimating main effects consists of the foldover pairs only; that is, all k attributes appear at different levels in the two options in each choice set. For the optimal designs

$$\det(C_{\text{opt},M}) = \left[\frac{1}{2^k}\right]^k.$$

Table 4.5 The 7 Competing Designs for Main Effects Only for Pairs with $k = 3$

a_1	a_2	a_3	N	$12a_1 + 12a_2 + 4a_3$	$\det(C_M)$
$\frac{1}{12}$	0	0	12	$12\frac{1}{12} + 0 + 0 = 1$	$\frac{1}{8}(\frac{1}{12})^3 = 7.234 \times 10^{-5}$
0	$\frac{1}{12}$	0	12	$0 + 12\frac{1}{12} + 0 = 1$	$\frac{1}{8}(\frac{1}{6})^3 = 5.787 \times 10^{-4}$
0	0	$\frac{1}{4}$	4	$0 + 0 + 4\frac{1}{4} = 1$	$\frac{1}{8}(\frac{1}{4})^3 = 1.953 \times 10^{-3}$
0	$\frac{1}{16}$	$\frac{1}{16}$	16	$0 + 12\frac{1}{16} + 4\frac{1}{16} = 1$	$\frac{1}{8}(\frac{3}{16})^3 = 8.240 \times 10^{-4}$
$\frac{1}{16}$	0	$\frac{1}{16}$	16	$12\frac{1}{16} + 0 + 4\frac{1}{16} = 1$	$\frac{1}{8}(\frac{1}{8})^3 = 2.441 \times 10^{-4}$
$\frac{1}{24}$	$\frac{1}{24}$	0	24	$12\frac{1}{24} + 12\frac{1}{24} + 0 = 1$	$\frac{1}{8}(\frac{1}{8})^3 = 2.441 \times 10^{-4}$
$\frac{1}{28}$	$\frac{1}{28}$	$\frac{1}{28}$	28	$12\frac{1}{28} + 12\frac{1}{28} + 4\frac{1}{28} = 1$	$\frac{1}{8}(\frac{1}{7})^3 = 3.644 \times 10^{-4}$

Proof. To find the D-optimal design, we must maximize

$$\det(C_M) = 2^{-k}\left[\sum_{v=1}^{k} a_v \binom{k-1}{v-1}\right]^k$$

subject to the constraint $2^{k-1}\sum_{v=1}^{k}\binom{k}{v}a_v = 1$. Rearranging this constraint, we get $2^{k-1}a_k = 1 - 2^{k-1}\sum_{v=1}^{k-1}\binom{k}{v}a_v$. Substituting for a_k into the expression for $\det(C_{opt,M})$, we get

$$2^{-k}\left[\sum_{v=1}^{k-1}\left[\binom{k-1}{v-1}a_v\right] + \frac{1}{2^{k-1}} - \sum_{v=1}^{k-1}\left[\binom{k}{v}a_v\right]\right]^k$$

which can be rearranged to give

$$2^{-k}\left[\frac{1}{2^{k-1}} - \sum_{v=1}^{k-1}\left[\binom{k-1}{v}a_v\right]\right]^k.$$

Consider this expression; it is clear that the maximum value of $\det(C_M)$ is attained when $a_v = 0$, $v \neq k$ and $a_k = 1/2^{k-1}$. □

For any design that we construct we can calculate the D-efficiency of that design, relative to the optimal design, using the expression

$$\text{Eff}_D = \left(\frac{\det(C_M)}{\det(C_{\text{opt},M})}\right)^{1/p}$$

where $p = k$.

■ EXAMPLE 4.1.10.
When $k = 3$, we see from Table 4.5, that $\det(C_{opt,M}) = 1.953 \times 10^{-3}$ and so the design with $i_1 = i_3 = 0$ and $i_2 = 1$ has

$$\text{Eff}_D = 100\left(\frac{5.787 \times 10^{-4}}{1.953 \times 10^{-3}}\right)^{1/3} = 66.6679\%.$$

□

■ **THEOREM 4.1.2.**
The A-optimal paired comparison design for estimating main effects consists of the foldover pairs only. For the optimal designs

$$\operatorname{tr}(C^{-1}_{\text{opt},M}) = k2^k.$$

Proof. To find the A-optimal design, we must minimize $\operatorname{tr}(C_M^{-1})$ subject to the constraint $2^{k-1}\sum_{v=1}^{k}\binom{k}{v}a_v = 1$. Because C_M is diagonal and all entries are equal, this is equivalent to maximizing $\operatorname{tr}(C_M)$. Since $\operatorname{tr}(C_M) = \frac{k}{2}\sum_{v=1}^{k}\binom{k-1}{v-1}a_v$, this must be maximized, again subject to the constraint. As above, this maximum is obtained when $a_v = 1/2^{k-1}$ and all other a_v are 0. □

We can also calculate the A-efficiency of any design, relative to the optimal design, using the expression

$$\text{Eff}_A = \left(\frac{\operatorname{tr}(C^{-1}_{\text{opt},M})}{\operatorname{tr}(C_M^{-1})}\right).$$

Thus we find that the A- and D-optimal designs coincide for the estimation of main effects only.

4.1.4 The Model for Main Effects and Two-factor Interactions

We evaluate the information matrix associated with the estimation of main effects and two-factor interactions so that we can determine the A- and D-optimal designs.

As before, we let $B_{2^k,M}$ be the rows of B_{2^k} that correspond to main effects and we now let $B_{2^k,T}$ be the rows of B_{2^k} that correspond to two-factor interactions. Thus the matrix associated with main effects and two-factor interactions is the concatenation of these matrices and we denote it by $B_{2^k,MT}$. We want to get a neat expression for $B_{2^k,T}D_{k,v}$.

■ **EXAMPLE 4.1.11.**
Let $k = 3$. Then

$$B_{2^3,T} = \frac{1}{\sqrt{8}}\begin{bmatrix} 1 & 1 & -1 & -1 & -1 & -1 & 1 & 1 \\ 1 & -1 & 1 & -1 & -1 & 1 & -1 & 1 \\ 1 & -1 & -1 & 1 & 1 & -1 & -1 & 1 \end{bmatrix}$$

and, using the results from Example 4.1.8, we see that

$$B_{2^3,T}D_{3,1} = -B_{2^3,T},\ B_{2^3,T}D_{3,2} = -B_{3,T},\ \text{and } B_{2^3,T}D_{3,3} = B_{2^3,T}.$$ □

These results are extended in the next lemma.

■ **LEMMA 4.1.5.**

$$B_{2^k,T}D_{k,v} = \left[\binom{k-2}{v} - 2\binom{k-2}{v-1} + \binom{k-2}{v-2}\right] B_{2^k,T}$$

for all $k \geq 3$ for all allowable v.

Proof. By straight-forward multiplication, we know that the result holds for $k = 3$ for all allowable v.

Assume that the result is true for k and consider $k+1$. Then

$$\begin{aligned} B_{2^{k+1},T}D_{k+1,v} &= \frac{1}{\sqrt{2}}\begin{bmatrix} B_{2^k,T} & B_{2^k,T} \\ -B_{2^k,M} & B_{2^k,M} \end{bmatrix}\begin{bmatrix} D_{k,v} & D_{k,v-1} \\ D_{k,v-1} & D_{k,v} \end{bmatrix} \\ &= \frac{1}{\sqrt{2}}\begin{bmatrix} B_{2^k,T}(D_{k,v}+D_{k,v-1}) & B_{2^k,T}(D_{k,v}+D_{k,v-1}) \\ B_{2^k,M}(D_{k,v-1}-D_{k,v}) & B_{2^k,M}(D_{k,v}-D_{k,v-1}) \end{bmatrix}. \end{aligned}$$

Using the results of Lemma 4.1.3, we have that the (2,1) position is

$$\left[\binom{k-1}{v-1} - \binom{k-1}{v-2} - \binom{k-1}{v} + \binom{k-1}{v-1}\right] B_{2^k,M}$$

$$= -\left[\binom{k-2}{v} - 2\binom{k-2}{v-1} + \binom{k-2}{v-2}\right] B_{2^k,M},$$

as required. The (2,2) position is the negative of the (2,1) position.

The (1,1) (and the (1,2)) positions are given by

$$\begin{aligned} B_{2^k,T}(D_{k,v}+D_{k,v-1}) &= \left\{\left[\binom{k-2}{v} - 2\binom{k-2}{v-1} + \binom{k-2}{v-2}\right]\right. \\ &\quad \left. + \left[\binom{k-2}{v-1} - 2\binom{k-2}{v-2} + \binom{k-2}{v-3}\right]\right\} B_{2^k,T} \end{aligned}$$

by the induction hypothesis. Observing that

$$\binom{k-1}{v} = \binom{k-2}{v} + \binom{k-2}{v-1},$$

$$\binom{k-1}{v-1} = \binom{k-2}{v-1} + \binom{k-2}{v-2},$$

and

$$\binom{k-1}{v-2} = \binom{k-2}{v-2} + \binom{k-2}{v-3},$$

we see that we have established the result. □

Using this result, we can evaluate the information matrix for main effects plus two-factor interactions. Note that we now have $B_h = B_{2^k,MT}$.

■ LEMMA 4.1.6.

Under the null hypothesis, the information matrix for main effects plus two-factor interactions is given by

$$C_{MT} = \begin{bmatrix} \frac{1}{2}\sum_{v=1}^{k}\binom{k-1}{v-1}a_v I_k & \mathbf{0} \\ \mathbf{0} & \sum_{v=1}^{k}\binom{k-2}{v-1}a_v I_{k(k-1)/2} \end{bmatrix}.$$

Proof. Using the definition for the information matrix and Lemmas 4.1.3 and 4.1.4, we have that

$$\begin{aligned} C_{MT} &= B_{2^k,MT}\Lambda B'_{2^k,MT} \\ &= \begin{bmatrix} B_{2^k,M}\Lambda B'_{2^k,M} & B_{2^k,M}\Lambda B'_{2^k,T} \\ B_{2^k,T}\Lambda B'_{2^k,M} & B_{2^k,T}\Lambda B'_{2^k,T} \end{bmatrix} \\ &= \begin{bmatrix} \frac{1}{2}\sum_{v=1}^{k}\binom{k-1}{v-1}a_v I_k & \mathbf{0} \\ \mathbf{0} & B_{2^k,T}\Lambda B'_{2^k,T} \end{bmatrix}. \end{aligned}$$

Now

$$\begin{aligned} B_{2^k,T}\Lambda &= \frac{1}{4}B_{2^k,T}\left[\sum_{v=1}^{k}\binom{k}{v}a_v I_{2^k} - \sum_{v=1}^{k} a_v D_{k,v}\right] \\ &= \sum_{v=1}^{k}\binom{k-2}{v-1}a_v B_{2^k,T}. \end{aligned}$$

Thus $B_{2^k,T}\Lambda B'_{2^k,T} = \sum_{v=1}^{k}\binom{k-2}{v-1}a_v I_{k(k-1)/2}$. □

To find the D-optimal design, we need to maximize

$$\det(C_{MT}) = \left[\frac{1}{2}\sum_{v=1}^{k}\binom{k-1}{v-1}a_v\right]^k \times \left[\sum_{v=1}^{k}\binom{k-2}{v-1}a_v\right]^{k(k-1)/2}$$

subject to the constraint that $2^{k-1}\sum_v \binom{k}{v}a_v = 1$. We look at a small example before proving a general result.

■ **EXAMPLE 4.1.12.**
Let $k = 3$. Then

$$\det(C_{MT}) = \left[\frac{1}{2}\left[\binom{2}{0}a_1 + \binom{2}{1}a_2 + \binom{2}{2}a_3\right]\right]^3 \times \left[\binom{1}{0}a_1 + \binom{1}{1}a_2 + \binom{1}{2}a_3\right]^3,$$

and we want to maximize this, subject to the constraint $4(3a_1 + 3a_2 + a_3) = 1$, the same constraint that we had when considering the estimation of main effects only. Simplifying, we see that we want to maximize

$$\frac{1}{8}\left[a_1 + 2a_2 + a_3\right]^3 \times \left[a_1 + a_2\right]^3$$

subject to the constraint. We can calculate the possible values of N and the corresponding values of $\det(C_{MT})$. We know there are $\binom{8}{2} = 28$ pairs of distinct treatment combinations; 4 have three attributes different, 12 have two attributes different and 12 have one attribute different. Since we are including all pairs with a given number of attributes different, or none of them, there are 7 possible sets of pairs to consider. These are given in Table 4.6 together with the corresponding values of N and $\det(C_{MT})$. We see that the first design consists of all the pairs with only one attribute different, the second design consists of all the pairs with 2 attributes different, and so on until the final design consists of all 28 pairs. From the table, we see that the optimal design for estimating main effects and two-factor interactions has the 12 pairs with two attributes different. □

■ **THEOREM 4.1.3.**
The D-optimal design for testing main effects and two-factor interactions when all other effects are assumed to be zero is given by

$$a_v = \begin{cases} \left\{2^{k-1}\binom{k}{(k+1)/2}\right\}^{-1} & v = (k+1)/2, \quad \textit{if } k \textit{ is odd} \\ 0 & \textit{otherwise,} \end{cases}$$

and

$$a_v = \begin{cases} \left\{2^{k-1}\binom{k+1}{k/2}\right\}^{-1} & v = k/2, k/2+1, \quad \textit{if } k \textit{ is even} \\ 0 & \textit{otherwise.} \end{cases}$$

Table 4.6 The 7 Competing Designs for Main Effects and Two-Factor Interactions for Pairs with $k = 3$

a_1	a_2	a_3	N	$\det(C_{MT})$	$\mathrm{tr}(C_{MT}^{-1})$
$\frac{1}{12}$	0	0	12	$(\frac{1}{24})^3(\frac{1}{12})^3 = 4.1862 \times 10^{-8}$	108
0	$\frac{1}{12}$	0	12	$(\frac{1}{12})^3(\frac{1}{12})^3 = 3.3490 \times 10^{-7}$	72
0	0	$\frac{1}{4}$	4	$(\frac{1}{8})^3 0^3 = 0$	not defined
0	$\frac{1}{16}$	$\frac{1}{16}$	16	$(\frac{3}{32})^3(\frac{1}{16})^3 = 2.0117 \times 10^{-7}$	80
$\frac{1}{16}$	0	$\frac{1}{16}$	16	$(\frac{1}{16})^3(\frac{1}{16})^3 = 5.9605 \times 10^{-8}$	96
$\frac{1}{24}$	$\frac{1}{24}$	0	24	$(\frac{1}{16})^3(\frac{1}{12})^3 = 1.4129 \times 10^{-7}$	84
$\frac{1}{28}$	$\frac{1}{28}$	$\frac{1}{28}$	28	$(\frac{1}{14})^3(\frac{1}{14})^3 = 1.3281 \times 10^{-7}$	84

The determinant of the optimal design is given by

$$\det(C_{\mathrm{opt},MT}) = \begin{cases} \left\{\frac{k+1}{2^{k+1}k}\right\}^{k+k(k-1)/2} & \text{if } k \text{ is odd,} \\ \left\{\frac{k+2}{2^{k+1}(k+1)}\right\}^{k+k(k-1)/2} & \text{if } k \text{ is even.} \end{cases}$$

Proof. Let

$$x_v = 2^{k-1}a_v, \quad W = \sum_{v=1}^{k}\binom{k-1}{v-1}x_v \quad \text{and} \quad Z = \sum_{v=1}^{k}\binom{k-2}{v-1}x_v.$$

Thus we want to maximize $f = WZ^{(k-1)/2}$, subject to the constraint that $\sum_{v=1}^{k}\binom{k}{v}x_v = 1$. Substituting for $x_k = 1 - \sum_{v=1}^{k-1}\binom{k}{v}x_v$ in W gives

$$W = 1 - \sum_{v=1}^{k-1}\binom{k-1}{v}x_v.$$

Thus f is a function of $(k-1)$ variables and needs to be maximized over the region described by the inequalities $\sum_{v=1}^{k-1}\binom{k}{v}x_v \leq 1$ with $x_v \geq 0$ for $v = 1, \ldots, k-1$. In this region, we note that $f \geq 0$.

Any local extreme values of f will be found by solving the system of equations

$$\frac{\partial f}{\partial x_v} = 0, \quad v = 1, \ldots, (k-1).$$

For $k \geq 3$ this gives

$$\left\{\frac{k-1}{2}\binom{k-2}{v-1}W - \binom{k-1}{v}Z\right\} = 0.$$

Hence

$$\binom{k-1}{v}Z = \frac{k-1}{2}\binom{k-2}{v-1}W = \frac{v}{2}\binom{k-1}{v}W, \quad v = 1, \ldots, (k-1)$$

and so we have that

$$2Z = vW, \quad v = 1, \ldots, (k-1).$$

Either these equations are inconsistent or $W = 0 = Z$, and hence $f = 0$ at all local extrema; thus the maximum value of f will occur on the boundary of the region of allowed values of x_v in $\Re^{k-1}$. (We use $\Re$ to denote the set of real numbers and $\Re^{(k-1)}$ to denote the set of (k-1)-dimensional vectors over the real numbers. So $\Re^2$ is the set of pairs over the real numbers, familiar from the two-dimensional graphs of high school.)

There are k subspaces of dimension $k-2$; they are given by the equations

$$x_v = 0, v = 1, \ldots, (k-1) \text{ and } \sum_{v=1}^{k-1} \binom{k}{v} x_v = 1.$$

For the first $k-1$ of these subspaces, the above analysis still holds; we simply delete the mth equation from the system and put $x_m = 0$ in the remaining equations. We either obtain inconsistent equations or $W = 0 = Z$; thus $f = 0$ at all local extrema of f in these subspaces, and so the maximum allowed value of f will occur on the boundaries of the region of allowed values of x_v in these subspaces of $\Re^{(k-1)}$.

In the subspace given by

$$\sum_{v=1}^{k-1} \binom{k}{v} x_v = 1,$$

we use Lagrange multipliers to locate the extreme values of f; this gives

$$\binom{k-1}{v}\left\{-Z + \frac{v}{2}W\right\} = \lambda\binom{k}{v}.$$

Eliminating λ between two of these equations gives

$$\{(g+v-k)W - 2Z\}\,\frac{g-v}{k}\binom{k}{v}\binom{k}{g} = 0.$$

This gives

$$2Z = (g+v-k)W, \quad v = 1, \ldots, (k-1), \quad v \neq g, \quad v \neq k-g.$$

Either these equations are inconsistent or $W = Z = 0$, and so $f = 0$ provided $k \geq 4$.

Thus we find that $f = 0$ at all local extrema of f in this subspace; so the maximum allowed value of f will occur on the boundaries of the region of allowed values of x_v in this subspace of $\Re^{k-1}$. We continue in this way, projecting the region of allowed values of x_v onto subspaces of lower and lower dimension. It is only when we reach the one-dimensional subspaces (the edges of the region of the allowed values of x_v) that we obtain just one equation to be solved to locate extreme values of f in this subspace; in general, $f \neq 0$ at these points.

There are two types of edges that need to be considered:

1. those along the coordinate axis with

$$x_v = 0, v \neq g, \text{and} 0 \leq \binom{k}{g} x_g \leq 1;$$

2. those bounding the constraint surface with

$$\binom{k}{g}x_g + \binom{k}{h}x_h = 1,\ g \neq h,\ x_v = 0,\ v \neq g, h,$$

$$0 \leq \binom{k}{g}x_g \leq 1, \text{ and } 0 \leq \binom{k}{h}x_h \leq 1.$$

Along the coordinate axes,

$$W = 1 - \binom{k-1}{g}x_g,\ Z = \binom{k-2}{g-1}x_g \text{ and } \frac{\partial f}{\partial x_g} = 0$$

requires

$$2Z = gW,$$

which gives

$$x_g = \frac{g}{k+1}\binom{k-2}{g-1}^{-1} = \frac{k(k-1)}{(k+1)(k-g)}\binom{k}{g}^{-1}.$$

It is only when $g = 1$ that this value of x_g lies in the interval of allowed values of x_g; in this case

$$W = \frac{2}{k+1} \text{ and } Z = \frac{1}{k+1}.$$

For all other values of g, the maximum value of f on $0 \leq \binom{k}{g}x_g \leq 1$ occurs at $x_g = \binom{k}{g}^{-1}$, where

$$W = \frac{g}{k} \text{ and } Z = \frac{g(k-g)}{k(k-1)}.$$

For $2 \leq g \leq k-1$ with $k \geq 3$ these values of W and Z are larger than those found above for $g = 1$. To determine which value of g maximizes f amongst these alternatives, we use these expressions for W and Z in f and set $\frac{df}{dg} = 0$. This gives a maximum when $g = \frac{k+1}{2}$, provided this is an integer. If (k+1)/2 is not an integer, we shall show that the maximum allowed value of f occurs along the edge joining $g = k/2$ and $g = k/2 + 1$.

It remains to consider the behavior of f along the edges of the constraint surface; in particular, we consider what happens along the edge given by

$$\binom{k}{g}x_g + \binom{k}{h}x_h = 1,\ g \neq h,\ x_v = 0,\ v \neq g, h,$$

$$0 \leq \binom{k}{g}x_g \leq 1, \text{ and } 0 \leq \binom{k}{h}x_h \leq 1.$$

Along this edge

$$W = \frac{h}{k} + \frac{g-h}{k}\binom{k}{g}x_g$$

and

$$Z = \frac{h(k-h)}{k(k-1)} + \frac{(h-g)(h+g-k)}{k(k-1)}\binom{k}{g}x_g.$$

Setting

$$\frac{\partial f}{\partial x_g} = 0$$

gives

$$(g+h-k)W = 2Z.$$

If $g+h-k=0$, Z is constant and the equation is inconsistent; thus the maximum allowed value of f will occur at the endpoint of the interval (that is, at one of the vertices studied above). If $g+h-k \neq 0$, we solve for x_g and get

$$\binom{k}{g} x_g = \frac{n[(k+1)(k-h) - g(k-1)]}{(g-h)(g+h-k)(k+1)}.$$

This gives the location of a local maximum along the edge. At this local maximum,

$$W = \frac{2gh}{(g+h-k)k(k+1)} \quad \text{and} \quad Z = \frac{gh}{k(k+1)}.$$

The value of h which maximizes f for a particular value of g is obtained from $\frac{\partial f}{\partial h} = 0$; this gives

$$h = \frac{k+1}{k-1}(k-g), \quad W = \frac{g}{k} \quad \text{and} \quad Z = \frac{g(k-g)}{k(k-1)}.$$

The value of g which gives the overall maximum allowed value of f is obtained from $\frac{\partial f}{\partial g} = 0$; this gives $g = \frac{k+1}{2}$. In this case, $h = g$ which is not allowed along an edge. If $g \neq h$, the situation closest to equality holds when $g = h+1 = \frac{k+1}{k-1}(k-g)+1$; this gives $g = \frac{k}{2} + 1 - \frac{1}{2k}$ and $h = \frac{k}{2} - \frac{1}{2k}$. When k is an even integer, this result gives values for g and h which are close to $k/2+1$ and $k/2$ respectively. Along the edge where $g = \frac{k}{2}+1$ and $h = \frac{k}{2}$ (for k even), the maximum value of f occurs where $x_{k/2} = x_{k/2+1} = \left[\binom{k+1}{k/2}\right]^{-1}$.

We now determine the maximum value of the determinant at these a_v values. For k even, only two values of v will give the maximum determinant. These are $v = k/2$ and $v = k/2+1$, where

$$a_{k/2} = a_{k/2+1} = \left[2^{k-1}\binom{k+1}{k/2}\right]^{-1}$$

and all other $a_v = 0$. Then

$$\begin{aligned}
\det(C_{MT}) &= \left[\frac{1}{2}\sum_{v=k/2,k/2+1}\binom{k-1}{v-1}a_v\right]^k \times \left[\sum_{v=k/2,k/2+1}^{k}\binom{k-2}{i-1}a_v\right]^{k(k-1)/2} \\
&= \left[\frac{1}{2^k}\binom{k+1}{k/2}^{-1}\left[\binom{k-1}{k/2-1}+\binom{k-1}{k/2}\right]\right]^k \\
&\quad \times \left[\frac{1}{2^{k-1}}\binom{k+1}{k/2}^{-1}\left[\binom{k-2}{k/2-1}+\binom{k-2}{k/2}\right]\right]^{k(k-1)/2}.
\end{aligned}$$

Now $\binom{k-1}{k/2-1}+\binom{k-1}{k/2} = \binom{k}{k/2}$, $\binom{k-2}{k/2-1}+\binom{k-2}{k/2} = \binom{k-1}{k/2}$,

$$\frac{\binom{k}{k/2}}{\binom{k+1}{k/2}} = \frac{k+2}{2(k+1)} \quad \text{and} \quad \frac{\binom{k-1}{k/2}}{\binom{k+1}{k/2}} = \frac{k+2}{4(k+1)}.$$

Therefore, for k even,

$$\det(C_{MT}) = \left(\frac{(k+2)}{(2^{k+1}(k+1))}\right)^{k+k(k-1)/2}.$$

For k odd, the only value of v that will give the maximum determinant is $v = (k+1)/2$, where

$$a_{(k+1)/2} = \left[2^{k-1}\binom{k}{(k+1)/2}\right]^{-1}$$

and all other $a_v = 0$. Then

$$\begin{aligned}\det(C_{MT}) &= \left[\frac{1}{2^k}\binom{k}{(k+1)/2}^{-1}\binom{k-1}{(k+1)/2-1}\right]^{k} \\ &\quad \times \left[\frac{1}{2^{k-1}}\binom{k}{(k+1)/2}^{-1}\binom{k-2}{(k+1)/2-1}\right]^{k(k-1)/2}.\end{aligned}$$

Now

$$\frac{\binom{k-1}{(k+1)/2-1}}{\binom{k}{(k+1)/2}} = \frac{\binom{k-1}{(k-1)/2}}{\binom{k}{(k+1)/2}} = \frac{k+1}{2k}$$

and

$$\frac{\binom{k-2}{(k+1)/2-1}}{\binom{k}{(k+1)/2}} = \frac{\binom{k-2}{(k-1)/2}}{\binom{k}{(k+1)/2}} = \frac{k+1}{4k}.$$

Therefore, for k odd,

$$\det(C_{MT}) = \left(\frac{(k+1)}{2^{k+1}k}\right)^{k+k(k-1)/2}. \qquad \square$$

When $k = 3$, this result says that the pairs to use are those with $(k+1)/2 = 4/2 = 2$ attributes different, just as we found in Example 4.1.12.

For any design that we construct we can calculate the D-efficiency of that design, relative to the optimal design, using the expression

$$\text{Eff}_D = \left(\frac{\det(C_{MT})}{\det(C_{\text{opt},MT})}\right)^{1/p},$$

where $p = k + k(k-1)/2$.

We now want to establish a similar result for A-optimal designs. We begin by looking at an example.

■ EXAMPLE 4.1.13.

Let $k = 3$. We know that

$$C_{MT} = \begin{bmatrix} \frac{1}{2}(a_1 + 2a_2 + a_3)I_3 & \mathbf{0} \\ \mathbf{0} & (a_1 + a_2)I_3 \end{bmatrix}.$$

So $\text{tr}(C_{MT}^{-1}) = 6(a_1 + 2a_2 + a_3)^{-1} + 3(a_1 + a_2)^{-1}$. These values are given in Table 4.6, and we see that the A-optimal design has the 12 pairs with two attributes different. □

The A-optimal design in the previous example is the same as the D-optimal design when $k = 3$. In the next result, we determine the A-optimal design by minimizing $\text{tr}(C_{MT}^{-1})$ and find that the A-optimal design is always the same as the D-optimal design for the same value of k.

THEOREM 4.1.4.

The A-optimal design for testing main effects and two-factor interactions, when all other interaction effects are assumed to be zero, is given by

$$a_v = \begin{cases} \left\{2^{k-1}\binom{k}{(k+1)/2}\right\}^{-1}, & v = (k+1)/2, \quad \text{if } k \text{ is odd} \\ 0, & \text{otherwise,} \end{cases}$$

and

$$a_v = \begin{cases} \left\{2^{k-1}\binom{k+1}{k/2}\right\}^{-1}, & i = k/2, k/2+1, \quad \text{if } k \text{ is even} \\ 0, & \text{otherwise.} \end{cases}$$

The trace of the optimal design is given by

$$\text{tr}(C_{\text{opt},MT}^{-1}) = \begin{cases} \left\{\frac{2^{k+1}k}{k+1}\right\} \times (k + k(k-1)/2), & \text{if } k \text{ is odd,} \\ \left\{\frac{2^{k+1}(k+1)}{k+2}\right\} \times (k + k(k-1)/2), & \text{if } k \text{ is even.} \end{cases}$$

Proof. Let

$$x_v = 2^{k-1}a_v, \quad W = \sum_{v=1}^{k}\binom{k-1}{v-1}x_v, \quad \text{and} \quad Z = \sum_{v=1}^{k-1}\binom{k-2}{v-1}x_v.$$

Thus we want to minimize

$$f = W^{-1} + \frac{(k-1)}{4}Z^{-1}$$

subject to the constraint

$$\sum_{v=1}^{k}\binom{k}{v}x_v = 1.$$

We use the constraint to obtain

$$W = 1 - \sum_{v=1}^{k-1}\binom{k-1}{v}x_v.$$

In this way, f is now a function of $(k-1)$ variables which needs to be minimized over the region described by the inequalities $\sum_{v=1}^{k-1}\binom{k}{v}x_v \leq 1$ with $x_v \geq 0$ for $v = 1, 2, \ldots, (k-1)$. In this region $f \geq 0$ because $x_v \geq 0 \; \forall \; v$.

Any local extreme value of f will be found by solving the system of equations

$$\frac{\partial f}{\partial x_v} = 0, \quad v = 1, 2, \ldots, (k-1).$$

Thus

$$\binom{k-1}{v} W^{-2} = \frac{(k-1)}{4}\binom{k-2}{v-1} Z^{-2},$$

which we can re-write as

$$Z^2 = \frac{v}{4} W^2.$$

As both $W > 0$ and $Z > 0$, we require that

$$Z = \frac{\sqrt{v}}{2} W \ \forall \ v = 1, 2, \ldots, (k-1).$$

Thus we have obtained a system of inconsistent equations which has no solution; so the minimum allowed value of f will occur on the boundary of the region of allowed values of x_v in $\Re^{k-1}$.

Along the coordinate axis with $x_v = 0, v \neq 0$, and $0 \leq \binom{k}{g} x_g \leq 1$, we have that

$$\begin{aligned} f &= \frac{1}{1 - \binom{k-1}{g} x_g} + \frac{k-1}{4\binom{k-2}{g-1} x_g}, \\ \frac{\partial f}{\partial x_g} &= \frac{\binom{k-1}{g}}{\left[1 - \binom{k-1}{g} x_g\right]^2} - \frac{(k-1)\binom{k-2}{g-1}}{4\left[\binom{k-2}{g-1} x_g\right]^2} \\ &= 0 \end{aligned}$$

when

$$\binom{k-2}{g-1} x_g = \frac{\sqrt{g}}{2}\left[1 - \binom{k-1}{g} x_g\right].$$

Thus we see that

$$\binom{k-1}{g} x_g = [1 + \frac{\sqrt{g}}{2(k-1)}]^{-1}.$$

However this value of x_g lies outside the interval of allowed values as

$$\frac{k}{k-g}[1 + \frac{\sqrt{g}}{2(k-1)}]^{-1} \geq 1 \quad \text{for all allowed } g.$$

Thus the extreme values of f will occur at $\binom{k}{g} x_g = 0$ and $\binom{k}{g} x_g = 1$; the minimum value occurs at

$$\binom{k}{g} x_g = 1 \text{ where } f = \frac{k}{g}[1 + \frac{(k-1)^2}{4(k-g)}].$$

To find the value of g which minimizes f amongst these alternatives, we require

$$\frac{df}{dg} = 0,$$

which gives

$$\frac{k}{g}\left\{\frac{(k-1)^2}{4(k-g)^2} - \frac{1}{g}\left[1 + \frac{(k-1)^2}{4(k-g)}\right]\right\} = 0.$$

So we have

$$g = \frac{k+1}{2}, \; k\frac{k+1}{2}.$$

However, the second root lies outside the interval of allowed values $[1, k-1]$.

Thus, provided that k is odd, the minimum value of f along the coordinate axes

$$x_v = 0, \; v \neq g, \; 0 \leq \binom{k}{g} x_g \leq 1,$$

occurs at

$$g = \frac{k+1}{2}$$

where

$$f = k.$$

When k is even, we find that

$$g = \frac{k}{2}$$

gives

$$f = k + \frac{1}{k}$$

while

$$g = \frac{k}{2} + 1$$

gives

$$f = k + \frac{1}{k - 4/k} > k + \frac{1}{k}.$$

Along the edge

$$\binom{k}{g} x_g + \binom{k}{h} x_h = 1, \quad g \neq h$$

with

$$0 \leq \binom{k}{g} x_g \leq 1, \quad 0 \leq \binom{k}{h} x_h \leq 1, \quad x_v = 0, \; v \neq g, h$$

we have that

$$W = \frac{h}{k} + \frac{g-h}{k} \binom{k}{g} x_g$$

and

$$Z = \frac{h(k-h)}{k(k-1)} + \frac{(h-g)(h+g-k)}{k(k-1)} \binom{k}{g} x_g.$$

We find that

$$\frac{df}{dx_g} = 0$$

when

$$Z = \frac{\sqrt{h+g-k}}{2} W,$$

provided that

$$h + g - k \neq 0.$$

When

$$h + g - k = 0,$$

$$\frac{df}{dx_g} \neq 0;$$

thus f will be minimized at a vertex studied previously. When

$$h + g - k \neq 0,$$

$$\binom{k}{g} x_g = \frac{h}{h-g} \left[\frac{2(k-h) - (k-1)\sqrt{h+g-k}}{2(h+g-k) + (k-1)\sqrt{h+g-k}} \right],$$

which gives

$$\begin{aligned} W &= \frac{2gh}{k} \left[2(g+h-k) + (k-1)\sqrt{h+g-k} \right]^{-1}, \\ Z &= \frac{\sqrt{h+g-k}}{2} W, \end{aligned}$$

and

$$f = \frac{k}{4gh} \left[4(g+h-k) + 4(k-1)\sqrt{g+h-k} + (k-1)^2 \right].$$

To find the value of h which minimizes f for a particular value of g, we require $\frac{\partial f}{\partial h} = 0$. Thus

$$0 = -\frac{f}{h} + \frac{k}{4gh} \left[4 + \frac{2(k-1)}{\sqrt{g+h-k}} \right],$$

which gives

$$\sqrt{g+h-k} = \frac{2(k-g)}{k-1} \quad \text{or} \quad h = \left[\frac{2(k-g)}{k-1} \right]^2 + (k-g).$$

To find the pair of numbers (g, h) which minimizes f for a particular value of k, we attempt to solve

$$\frac{\partial f}{\partial h} = 0 = \frac{\partial f}{\partial g};$$

this gives

$$g = h = \frac{k+1}{2},$$

which is not allowed as $g \neq h$, but agrees with the earlier result for the location of the minimum value of f along a coordinate axis.

When k is even, $(\frac{k}{2}, \frac{k}{2}+1)$ is the pair of integers closest to $(\frac{k+1}{2}, \frac{k+1}{2})$; indeed substituting $g = \frac{k}{2}$ in the expression for h gives $h = \frac{k}{2} + 1 + \frac{1}{(k-1)^2} \simeq \frac{k}{2} + 1$ and the corresponding value for f is

$$f = \frac{(k+1)^2}{k+2} = k + \frac{1}{k+2} < k + \frac{1}{k}.$$

Thus, when k is even, $\text{tr}(C^{-1})$ is minimized when $g = \frac{k}{2}$ and $h = \frac{k}{2} + 1$, giving the result in the statement of the theorem.

To determine the maximum value of the trace of $C^{-1}_{\text{opt},MT}$, we substitute the a_v values into the expression for C_{MT} in Lemma 4.1.6 and invert. C^{-1}_{MT} is a diagonal matrix and the $(k + k(k-1)/2)$ diagonal entries are given by

$$\begin{array}{ll} \frac{2^{k+1}k}{k+1}, & \text{if } k \text{ is odd,} \\ \frac{2^{k+1}(k+1)}{k+2}, & \text{if } k \text{ is even.} \end{array}$$

Then, if k is odd,

$$\text{tr}(C^{-1}_{\text{opt},MT}) = \left\{\frac{2^{k+1}k}{k+1}\right\} \times (k + k(k-1)/2)$$

and, if k is even,

$$\text{tr}(C^{-1}_{\text{opt},MT}) = \left\{\frac{2^{k+1}(k+1)}{k+2}\right\} \times (k + k(k-1)/2).$$

□

We can calculate the A-efficiency of any design, relative to the optimal design, using the expression

$$\text{Eff}_A = \left(\frac{\text{tr}(C^{-1}_{\text{opt},MT})}{\text{tr}(C^{-1}_{MT})}\right).$$

Hence we see that the D- and A-optimal designs for estimating main effects and two-factor interactions coincide.

■ **EXAMPLE 4.1.14.**

If $k = 4$, then the D- and A-optimal design consists of the 80 pairs with two and three attributes different. So $a_1 = a_4 = 0$ and $a_2 = a_3 = \frac{1}{80}$. □

4.1.5 Exercises

1. If $k = 4$, give the set of pairs with $v = 1$, $v = 2$, $v = 3$, and $v = 4$ attributes different.

2. Give $D_{3,3}$. Verify the recursive formula given for the $D_{k,v}$ for $k = 4$ for $v = 1, 2, 3$, and 4.

3. For $k = 3$ give B_{2^3} and verify that $B_{2^3} = \frac{1}{\sqrt{8}}\begin{bmatrix} -\mathbf{j}'_{2^2} & \mathbf{j}'_{2^2} \\ 2B_{2^2} & 2B_{2^2} \\ -2B_{2^2} & 2B_{2^2} \end{bmatrix}$.

4. Show that $B_{2^3,M}D_{3,i} = \left[\binom{2}{i} - \binom{2}{i-1}\right] B_{2^3,M}$ for $i = 0, 1, 2$, and 3.

5. (a) Let $k = 3$. Show that $C_{hr} = 0$ for the estimation of main effects when B_a contains the two-factor interaction contrasts.

   (b) Let $k = 3$. Show that $C_{hr} = 0$ for the estimation of main effects when B_a contains the three-factor interaction contrast only.

6. Suppose that $k = 5$.

(a) Using Theorem 4.1.1, give the D-optimal set of pairs from the complete factorial for estimating main effects.

(b) Calculate the corresponding C matrix.

7. Suppose that $k = 5$.

(a) Using Theorem 4.1.3, give the D-optimal pairs for estimating main effects plus two-factor interactions.

(b) Calculate the corresponding C matrix.

(c) Compare this to the C matrix of the previous exercise. Comment.

8. Suppose that $k = 2$.

(a) Give the optimal pairs for estimating main effects.

(b) Give the optimal pairs for estimating main effects and the two-factor interaction.

(c) Use the ideas developed in this section to find the optimal pairs for estimating the two-factor interaction only. Comment.

9. Let $k = 4$. There are 15 possible sets of pairs to consider. Draw up a table like Table 4.6 for $k = 4$, and hence confirm the results in Theorems 4.1.3 and 4.1.4.

4.2 SMALL OPTIMAL AND NEAR-OPTIMAL DESIGNS FOR PAIRS

In this section we show how a fractional factorial design can be used to provide the treatment combinations for an optimal paired comparison design. We begin by deriving the information matrix when the treatment combinations come from a fractional factorial design.

4.2.1 The Derivation of the Λ Matrix

The Λ matrix is defined in the same way as in Section 4.1.1 but because we are working with a fractional factorial design it is possible that some of the rows and columns of Λ will be 0. Similarly the contrast matrix, B_h, will be the same as before, being a contrast matrix for main effects or for main effects plus two-factor interactions, depending on what effects are of interest. As before the information matrix C for the effects of interest is given by $C = B_h \Lambda B_h'$ but now we need to assume that all contrasts other than the ones we want to estimate must be 0 because fractional factorial designs are constructed assuming that higher order interaction effects are 0. So B_a includes all the contrasts that are not in B_h and B_r is empty.

We illustrate these comments with a small example.

■ EXAMPLE 4.2.1.

Let $k = 4$, and consider the 4 pairs in Design 1 in Table 4.7. Then there are 8 treatment combinations that do not appear in the choice experiment and so 8 rows and columns of Λ

are equal to 0, as we can see below.

$$\Lambda = \frac{1}{16}\left[\begin{array}{cccccccccccccccc}
1 & 0 & 0 & 0 & 0 & 0 & 0 & 0 & 0 & 0 & 0 & 0 & 0 & 0 & 0 & -1 \\
0 & 0 & 0 & 0 & 0 & 0 & 0 & 0 & 0 & 0 & 0 & 0 & 0 & 0 & 0 & 0 \\
0 & 0 & 0 & 0 & 0 & 0 & 0 & 0 & 0 & 0 & 0 & 0 & 0 & 0 & 0 & 0 \\
0 & 0 & 0 & 1 & 0 & 0 & 0 & 0 & 0 & 0 & 0 & 0 & -1 & 0 & 0 & 0 \\
0 & 0 & 0 & 0 & 0 & 0 & 0 & 0 & 0 & 0 & 0 & 0 & 0 & 0 & 0 & 0 \\
0 & 0 & 0 & 0 & 0 & 1 & 0 & 0 & 0 & 0 & -1 & 0 & 0 & 0 & 0 & 0 \\
0 & 0 & 0 & 0 & 0 & 0 & 1 & 0 & 0 & -1 & 0 & 0 & 0 & 0 & 0 & 0 \\
0 & 0 & 0 & 0 & 0 & 0 & 0 & 0 & 0 & 0 & 0 & 0 & 0 & 0 & 0 & 0 \\
0 & 0 & 0 & 0 & 0 & 0 & 0 & 0 & 0 & 0 & 0 & 0 & 0 & 0 & 0 & 0 \\
0 & 0 & 0 & 0 & 0 & 0 & -1 & 0 & 0 & 1 & 0 & 0 & 0 & 0 & 0 & 0 \\
0 & 0 & 0 & 0 & 0 & -1 & 0 & 0 & 0 & 0 & 1 & 0 & 0 & 0 & 0 & 0 \\
0 & 0 & 0 & 0 & 0 & 0 & 0 & 0 & 0 & 0 & 0 & 0 & 0 & 0 & 0 & 0 \\
0 & 0 & 0 & -1 & 0 & 0 & 0 & 0 & 0 & 0 & 0 & 0 & 1 & 0 & 0 & 0 \\
0 & 0 & 0 & 0 & 0 & 0 & 0 & 0 & 0 & 0 & 0 & 0 & 0 & 0 & 0 & 0 \\
0 & 0 & 0 & 0 & 0 & 0 & 0 & 0 & 0 & 0 & 0 & 0 & 0 & 0 & 0 & 0 \\
-1 & 0 & 0 & 0 & 0 & 0 & 0 & 0 & 0 & 0 & 0 & 0 & 0 & 0 & 0 & 1
\end{array}\right].$$

Using $B_h = B_{2^4,M}$ and the Λ matrix above we see that

$$C_M = B_{2^4,M}\Lambda B'_{2^4,M} = \frac{1}{16}I_4.$$

This is the same C matrix obtained by using the foldover pairs from the complete factorial. Thus the 4 pairs in Design 1 in Table 4.7 give an optimal choice experiment for estimating main effects. The 4 pairs in Design 2 in Table 4.7 are also optimal for estimating main effects. Within both of these designs, for the first attribute, the level is the same for all of the treatment combinations within an option. If this is a problem in a practical sense, then combining the two designs to get 8 pairs will ensure that both of the levels of the first attribute appear in both options. The combined design consists of all the foldover pairs and hence is optimal too. □

Table 4.7 Two Designs with Four Pairs for $k = 4$ Binary Attributes

Option A	Option B	Option A	Option B
0000	1111	1110	0001
0011	1100	1101	0010
0101	1010	1011	0100
0110	1001	1000	0111
Design 1		Design 2	

4.2.2 The Model for Main Effects Only

In Section 4.1.3 we showed that the optimal pairs for the estimation of main effects in a forced choice stated preference experiment were all the foldover pairs. This suggests that an optimal choice experiment might be obtainable from a fractional factorial by taking the foldover pairs. From Section 2.2.1, we know that that all contrasts for main effects can be independently estimated in a resolution 3 fraction provided that we assume that all the other contrasts are 0. That is, we assume that B_a will contain all the contrasts except those in $B_{2^k,M}$. This suggests that a resolution 3 fraction could be the starting design for the foldover pairs in a construction for an optimal design with fewer pairs. The only additional constraint that we need to impose is that the resolution 3 fraction must be regular. The following construction, without the regularity constraint, is found as Option 3 in Appendix A5 of Louviere et al. (2000), although no formal proof of the properties of these designs is given there.

■ CONSTRUCTION 4.2.1.

To construct a set of pairs to compare products described by k binary attributes, first construct a regular orthogonal main effect plan with k binary attributes. From each row of this OMEP obtain a choice pair by pairing the row with its foldover. If any pair appears twice then the duplicate choice set is omitted. Thus each treatment combination, and each pair, appears only once in the final set of choice pairs. The design has a diagonal information matrix C and a D-efficiency of 100% for estimating main effects.

Proof. Assume that all the equations that define the OMEP have an even number of non-zero coefficients. Let

$$\sum_i \eta_i x_i = 0$$

be one of these equations. Then

$$\sum_i \eta_i = 0.$$

If $(a_1, a_2, \ldots, a_k)$ is a solution of this equation then

$$\sum_i \eta_i a_i = 0.$$

Hence

$$\sum_i \eta_i (1 - a_i) = \sum_i \eta_i - \sum_i \eta_i a_i = 0.$$

Thus for each treatment combination in the OMEP its foldover also appears.

Let the levels for the attributes be -1 and 1 and let A denote the $N \times k$ array for the OMEP. Since for each treatment combination that is in A, its foldover is also in A, we can represent A as $\begin{bmatrix} A_1 \\ -A_1 \end{bmatrix}$. Thus

$$A'A = NI_k = 2A_1'A_1.$$

We can write the pairs as $(A_1, -A_1)$. We will order the treatment combinations by writing all the treatment combinations in A_1 first, then all the treatment combinations in $-A_1$ and then the remaining treatment combinations in any order. Then the B matrix for main effects is

$$B_{2^k,M} = \frac{1}{\sqrt{2^k}}(A_1', -A_1', B_{\bar{A}}),$$

where $B_{\bar{A}}$ contains the coefficients of the main effect contrasts for the treatment combinations that are not in A. Using the same ordering for the treatment combinations, we have that

$$\Lambda = \frac{1}{2N}\begin{bmatrix} I & -I & \mathbf{0} \\ -I & I & \mathbf{0} \\ \mathbf{0} & \mathbf{0} & \mathbf{0} \end{bmatrix}.$$

Then we get

$$B_{2^k,M} \Lambda B_{2^k,M}' = \frac{1}{2^k}\frac{1}{4N}A_1'A_1 = \frac{1}{2^k}I_k.$$

Thus

$$\det(C_M) = \left[\frac{1}{2^k}\right]^k,$$

which is equal to $\det(C_{\text{opt},M})$ in Theorem 4.1.1, and so this design has a D-efficiency of 100%.

Next, suppose that the set of binary equations that define the OMEP, A, has at least one equation with an odd number of non-zero coefficients. Thus there are no foldover pairs in the experiment and the pairs in the choice experiment are given by $(A, -A)$; the argument then proceeds as above. □

■ EXAMPLE 4.2.2.
Let $k = 4$, and consider the OMEP in Table 2.7. The pairs derived from Construction 4.2.1 and this OMEP are given as Design 1 in Table 4.7. As we have remarked previously, there are only 8 of the 16 possible treatment combinations involved in these pairs, but this design is as efficient as the design based on all 8 foldover pairs. If the same construction is used on the OMEP given in Table 2.5, we get the same design as we get when we take the foldover pairs in the complete factorial. □

As initially described, Construction 4.2.1 did not include the restriction that the OMEP used in the construction be regular. However, Example 4.2.3 shows that, without that restriction, the pairs that result may not have a diagonal information matrix or be 100% efficient; see Exercise 4.2.5.2.

■ EXAMPLE 4.2.3.
Let $k = 6$, and consider the OMEP in Table 4.8(a). Observe that it has one foldover pair of treatment combinations and that it is not regular; see Exercise 4.2.5.2. Applying Construction 4.2.1 to this OMEP gives 11 distinct pairs with information matrix

$$C_M = \frac{1}{2^6 \times 44}(48I - 4J)$$

and a D-efficiency of 97.2%. □

Construction 4.2.1 can be extended to the union of regular designs. For example the design in Table 4.8(b) is the union of a 2^{4-1} and two copies of a 2^{4-2}. Thus it has 8 treatment combinations in which the foldover occurs in the design and four which do not (and which are repeated). However, Construction 4.2.1 applied to this design gives a set of pairs that is 100% efficient.

4.2.3 The Model for Main Effects and Two-Factor Interactions

We begin by recalling that a resolution 5 fractional factorial design allows for the independent estimation of all main effects and all two-factor interactions in the ordinary least squares setting. Also recall that when deriving pairs from the complete factorial the optimal pairs for estimation of all main effects plus two-factor interactions are all those pairs in which either $(k+1)/2$ attributes are different (k odd) or $k/2$ and $k/2+1$ attributes are different (k even). These ideas are exploited in the remainder of this section to give sets of generators to use to define the choice sets. Although the resulting designs are near-optimal, it is not yet possible to give a definitive construction method for optimal pairs for the estimation of main effects plus two-factor interactions except when starting with the complete factorial.

Table 4.8 Non-regular OMEPs of Resolution 3

0	0	0	0	0	0
1	1	1	1	1	1
0	1	0	1	1	0
1	0	1	0	1	0
0	1	1	1	0	0
0	0	0	1	1	1
0	0	1	0	1	1
1	0	0	1	0	1
1	1	0	0	1	0
1	1	0	0	0	1
0	1	1	0	0	1
1	0	1	1	0	0

(a) $k = 6$

0	0	0	0
1	1	1	1
0	0	0	1
1	1	1	0
0	1	0	0
1	0	1	1
0	1	0	1
1	0	1	0
1	0	0	0
1	0	0	0
0	0	1	1
0	0	1	1
1	1	0	1
1	1	0	1
0	1	1	0
0	1	1	0

(b) $k = 4$

Consider a regular fractional factorial design F of resolution 5. Choose any treatment combination not in the fraction, **e** say, and form pairs by pairing $\mathbf{f} \in F$ with **f+e**, where the addition is done component-wise modulo 2. We will write the complete set of pairs as $(F, F + \mathbf{e})$. We refer to **e** as the *generator* of the pairs.

To evaluate the information matrix of these pairs easily, we need to define two incidence matrices, $D_{M,\mathbf{e}}$ and $D_{T,\mathbf{e}}$. We let X and Z be any two attributes in the experiment. We define a diagonal matrix $D_{M,\mathbf{e}}$ by $(D_{M,\mathbf{e}})_{XX} = 1$ if $\mathbf{e}_X = 0$ and $(D_{M,\mathbf{e}})_{XX} = -1$ if $\mathbf{e}_X = 1$, where the attributes label the rows and columns of $D_{M,\mathbf{e}}$. We define a diagonal matrix $D_{T,\mathbf{e}}$ of size $k(k-1)/2$ by $(D_{T,\mathbf{e}})_{XZ,XZ} = 1$ if $\mathbf{e}_X = \mathbf{e}_Z$ and $(D_{T,\mathbf{e}})_{XZ,XZ} = -1$ if $\mathbf{e}_X \neq \mathbf{e}_Z$ (where we label the rows and columns of $D_{T,\mathbf{e}}$ by the unordered pairs of distinct attributes).

■ EXAMPLE 4.2.4.

Let $k = 5$, and let F be the solutions to $x_1 + x_2 + x_3 + x_4 + x_5 = 0$. So F contains the treatment combinations

00000 00011 00101 00110 01001 01010 01100 01111
10001 10010 10100 10111 11000 11011 11101 11110.

Let $\mathbf{e} = (00111)$. Then the treatment combinations in $F + \mathbf{e}$ are given by

00111 00100 00010 00001 01110 01101 01011 01000
10110 10101 10011 10001 11111 11100 11010 11001,

$$D_{M,\mathbf{e}} = \begin{bmatrix} 1 & 0 & 0 & 0 & 0 \\ 0 & 1 & 0 & 0 & 0 \\ 0 & 0 & -1 & 0 & 0 \\ 0 & 0 & 0 & -1 & 0 \\ 0 & 0 & 0 & 0 & -1 \end{bmatrix},$$

and

$$D_{T,\mathbf{e}} = \begin{bmatrix} 1 & 0 & 0 & 0 & 0 & 0 & 0 & 0 & 0 & 0 \\ 0 & -1 & 0 & 0 & 0 & 0 & 0 & 0 & 0 & 0 \\ 0 & 0 & -1 & 0 & 0 & 0 & 0 & 0 & 0 & 0 \\ 0 & 0 & 0 & -1 & 0 & 0 & 0 & 0 & 0 & 0 \\ 0 & 0 & 0 & 0 & -1 & 0 & 0 & 0 & 0 & 0 \\ 0 & 0 & 0 & 0 & 0 & -1 & 0 & 0 & 0 & 0 \\ 0 & 0 & 0 & 0 & 0 & 0 & -1 & 0 & 0 & 0 \\ 0 & 0 & 0 & 0 & 0 & 0 & 0 & 1 & 0 & 0 \\ 0 & 0 & 0 & 0 & 0 & 0 & 0 & 0 & 1 & 0 \\ 0 & 0 & 0 & 0 & 0 & 0 & 0 & 0 & 0 & 1 \end{bmatrix}.$$

The treatment combinations in the pairs $(F, F+\mathbf{e})$ include all the treatment combinations in the complete factorial. If we write the treatment combinations in F in the order we gave them above, then the contrast, within F, for the main effect of the first attribute is

$$(-1,-1,-1,-1,-1,-1,-1,-1,1,1,1,1,1,1,1,1).$$

If we then add $\mathbf{e}$ to each treatment combination in F we see that the contrast, within $F+\mathbf{e}$, for the main effect of the first attribute is exactly the same as that in F since the first entry of $\mathbf{e}$ is 0; so the first levels of the treatment combinations in F and $F+\mathbf{e}$ are the same. However, the contrast for the main effect of the third attribute is

$$(-1,-1,1,1,-1,-1,1,1,-1,-1,1,1,-1,-1,1,1)$$

in F and is

$$(1,1,-1,-1,1,1,-1,-1,1,1,-1,-1,1,1,-1,-1)$$

in $F+\mathbf{e}$. This is the negative of the contrast in F and happens because the third entry in $\mathbf{e}$ is 1.

Similarly, the coefficients for the interaction of the first two attributes are the same in F as they are in $F+\mathbf{e}$ since the addition of the generator $\mathbf{e}$ does not change the levels of these two attributes. The coefficients for the interaction of the last two attributes are the same in F as they are in $F+\mathbf{e}$ since the addition of the generator $\mathbf{e}$ changes the levels of both of these two attributes.

Thus we have that

$$B_{MT} = \begin{bmatrix} B_{M,F} & D_{M,\mathbf{e}}B_{M,F} \\ B_{T,F} & D_{T,\mathbf{e}}B_{T,F} \end{bmatrix},$$

where $B_{M,F}$ and $B_{T,F}$ are the matrices of the coefficients for main effects and two-factor interactions, respectively, in the fraction F only. Specifically, we have $\sqrt{32}B_{M,F}$ is

$$\begin{bmatrix} -1 & -1 & -1 & -1 & -1 & -1 & -1 & -1 & 1 & 1 & 1 & 1 & 1 & 1 & 1 & 1 \\ -1 & -1 & -1 & -1 & 1 & 1 & 1 & 1 & -1 & -1 & -1 & -1 & 1 & 1 & 1 & 1 \\ -1 & -1 & 1 & 1 & -1 & -1 & 1 & 1 & -1 & -1 & 1 & 1 & -1 & -1 & 1 & 1 \\ -1 & 1 & -1 & 1 & -1 & 1 & -1 & 1 & -1 & 1 & -1 & 1 & -1 & 1 & -1 & 1 \\ -1 & 1 & 1 & -1 & 1 & -1 & -1 & 1 & 1 & -1 & -1 & 1 & -1 & 1 & 1 & -1 \end{bmatrix}$$

and $\sqrt{32}B_{T,F}$ is

$$\begin{bmatrix}
1 & 1 & 1 & 1 & -1 & -1 & -1 & -1 & -1 & -1 & -1 & -1 & 1 & 1 & 1 & 1 \\
1 & 1 & -1 & -1 & 1 & 1 & -1 & -1 & -1 & -1 & 1 & 1 & -1 & -1 & 1 & 1 \\
1 & -1 & 1 & -1 & 1 & -1 & 1 & -1 & -1 & 1 & -1 & 1 & -1 & 1 & -1 & 1 \\
1 & -1 & -1 & 1 & -1 & 1 & 1 & -1 & 1 & -1 & -1 & 1 & -1 & 1 & 1 & -1 \\
1 & 1 & -1 & -1 & -1 & -1 & 1 & 1 & 1 & 1 & -1 & -1 & -1 & -1 & 1 & 1 \\
1 & -1 & 1 & -1 & -1 & 1 & -1 & 1 & 1 & -1 & 1 & -1 & -1 & 1 & -1 & 1 \\
1 & -1 & -1 & 1 & 1 & -1 & -1 & 1 & -1 & 1 & 1 & -1 & -1 & 1 & 1 & -1 \\
1 & -1 & -1 & 1 & 1 & -1 & -1 & 1 & 1 & -1 & -1 & 1 & 1 & -1 & -1 & 1 \\
1 & -1 & 1 & -1 & -1 & 1 & -1 & 1 & -1 & 1 & -1 & 1 & 1 & -1 & 1 & -1 \\
1 & 1 & -1 & -1 & -1 & -1 & 1 & 1 & -1 & -1 & 1 & 1 & 1 & 1 & -1 & -1
\end{bmatrix}.$$

Calculating C_{MT}, we get

$$C_{MT} = \frac{1}{32}\begin{bmatrix}
0 & 0 & 0 & 0 & 0 & 0 & 0 & 0 & 0 & 0 & 0 & 0 & 0 & 0 & 0 \\
0 & 0 & 0 & 0 & 0 & 0 & 0 & 0 & 0 & 0 & 0 & 0 & 0 & 0 & 0 \\
0 & 0 & 1 & 0 & 0 & 0 & 0 & 0 & 0 & 0 & 0 & 0 & 0 & 0 & 0 \\
0 & 0 & 0 & 1 & 0 & 0 & 0 & 0 & 0 & 0 & 0 & 0 & 0 & 0 & 0 \\
0 & 0 & 0 & 0 & 1 & 0 & 0 & 0 & 0 & 0 & 0 & 0 & 0 & 0 & 0 \\
0 & 0 & 0 & 0 & 0 & 0 & 0 & 0 & 0 & 0 & 0 & 0 & 0 & 0 & 0 \\
0 & 0 & 0 & 0 & 0 & 0 & 1 & 0 & 0 & 0 & 0 & 0 & 0 & 0 & 0 \\
0 & 0 & 0 & 0 & 0 & 0 & 0 & 1 & 0 & 0 & 0 & 0 & 0 & 0 & 0 \\
0 & 0 & 0 & 0 & 0 & 0 & 0 & 0 & 1 & 0 & 0 & 0 & 0 & 0 & 0 \\
0 & 0 & 0 & 0 & 0 & 0 & 0 & 0 & 0 & 1 & 0 & 0 & 0 & 0 & 0 \\
0 & 0 & 0 & 0 & 0 & 0 & 0 & 0 & 0 & 0 & 1 & 0 & 0 & 0 & 0 \\
0 & 0 & 0 & 0 & 0 & 0 & 0 & 0 & 0 & 0 & 0 & 1 & 0 & 0 & 0 \\
0 & 0 & 0 & 0 & 0 & 0 & 0 & 0 & 0 & 0 & 0 & 0 & 0 & 0 & 0 \\
0 & 0 & 0 & 0 & 0 & 0 & 0 & 0 & 0 & 0 & 0 & 0 & 0 & 0 & 0 \\
0 & 0 & 0 & 0 & 0 & 0 & 0 & 0 & 0 & 0 & 0 & 0 & 0 & 0 & 0
\end{bmatrix}.$$

Notice that the 0 entries on the diagonal in C_{MT} correspond to 1s on the diagonal in D_M (for the first 5 values) and D_T (for the remaining 10 values). □

The following result lets us say something about the C matrix in terms of the generator **e** in general.

■ **LEMMA 4.2.1.**

Consider the pairs $(F, F + \mathbf{e})$, where F is a fractional factorial design for k attributes and **e** *is any treatment combination not in F. Then the information matrix $C_{\mathbf{e}}$ is given by*

$$C_{\mathbf{e}} = \frac{1}{4 \times 2^k}\begin{bmatrix} 2I_k - 2D_{M,\mathbf{e}} & \mathbf{0} \\ \mathbf{0} & 2I_{k(k-1)/2} - 2D_{T,\mathbf{e}} \end{bmatrix}.$$

Proof. Let $B_{M,F}$ be the submatrix of the contrast matrix for main effects associated with the treatment combinations in F, and let $B_{T,F}$ be the submatrix of the contrast matrix for two-factor interactions associated with the treatment combinations in F. Then, assuming that there are k attributes and that there are N treatment combinations in F, we know that $B_{M,F}B'_{M,F} = \frac{N}{2^k}I_k$, $B_{T,F}B'_{T,F} = \frac{N}{2^k}I_{k(k-1)/2}$ and $B_{M,F}B'_{T,F} = 0$.

For convenience, we order the treatment combinations in the paired comparison experiment as $\mathbf{f}_1, \mathbf{f}_2, \ldots, \mathbf{f}_N$ (for some fixed but arbitrary order), the treatment combinations in

F, followed by $\mathbf{f}_1+\mathbf{e}, \mathbf{f}_2+\mathbf{e}, \ldots, \mathbf{f}_N+\mathbf{e}$ and then the remaining treatment combinations. Using this order to label the rows and columns of Λ we get

$$\Lambda_{\mathbf{e}} = \frac{1}{4N}\begin{bmatrix} I & -I & \mathbf{0} \\ -I & I & \mathbf{0} \\ \mathbf{0} & \mathbf{0} & \mathbf{0} \end{bmatrix}.$$

To calculate the information matrix $C_{\mathbf{e}}$, we need to calculate the B matrix for the treatment combinations in this order. If a particular attribute X has a 0 in $\mathbf{e}$, then the X contrast in $F+\mathbf{e}$ is the same as it is in F; if attribute X has a 1 in $\mathbf{e}$, then the X contrast in $F+\mathbf{e}$ is the negative of the one in F. So the matrix for main effect contrasts is given by

$$B_M = \begin{bmatrix} B_{M,F} & D_{M,\mathbf{e}}B_{M,F} & B_{M,\bar{A}} \end{bmatrix},$$

where $A = F \cup (F+\mathbf{e})$.

Similarly, consider two attributes X and Z. If $\mathbf{e}_X = \mathbf{e}_Z$, then the two-factor interaction contrast for the attributes X and Z is the same in $F+\mathbf{e}$ as it is in F. If $\mathbf{e}_X \neq \mathbf{e}_Z$ then the two-factor interaction contrast for attributes X and Z is the negative in $F+\mathbf{e}$ of the corresponding contrast in F. So the matrix for two-factor interaction contrasts is given by

$$B_T = \begin{bmatrix} B_{T,F} & D_{T,\mathbf{e}}B_{T,F} & B_{T,\bar{A}} \end{bmatrix}.$$

Then we get

$$\begin{aligned} C_{\mathbf{e}} &= \frac{1}{4N}\begin{bmatrix} B_{M,F} & D_{M,\mathbf{e}}B_{M,F} & B_{M,\bar{A}} \\ B_{T,F} & D_{T,\mathbf{e}}B_{T,F} & B_{T,\bar{A}} \end{bmatrix}\begin{bmatrix} I & -I & \mathbf{0} \\ -I & I & \mathbf{0} \\ \mathbf{0} & \mathbf{0} & \mathbf{0} \end{bmatrix}\begin{bmatrix} B'_{M,F} & B'_{T,F} \\ B'_{M,F}D'_{M,\mathbf{e}} & B'_{T,F}D'_{T,\mathbf{e}} \\ B'_{M,\bar{A}} & B'_{T,\bar{A}} \end{bmatrix} \\ &= \frac{1}{4N}\begin{bmatrix} B_{M,F} & D_{M,\mathbf{e}}B_{M,F} & B_{M,\bar{A}} \\ B_{T,F} & D_{T,\mathbf{e}}B_{T,F} & B_{T,\bar{A}} \end{bmatrix}\begin{bmatrix} B'_{M,F}-B'_{M,F}D'_{M,\mathbf{e}} & B'_{T,F}-B'_{T,F}D'_{T,\mathbf{e}} \\ B'_{M,F}D'_{M,\mathbf{e}}-B'_{M,F} & B'_{T,F}D'_{T,\mathbf{e}}-B'_{T,F} \\ \mathbf{0} & \mathbf{0} \end{bmatrix}. \end{aligned}$$

Now the (1,1) position of this matrix is given by

$$\begin{aligned} B_{M,F}B'_{M,F} - B_{M,F}B'_{M,F}D'_{M,\mathbf{e}} - D_{M,\mathbf{e}}B_{M,F}B'_{M,F} + D_{M,\mathbf{e}}B_{M,F}B'_{M,F}D'_{M,\mathbf{e}} \\ = \frac{N}{2^k}(I_k - D_{M,\mathbf{e}} - D_{M,\mathbf{e}} + I_k). \end{aligned}$$

Proceeding similarly with the other entries in $C_{\mathbf{e}}$ we get

$$C_{\mathbf{e}} = \frac{1}{4\times 2^k}\begin{bmatrix} 2I_k - 2D_{M,\mathbf{e}} & \mathbf{0} \\ \mathbf{0} & 2I_{k(k-1)/2} - 2D_{T,\mathbf{e}} \end{bmatrix},$$

as required. □

Thus $C_{\mathbf{e}}$ is diagonal, and the non-zero entries in $C_{\mathbf{e}}$ correspond to those positions in $\mathbf{e}$ where there is a 1 for the main effects part of $C_{\mathbf{e}}$, and to those positions in the two-factor interaction part where one attribute corresponds to a 1 and one to a 0 in $\mathbf{e}$. One can see this by considering the matrices calculated in Example 4.2.4.

The next result extends the previous result to two generators.

LEMMA 4.2.2.
Consider the pairs generated by $\mathbf{e}$ *and* $\mathbf{g}$ *where* $\mathbf{e}, \mathbf{g} \notin F$ *but where* $\mathbf{e} + \mathbf{g} \in F$. *Then the information matrix* $C_{\mathbf{e},\mathbf{g}}$ *is given by*

$$C_{\mathbf{e},\mathbf{g}} = \frac{1}{8 \times 2^k} \begin{bmatrix} 4I_k - 2D_{M,\mathbf{e}} - 2D_{M,\mathbf{g}} & \mathbf{0} \\ \mathbf{0} & 4I_{k(k-1)/2} - 2D_{T,\mathbf{e}} - 2D_{T,\mathbf{g}} \end{bmatrix}.$$

Proof. As $\mathbf{e} + \mathbf{g} \in F$, only treatment combinations in F and $F + \mathbf{e}$ have been used in the construction of the additional pairs. Hence we can write

$$\Lambda_{\mathbf{e},\mathbf{g}} = \frac{1}{8N} \begin{bmatrix} 2I & -I - P & \mathbf{0} \\ -I - P & 2I & \mathbf{0} \\ \mathbf{0} & \mathbf{0} & \mathbf{0} \end{bmatrix},$$

where P is a permutation matrix that ensures that $\Lambda_{\mathbf{e},\mathbf{g}}$ contains the correct pairs. Consider $C_{\mathbf{e},\mathbf{g}} = B_{MT}\Lambda_{\mathbf{e},\mathbf{g}}B'_{MT}$ as a 2×2 block matrix. Then similar calculations to those in Lemma 4.2.1 give

$$(B_{MT}\Lambda_{\mathbf{e},\mathbf{g}}B'_{MT})_{12} = -D_{M,\mathbf{e}}B_{M,F}PB'_{T,F} - B_{M,F}PB'_{T,F}D'_{T,\mathbf{e}}.$$

If we define $D_{M,\mathbf{g}}$ and $D_{T,\mathbf{g}}$ for the generator $\mathbf{g}$ in the same way we defined $D_{M,\mathbf{e}}$ and $D_{T,\mathbf{e}}$ for the generator $\mathbf{e}$, then we have that

$$B_{M,F}P = D_{M,\mathbf{e}}D_{M,\mathbf{g}}B_{M,F}$$

since $B_{M,F}P$ is a permutation of the columns of $B_{M,F}$ and can be thought of as a permutation of the treatment combinations in F. P is a permutation of the treatment combinations in $F + \mathbf{e}$ so that the order of the treatment combinations corresponds to that of $F + \mathbf{g}$. Since we know $F = F + \mathbf{e} + \mathbf{g}$, we see that the contrast matrix for the treatment combinations in this order is

$$D_{M,\mathbf{e}+\mathbf{g}}B_{M,F} = D_{M,\mathbf{e}}D_{M,\mathbf{g}}B_{M,F};$$

this gives the result. Thus

$$(B_{MT}\Lambda_{\mathbf{e},\mathbf{g}}B'_{MT})_{12} = 0.$$

Again we find that

$$(B_{MT}\Lambda_{\mathbf{e},\mathbf{g}}B'_{MT})_{11} = \frac{1}{8 \times 2^k}(4I_k - 2D_{M,\mathbf{e}} - 2D_{M,\mathbf{g}}).$$

Finally, noting that

$$B_{T,F}P = D_{T,\mathbf{e}}D_{T,\mathbf{g}}B_{T,F},$$

we see that

$$(B_{MT}\Lambda_{\mathbf{e},\mathbf{g}}B'_{MT})_{22} = \frac{1}{8 \times 2^k}(4I_{k(k-1)/2} - 2D_{T,\mathbf{e}} - 2D_{T,\mathbf{g}}).$$

Hence we see that $C_{\mathbf{e},\mathbf{g}}$ is diagonal, and the effects that can be estimated are those that correspond to a non-zero entry in one of the generators (for main effects) and those that correspond to positions with unequal entries (for two-factor interactions). □

EXAMPLE 4.2.5.
Let $k = 5$ and use the F of Example 4.2.4. Let $\mathbf{e} = (00111)$ and let $\mathbf{g} = (11100)$. Then

$\mathbf{e}, \mathbf{g} \notin F$ but $\mathbf{e} + \mathbf{g} \in F$. We use the order of the treatment combinations in $F + \mathbf{e}$ to give the order to use when labelling the final 16 rows and columns of the Λ matrix. Since the entries in $F + \mathbf{g}$ are

11100 11111 11001 11010 10101 10110 10000 10011
01101 01110 01000 01011 00100 00111 00001 00010

we see that the P matrix for these pairs is given by

$$P = \begin{bmatrix} 0&0&0&0&0&0&0&0&0&0&0&0&0&1&0&0 \\ 0&0&0&0&0&0&0&0&0&0&0&0&1&0&0&0 \\ 0&0&0&0&0&0&0&0&0&0&0&0&0&0&0&1 \\ 0&0&0&0&0&0&0&0&0&0&0&0&0&0&1&0 \\ 0&0&0&0&0&0&0&0&0&1&0&0&0&0&0&0 \\ 0&0&0&0&0&0&0&0&1&0&0&0&0&0&0&0 \\ 0&0&0&0&0&0&0&0&0&0&0&1&0&0&0&0 \\ 0&0&0&0&0&0&0&0&0&0&1&0&0&0&0&0 \\ 0&0&0&0&0&1&0&0&0&0&0&0&0&0&0&0 \\ 0&0&0&0&1&0&0&0&0&0&0&0&0&0&0&0 \\ 0&0&0&0&0&0&0&1&0&0&0&0&0&0&0&0 \\ 0&0&0&0&0&0&1&0&0&0&0&0&0&0&0&0 \\ 0&1&0&0&0&0&0&0&0&0&0&0&0&0&0&0 \\ 1&0&0&0&0&0&0&0&0&0&0&0&0&0&0&0 \\ 0&0&0&1&0&0&0&0&0&0&0&0&0&0&0&0 \\ 0&0&1&0&0&0&0&0&0&0&0&0&0&0&0&0 \end{bmatrix}.$$

□

■ LEMMA 4.2.3.

Consider the pairs generated by $\mathbf{e}$ *and* $\mathbf{g}$, *where* $\mathbf{e}, \mathbf{g} \notin F$ *and where* $\mathbf{e} + \mathbf{g} \notin F$. *The information matrix,* $C_{\mathbf{e},\mathbf{g}}$ *is given by*

$$C_{\mathbf{e},\mathbf{g}} = \frac{1}{8 \times 2^k} \begin{bmatrix} 4I_k - 2D_{M,\mathbf{e}} - 2D_{M,\mathbf{g}} & \mathbf{0} \\ \mathbf{0} & 4I_{k(k-1)/2} - 2D_{T,\mathbf{e}} - 2D_{T,\mathbf{g}} \end{bmatrix}.$$

Proof. We have pairs $(F, F+\mathbf{e})$ and $(F, F+\mathbf{g})$. Thus there are $2N$ pairs; N treatment combinations are in 2 pairs each and $2N$ treatment combinations are in one pair each. From the discussion above, we know that

$$B_M = \begin{bmatrix} B_{M,F} & D_{M,\mathbf{e}}B_{M,F} & D_{M,\mathbf{g}}B_{M,F} & B_{M,\bar{A}} \\ B_{T,F} & D_{T,\mathbf{e}}B_{T,F} & D_{T,\mathbf{g}}B_{T,F} & B_{T,\bar{A}} \end{bmatrix},$$

where $A = F \cup (F+\mathbf{e}) \cup (F+\mathbf{g})$, and

$$\Lambda = \frac{1}{8N} \begin{bmatrix} 2I & -I & -I & \mathbf{0} \\ -I & I & \mathbf{0} & \mathbf{0} \\ -I & \mathbf{0} & I & \mathbf{0} \\ \mathbf{0} & \mathbf{0} & \mathbf{0} & \mathbf{0} \end{bmatrix}.$$

Evaluating the information matrix, we get

$$C_{\mathbf{e},\mathbf{g}} = \frac{1}{2^k} \begin{bmatrix} 4I_k - 2D_{M,\mathbf{e}} - 2D_{M,\mathbf{g}} & \mathbf{0} \\ \mathbf{0} & 4I_{k(k-1)/2} - 2D_{T,\mathbf{e}} - 2D_{T,\mathbf{g}} \end{bmatrix}.$$

□

Finally, we need to consider generators that come from F. In this case, we can use such a generator to define a fraction of F and use one of the results that we have given above on that smaller fraction. The next examples show that the smaller fraction need not be uniquely determined and that it does not need to be of resolution 5 provided that all the treatment combinations that eventually appear correspond to a design of resolution 5.

■ EXAMPLE 4.2.6.
Let $k = 3$. Then F must be the complete factorial since anything smaller will not be of resolution 5. Let $\mathbf{e} = (0, 1, 1)$. Then we get the 4 distinct pairs given in Table 4.9. If we use the entries in the first column to define the fraction F then we can use the results above to determine the $C_{\mathbf{e}}$ matrix. □

Table 4.9 The Pairs from the Complete Factorial when $k = 3$ and $\mathbf{e} = (0, 1, 1)$

Option A	Option B
000	011
001	010
100	111
101	110

■ EXAMPLE 4.2.7.
Suppose that we use the F from Example 4.2.4. Let $\mathbf{e} = (01111)$; then we can see that we get the 8 pairs given in Table 4.10. The problem is how to decide which treatment combinations should be in the new fraction. Since the new fraction need not be of resolution 5, we use $x_1 + x_2 + x_3 + x_4 + x_5 = 0$ and $x_3 + x_4 + x_5 = 0$ to define the fraction. These equations give rise to the treatment combinations 00000, 00011, 00101, 00110, 11110, 11101, 10100, 10111, and have resolution 2. The same pairs would arise from using the equations $x_1 + x_2 + x_3 + x_4 + x_5 = 0$ and $x_1 = 0$ to define the fraction. Again, we can use the results above to derive the $C_{\mathbf{e}}$ matrix. □

Table 4.10 The Pairs from the Fraction in Example 4.2.4 when $\mathbf{e} = (0, 1, 1)$

Option A	Option B
00000	01111
00011	01100
00101	01010
00110	01001
10001	11110
10010	11101
10100	11011
10111	11000

Thus whether or not the generators are in F, or whether their sum is in F, does not have any bearing on the estimability properties of main effects and two-factor interactions. We can generalize the results in Lemmas 4.2.1, 4.2.2, and 4.2.3 to get the following result about the properties of a set of generators required to be able to estimate all main effects and two-factor interactions.

■ **THEOREM 4.2.1.**
Consider a set of generators such that

- *for each attribute there is at least one generator with a 1 in the corresponding position, and*
- *for any two attributes there is at least one generator in which the corresponding positions have a 0 and a 1.*

Then all main effects and two-factor interactions will be estimable from the pairs generated by this set of generators.

Proof. From Lemmas 4.2.2 and 4.2.3, we see that diagonal entries in C will be 0 only if all the D matrices have entries of 1 in some position. For main effects, this happens only if none of the generators has a 1 for that attribute. For two-factor interactions, this happens only if all of the generators have the same entry for two particular attributes. But the properties of the generators given in the statement of the theorem preclude these situations from arising. □

We will define an *estimable set of generators* to be a set of generators that satisfies the conditions of Theorem 4.2.1.

In Construction 4.2.1, we have shown that using a regular OMEP and the single generator $(1, 1, \ldots, 1)$ gives designs that are 100% efficient for estimating main effects. Thus we only consider sets of pairs for estimating main effects and two-factor interactions in the remainder of this section.

We would like to find a minimum set of generators from which all main effects and two-factor interactions can be estimated. For the estimation of main effects and two-factor interactions in the complete factorial, generators of weight $(k+1)/2$ have been shown to be optimal for odd k. For even k, generators of weights $k/2$ and $k/2+1$ have been shown to be optimal. We choose generators with these weights below, although we stress that we do not know that these weights are optimal in this setting.

In the next result, we give a recursive construction for sets of generators with $(k+1)/2$ non-zero entries if k is odd and $k/2$ non-zero entries if k is even. A similar result appears in Roberts (2000).

■ **LEMMA 4.2.4.**
If $2^m \leq k < 2^{m+1}$, *then there is an estimable set with* $m+1$ *generators.*

Proof. The proof proceeds recursively once we have the first two cases. When $k = 2$, use the generators (1,0) and (0,1). When $k = 3$, use the generators (1, 1, 0) and (0, 1, 1).

For the recursive construction, it is advantageous to write each generator as two sets, those positions in which the generator contains a 1 and those positions in which the generator contains a 0. Hence we get the partitions $[\{(1),(2)\}, \{(2),(1)\}]$ for $k = 2$ and $[\{(1,2), (3)\}, \{(2,3), (1)\}]$ for $k = 3$.

Consider $k = 2k_1 + 1$. We write the first partition as

$$\{(1, 2, \ldots, k_1 + 1), (k_1 + 2, k_1 + 3, \ldots, 2k_1 + 1)\}.$$

We then partition the sets of size k_1 and $k_1 + 1$ and take the union of the first set in the first partition of each and the union of the second set in the first partition of each to get the second partition for $k = 2k_1 + 1$. We continue in this way to get all the partitions for $k = 2k_1 + 1$. The only time that this might not work is when $k_1 + 1$ is a power of 2 since in that case $k_1 + 1$ has one more partition than k_1. However, using the initial generators given above, we see that the final generator for a power of 2 is just the foldover of the second last generator. This generator is required only so that main effects for the first $t + 1$ attributes can be estimated, not for the estimation of two-factor interactions, and all the main effects for the first $t + 1$ attributes can be estimated from the first generator. Hence we can ignore this generator when doing the recursive construction. This completes the construction for odd k.

Consider $k = 2k_1$. Do exactly the same construction as for odd k, using an initial partition of

$$\{(1, 2, \ldots, k_1), (k_1 + 1, k_1 + 2, \ldots, 2k_1)\}.$$

This proof is completed by noting that the set of generators satisfies the conditions of Theorem 4.2.1. □

■ EXAMPLE 4.2.8.

Let $k = 8$. Then the first partition is {(1,2,3,4), (5,6,7,8)}. For the first set, {1,2,3,4}, the partitions are {(1,2),(3,4)}, {(1,3),(2,4)}, and {(2,4),(1,3)}. For the second set, the partitions are {(5,6),(7,8)}, {(5,7),(6,8)} and {(6,8),(5,7)}. We combine these to get

$$[\{(1,2,5,6),(3,4,7,8)\},\{(1,3,5,7),(2,4,6,8)\},\{(2,4,6,8),(1,3,5,7)\}].$$

The final partition is only required to ensure that the main effect of attribute 8 can be estimated. Now construct the partitions for $k = 15$. We get

[{(1,2,3,4,5,6,7,8), (9,A,B,C,D,E,F)}, {(1,2,3,4,9,A,B,C), (5,6,7,8,D,E,F)},
{(1,2,5,6,9,A,D,E), (3,4,7,8,B,C,F)}, {1,3,5,7,9,B,E,F), (2,4,6,8,A,C,D)}],

and the final partition from $k = 8$ is not required since the main effect of attribute 8 can be estimated because there is a 1 in position 8 in the first generator. □

The efficiency of the designs that result from Lemma 4.2.4 depends on the particular resolution 5 design that is used as the starting design. For example when $k = 8$ using the fraction given by $I \equiv ABCDE \equiv DEFGH$, where we use this notation to represent the solutions to both the equations $x_1 + x_2 + x_3 + x_4 + x_5 = 0$ and $x_4 + x_5 + x_6 + x_7 + x_8 = 0$, and the generators from Lemma 4.2.4, gives 256 pairs that are 92.96% efficient. If we use the fraction given by $I \equiv ABCDEF \equiv DEFGH$ we get 224 pairs that are 86.51% efficient; see Table 4.12.

While Lemma 4.2.4 gives one set of generators for each value of k, it is often possible to get sets of generators that are better than these. For odd k, for example, it is possible to use a balanced incomplete block design (BIBD) (defined in Section 2.4) to give a set of generators in which each main effect and each two-factor interaction effect is estimated using the same number of generators. This idea is illustrated in the following example.

■ EXAMPLE 4.2.9.

The blocks in Table 4.11 form a BIBD on 7 items with 7 blocks each of size 3 and with each item appearing in 3 blocks. There is a unique block which contains every pair of items hence $\lambda = 1$.

Now suppose that we use the attributes as the items and let each block define a generator by placing a 1 for those attributes that appear in the block and a 0 for those attributes that do not appear in the block. Thus we would get the generator 1110000 from the first block, 1001100 from the second block and so on. The properties of the BIBD guarantee that the construction gives an estimable set of generators. □

Table 4.11 The Blocks a (7,3,1) BIBD

1	2	3
1	4	5
1	6	7
2	4	6
2	5	7
3	4	7
3	5	6

We can now use these BIBDs to give sets of generators. The pairs that result from Lemma 4.2.5 are often very efficient and relatively small. However there does not appear to be a general expression for the efficiency of these designs.

■ **LEMMA 4.2.5.**

1. *The blocks of a* $(4t+3, 2t+2, t+1)$ *SBIBD can be used to give* $4t+3$ *generators, each with weight* $2t+2$*, such that each main effect and each two-factor interaction can be estimated from* $2t+2$ *of the generators.*
2. *The blocks of a* $(4t+1, 2(4t+1), 2(2t+1), 2t+1, t+1)$ *BIBD can be used to give* $2(4t+1)$ *generators, each with weight* $2t+1$*, such that each main effect can be estimated from* $4t+2$ *of the generators and each two-factor interaction can be estimated from* $6t+2$ *of the generators.*
3. *For even* k*, estimable sets with high efficiency can be found by deleting one treatment from the designs above.*

Proof. The results follow by counting the number of blocks with one specific treatment, for main effects, and the number of blocks with only one of two specific treatment combinations, for interaction effects. □

In the following table, we give some fractions and generators, the number of pairs in the choice set and the D-efficiency of the set of pairs. For each value of k, the generators that come from Lemma 4.2.4 are indicated by an * and those that come from Lemma 4.2.5 are indicated by †. In Table 4.12, *MR* refers to Mathon and Rosa (2006), the largest published table of BIBDs.

Table 4.12: D-Efficiency and Number of Pairs for Some Constant Difference Choice Pairs

k	F	Generators	Number of pairs	Efficiency (%)
3†	Complete	011, 101, 110	12	100
3*	Complete	011, 101	8	94.5
4	Complete	All vectors of weights 2 and 3	80	100
4	Complete	Omit any one weight 2 or weight 3 vector	72	99.6
4	Complete	Omit any two weight 2 vectors or any two weight 3 vectors	64	99.21

4	Complete	Six weight 2 vectors and any weight 3 vector	56	98.95
4†	Complete	1100, 1010, 1001 0110, 0101, 0011	48	99.03
4	Complete	1110, 1101, 1011, 0111	32	98.01
4*	Complete	1100, 1010, 0101	24	93.98
5†	$I \equiv ABCDE$	All weight 3 vectors	160	100
5	$I \equiv ABCDE$	Any nine weight 3 vectors	144	99.60
5	$I \equiv ABCDE$	Any eight weight 3 vectors	128	99.08
5	$I \equiv ABCDE$	11100, 10011, 10101, 11010, 01110, 00111, 11001	112	98.46
5	$I \equiv ABCDE$	11100, 10110, 10101, 11010, 11001, 10110	96	97.92
5	$I \equiv ABCDE$	11100, 10110, 10101, 11010, 11001	80	96.49
5	$I \equiv ABCDE$	11100, 10110, 10101, 11010	64	95.72
5*	$I \equiv ABCDE$	11100, 11010, 01101	48	91.32
6†	Complete	110100, 111010, 011101, 001110,100111, 010011, 101001	224	100
6	Complete	Any 6 of the generators above	192	98.98
6	Complete	Any 5 of the generators above	160	97.44
6	Complete	The four weight 3 vectors from above	128	96.61
6†	$I \equiv ABCDEF$	110100, 111010, 011101, 001110, 100111, 010011, 101001	176	99.46
6	$I \equiv ABCDEF$	111100, 001111, 100111, 111010, 111001, 010111 110011, 101101, 011110	144	98.06
6	$I \equiv ABCDEF$	111100, 001111, 100111, 111010, 111001, 010111 110011, 101101	128	97.42
6	$I \equiv ABCDEF$	111100, 001111, 100111, 111010, 111001, 010111 110011	112	96.48
6	$I \equiv ABCDEF$	111100, 001111, 100111, 111010, 111001, 010111	96	95.71
6	$I \equiv ABCDEF$	111100, 001111, 100111, 111010, 111001	80	93.75
6	$I \equiv ABCDEF$	111100, 001111, 100111, 111010	64	92.49
6*	$I \equiv ABCDEF$	111000, 001011, 100110	96	91.85

7†	$I \equiv ABCDEFG$	1110100, 0111010, 0011101, 1001110, 0100111, 1010011, 1101001	224	100
7	$I \equiv ABCDEFG$	Any 6 of these 7 vectors	192	98.98
7	$I \equiv ABCDEFG$	Any 5 of these 7 vectors	160	97.44
7*	$I \equiv ABCDEFG$	1111000, 1100110, 1010011	96	91.85
8*	$I \equiv ABCDE$ $\equiv DEFGH$	11110000, 11001100, 10101010, 01010101	256	92.96
8†	$I \equiv ABCDE$ $\equiv DEFGH$	Generators from first (9,18,10,5,5) in *MR*	1120	99.97
8†	$I \equiv ABCDE$ $\equiv DEFGH$	Generators from sixth (9,18,10,5,5) in *MR*	1120	99.88
8†	$I \equiv ABCDE$ $\equiv DEFGH$	Generators from 11th (9,18,10,5,5) in *MR*	1056	99.90
8*	$I \equiv ABCDEF$ $\equiv DEFGH$	11110000, 11001100, 10101010, 01010101	224	86.51
8†	$I \equiv ABCDEF$ $\equiv DEFGH$	Generators from first (9,18,10,5,5) in *MR*	992	99.81
8†	$I \equiv ABCDEF$ $\equiv DEFGH$	Generators from sixth (9,18,10,5,5) in *MR*	960	99.68
8†	$I \equiv ABCDEF$ $\equiv DEFGH$	Generators from 11th (9,18,10,5,5) in *MR*	1056	99.9

4.2.4 Dominating Options

A binary attribute can have ordered levels in the sense that all respondents will prefer one level to the other level. For instance, everyone prefers to have wider rather than narrower seats on a plane, all other things being equal. So if all the attributes in an experiment have ordered levels, then, since foldover pairs are optimal for the estimation of main effects, it is clear that $111\ldots1$ will be preferred to $000\ldots0$ by all respondents. So we say that $111\ldots1$ *dominates* $000\ldots0$. When estimating main effects the only way that we can avoid having a pair with a dominating option is to start with a fractional factorial that does not contain either $111\ldots1$ or $000\ldots0$. We saw how to do this in Section 2.3.9.

If we want to estimate main effects and two-factor interactions optimally then we want to have all pairs in which about half the attributes are different. So again we would need to avoid having either $000\ldots0$ or $111\ldots1$ in any choice set. But there are other choice sets that can have an option which dominates. For example, if $k = 7$, then pairs with $(k+1)/2 = 4$ attributes different form the optimal design. So the pair (1000000, 1111100) would be in the choice experiment and the second option dominates the first. For $k > 5$ we can choose fractions that do not contain either $000\ldots0$ or $111\ldots1$, but it is not usually possible to choose a fraction so that all pairs with about half the attributes different do not contain a dominating option.

4.2.5 Exercises

1. Let $k = 5$.
   (a) Using the results in Table 2.9, or otherwise, construct a regular fractional factorial design of resolution 3.
   (b) Use Construction 4.2.1 to construct a set of pairs.
   (c) Verify that the resulting pairs are optimal for the estimation of main effects.
2. We have said that Construction 4.2.1 can be extended to the union of regular designs.
   (a) Show that the first 8 rows of the design in Table 4.8(b) is a regular 2^{4-1}. Show that the final 8 rows consist of two copies of 4 rows that form a regular 2^{4-2}.
   (b) Apply Construction 4.2.1 to this design and confirm that the resulting pairs are 100% efficient.
   (c) Show that the design in Table 4.8(a) is not regular.
3. Let $k = 5$ and let F be the solutions to $x_1 + x_2 + x_3 + x_4 + x_5 = 1$. Let $\mathbf{e} = (10011)$.
   (a) Find the B matrix for the pairs $(F, F + \mathbf{e})$.
   (b) Find the information matrix C_e for these pairs.
   (c) Comment on which effects can be estimated and which effects can not be estimated.
   (d) Can you find some other generators so that the resulting set of pairs allows all main effects plus two-factor interactions to be estimated?
4. Confirm the results in Table 4.12 when $k = 4$ and there are 48 pairs and 24 pairs.

4.3 REFERENCES AND COMMENTS

Most of the results in this chapter originally appeared in Street et al. (2001) and Street and Burgess (2004a). Readers can find software to construct choice sets from an initial factorial design and sets of generators, as well as calculate the corresponding information matrix and variance-covariance matrix, at http://maths.science.uts.edu.au/maths/wiki/SPExpts.

Comparisons of pairs of items have been used to study choices for nearly 150 years. There is a detailed account of a paired comparison experiment, allowing for order effects, in Fechner (1860), while MacKay (1988) notes that Thorndike (1910) "used paired comparisons to test hypotheses in discrimination". The idea of paired comparisons arose from consideration of problems where there are t items (products, individuals, options) that are to be ranked, but no natural measurement scale, such as height, is available to accomplish this objective. Such situations arise naturally when trying to rate players in tournaments or when trying to rank preferences for products in particular product categories or when trying to rank the quality of objects like wines or beers.

An extensive discussion of the literature on paired comparisons up until 1988 appears in David (1988). He discusses estimation, including Bayesian approaches, dealing with ties, models in which the order of presentation is taken into account and the early work on dealing with a factorial treatment structure.

This was considered in the 2×2 case by Abelson and Bradley (1954) and was extended to items with more than two binary attributes by Bradley and El-Helbawy (1976). These authors also provided a method to test for contrasts of the parameters, thus making it easy to extend the results to finding optimal paired comparison designs for particular effects of interest and for particular classes of competing designs; see El-Helbawy and Bradley (1978), El-Helbawy and Ahmed (1984), van Berkum (1987b), El-Helbawy et al. (1994) and Street et al. (2001).

Designs which use only some of the pairs are clearly essential if the number of items, t, is large and such designs appear to have been investigated first by McCormick and Bachus (1952) and McCormick and Roberts (1952); see David (1988) for a more extensive discussion.

The designs in Table 4.12 for $k = 4$ with 48 pairs and with 32 pairs are equivalent to designs given in Chapter 5 of van Berkum (1987a). The method of construction is quite different, however, as he focuses on finding sets of pairs within one or more fractions and we specify the fraction and one or more generators for the pairs, where the generators need not come from the fraction. For $k = 5$ for example, van Berkum's design with 80 pairs has an efficiency of 84% compared to our design with 80 pairs with an efficiency of 96.5%.

CHAPTER 5

LARGER CHOICE SET SIZES FOR BINARY ATTRIBUTES

So far we have only considered stated preference designs in which all choice sets have two options each. These designs are certainly frequently used in practice but, as we have seen, they can require a very large number of choice sets. In many cases, however, choice sets with three or more options in them would be just as acceptable. If it is possible to use larger choice sets, then often fewer choice sets are needed in total to give the same accuracy as a paired comparison design. The optimal number of options in a choice set, as well as choice experiments in which choice experiments may have choice sets of different sizes, are considered in Section 7.2. In this chapter, we investigate the form of optimal forced choice stated preference designs for k binary attributes when all choice sets have the same number of options, which is at least two (that is, $m \geq 2$), thus generalizing the results of the previous chapter.

We begin by considering an example based on a choice experiment used in Severin (2000).

■ **EXAMPLE 5.0.1.**

Suppose that we are interested in the effects of 16 attributes on the choice of holiday packages. The 16 attributes, together with the corresponding levels, are given in Table 5.1. Using these attributes we can describe $2^{16} = 65,536$ possible holiday packages or treatment combinations. One such holiday package is ($1200, Overseas, Qantas, 12 nights, No meals, No local tours, Peak, 4 star hotel, 3 hours, Museum, 100 yards, Pool, Friendly staff, Individual, No beach, Creative Holidays) which is equivalent to 1101110101100011 using the coded levels. A possible choice set from an experiment with choice sets of size 4 is given in Table 5.2. □

Table 5.1 Attributes and Levels for Holiday Packages

Attributes	Level 0	Level 1
Price	\$999	\$1200
Type of destination	Domestic	Overseas
Airline	Qantas	Ansett
Length of stay	7 nights	12 nights
Meal inclusion	Yes	No
Local tours availability	Yes	No
Season in destination	Peak	Off peak
Type of accommodation	2 star hotel	4 star hotel
Length of trip	3 hours	5 hours
Cultural activities	Historical sites	Museum
Distance from hotel to attractions	3 miles	100 yards
Swimming pool availability	Yes	No
Helpfulness of booking staff	Friendly	Unfriendly
Type of holiday	Individual	Organized tour
Beach availability	Yes	No
Brand	Jetset	Creative Holidays

We have seen that when the choice sets are only of size 2 the number of attributes in which the levels in the two options are different is important in determining the optimal design. Thus we know that the optimal design for estimating main effects is one in which all the levels are different between the two options in each choice set. For optimal designs for estimating main effects plus two-factor interactions, about half the attributes need to have different levels. This leads us to generalize the idea of the number of attributes that differ between two options by defining a difference vector for a choice set of size m.

We then define the class of competing designs, derive a general expression for the information matrix Λ, and use that to determine the optimal designs for estimating main effects and main effects plus two-factor interactions. Initially we work with choice sets determined from the complete factorial, but in the final section of the chapter we give a construction to get smaller optimal and near-optimal choice experiments.

5.1 OPTIMAL DESIGNS FROM THE COMPLETE FACTORIAL

We know from our study of optimal pairs that results from the complete factorial make it possible to decide how good any other proposed design is, and also give us some ideas about how to find constructions based on fractional factorial designs. So the goal of this section is to establish the optimal designs for the estimation of main effects and main effects plus two-factor interactions from the complete factorial.

5.1.1 Difference Vectors

In a choice set of size m, there are $\binom{m}{2}$ pairs of options in the choice set. We record the number of attributes different for each pair in the choice set in a *difference vector* $\mathbf{v} = (d_1, d_2, \ldots, d_{m(m-1)/2})$, where $1 \leq d_i \leq k$. Note that $d_i \neq 0$, so no repeated options are allowed in a choice set. We define d_1 to be the number of attributes different (the

Table 5.2 One Choice Set for the Possible Holiday Packages

Attributes	Option A	Option B	Option C	Option D
Price	$999	$999	$1200	$1200
Type of destination	Domestic	Domestic	Overseas	Overseas
Airline	Qantas	Qantas	Ansett	Ansett
Length of stay	7 nights	7 nights	12 nights	12 nights
Meal inclusion	Yes	Yes	No	No
Local tours availability	Yes	Yes	No	No
Season in destination	Peak	Peak	Off peak	Off peak
Type of accommodation	2 star hotel	2 star hotel	4 star hotel	4 star hotel
Length of trip	3 hours	5 hours	3 hours	5 hours
Cultural activities	Historical sites	Museum	Historical sites	Museum
Distance to attractions	3 miles	100 yards	3 miles	100 yards
Swimming pool availability	Yes	No	Yes	No
Helpfulness of booking staff	Friendly	Unfriendly	Friendly	Unfriendly
Type of holiday	Individual	Organized tour	Individual	Organized tour
Beach availability	Yes	No	Yes	No
Brand	Jetset	Creative Holidays	Jetset	Creative Holidays

Suppose that you have already narrowed down your choice of holiday packages to the four alternatives above.
Which of these four would you choose? *(tick one only)*

Option A ○ Option B ○ Option C ○ Option D ○

difference) between the first and second treatment combinations in the choice set; d_2 is the difference between the first and third treatment combinations, and so on.

This order is fixed but arbitrary. Since we can write the treatment combinations in the choice set in any order, the order of the d_i in $\mathbf{v}$ is not important; so we assume that any difference vector has $d_1 \leq d_2 \leq \cdots \leq d_{m(m-1)/2}$. Thus the d_i are written in increasing order.

■ EXAMPLE 5.1.1.

Let $k = 3$ and $m = 3$. There are 8 treatment combinations in a 2^3 factorial. For the choice set (000,001,110), there is one attribute different when comparing 000 and 001 ($d_1 = 1$), there are two attributes different when comparing 000 and 110 ($d_2 = 2$) and there are three attributes different when comparing 001 and 110 ($d_3 = 3$). So the difference vector is $\mathbf{v} = (d_1, d_2, d_3) = (1, 2, 3)$.

$$\underbrace{0\overbrace{00,\ 00}^{d_1=1}\overbrace{1,\ 1}^{d_3=3}10}_{d_2=2}$$

The choice set $(011, 100, 101)$ has difference vector $\mathbf{v} = (d_1, d_2, d_3) = (3, 2, 1)$, but this difference vector is considered to be the same as $\mathbf{v} = (1, 2, 3)$ since we can reorder the options in the choice set. The $\binom{8}{3} = 56$ possible choice sets of size 4 (triples), and their corresponding difference vectors (in lexicographic order) are given in Table 5.3. We see that there are three distinct values of $\mathbf{v}$ that can arise: (1,1,2), (1,2,3) and (2,2,2). □

When we had choice sets of size $m = 2$, the optimal designs for estimating main effects had each attribute level different between the two options. In the next result, we establish how "different" the options in a choice set with m options can be by finding an upper bound for the sum of the entries in the difference vector.

■ LEMMA 5.1.1.

For a particular difference vector $\mathbf{v}$*, for a given* m *and* $k \geq a$*, where* $2^{a-1} < m \leq 2^a$*, the least upper bound for the sum of the differences is*

$$S = \begin{cases} (m^2 - 1)k/4 & \text{for } m \text{ odd,} \\ m^2k/4 & \text{for } m \text{ even.} \end{cases}$$

Proof. Write the treatment combinations in the choice set as the rows of an $m \times k$ array. Then, for each column of length m, the maximum contribution to $\sum_{i=1}^{m(m-1)/2} d_i$ comes by having half the entries 1 and half 0 if m is even, or $(m-1)/2$ entries 1 and $(m+1)/2$ entries 0 (or the other way round) if m is odd. To get m distinct rows, we must have at least a columns where $2^{a-1} < m \leq 2^a$. Thus we get m distinct rows of a columns with the maximum difference by writing down the rows in foldover pairs. So we write

$$\begin{array}{ccccc} 0 & 0 & \ldots & 0 & 0 \\ 1 & 1 & \ldots & 1 & 1 \\ 0 & 0 & \ldots & 0 & 1 \\ 1 & 1 & \ldots & 1 & 0 \end{array}$$

and so on. If m is odd, we have $(m-1)/2$ foldover pairs and one extra row, which can be any treatment not already used. It does not matter which particular treatment combinations are used to construct the rows. Because the rows appear in foldover pairs, half the entries

Table 5.3 All Possible Triples when $k = 3$

Triple	v	Triple	v
(000,001,010)	(1,1,2)	(001,011,110)	(1,2,3)
(000,001,011)	(1,1,2)	(001,011,111)	(1,1,2)
(000,001,100)	(1,1,2)	(001,100,101)	(1,1,2)
(000,001,101)	(1,1,2)	(001,100,110)	(1,2,3)
(000,001,110)	(1,2,3)	(001,100,111)	(2,2,2)
(000,001,111)	(1,2,3)	(001,101,110)	(1,2,3)
(000,010,011)	(1,1,2)	(001,101,111)	(1,1,2)
(000,010,100)	(1,1,2)	(001,110,111)	(1,2,3)
(000,010,101)	(1,2,3)	(010,011,100)	(1,2,3)
(000,010,110)	(1,1,2)	(010,011,101)	(1,2,3)
(000,010,111)	(1,2,3)	(010,011,110)	(1,1,2)
(000,011,100)	(1,2,3)	(010,011,111)	(1,1,2)
(000,011,101)	(2,2,2)	(010,100,101)	(1,2,3)
(000,011,110)	(2,2,2)	(010,100,110)	(1,1,2)
(000,011,111)	(1,2,3)	(010,100,111)	(2,2,2)
(000,100,101)	(1,1,2)	(010,101,110)	(1,2,3)
(000,100,110)	(1,1,2)	(010,101,111)	(1,2,3)
(000,100,111)	(1,2,3)	(010,110,111)	(1,1,2)
(000,101,110)	(2,2,2)	(011,100,101)	(1,2,3)
(000,101,111)	(1,2,3)	(011,100,110)	(1,2,3)
(000,110,111)	(1,2,3)	(011,100,111)	(1,2,3)
(001,010,011)	(1,1,2)	(011,101,110)	(2,2,2)
(001,010,100)	(2,2,2)	(011,101,111)	(1,1,2)
(001,010,101)	(1,2,3)	(011,110,111)	(1,1,2)
(001,010,110)	(1,2,3)	(100,101,110)	(1,1,2)
(001,010,111)	(2,2,2)	(100,101,111)	(1,1,2)
(001,011,100)	(1,2,3)	(100,110,111)	(1,1,2)
(001,011,101)	(1,1,2)	(101,110,111)	(1,1,2)

are 1 and half are 0 (m even) and so $\sum_{i=1}^{m(m-1)/2} d_i$ is a maximum. Since these a columns guarantee that the rows are distinct, larger values of k can be obtained by writing down any columns of maximum difference for the remaining $k - a$ columns. The result follows. □

The treatment combinations given in Lemma 5.1.1 are not the only treatment combinations which achieve the bound; it is sufficient to have any foldover pairs in any order.

■ EXAMPLE 5.1.2.
Let $k = m = 3$. Then we have seen that there are three different difference vectors: (1,1,2) with sum 4, (1,2,3) with sum 6, and (2,2,2) with sum 6. The upper bound is $(3^2-1)3/4 = 6$. The value of a is 2 (since $2^{2-1} < 3 \leq 2^2$), and a set of rows constructed as in the lemma is

$$\begin{array}{cc} 0 & 0 \\ 1 & 1 \\ 0 & 1. \end{array}$$

These rows give the levels for the first two attributes for the three options in the choice set. To get a choice set which meets the bound, we can now adjoin any column with a 1 and two 0s, or a 0 and two 1s, for the levels of the third attribute. So we might adjoin 0, 0 and 1 and get $(000, 110, 011)$ which has difference vector $(2, 2, 2)$. Or we could adjoin 0, 1, and 0 and get $(000, 111, 010)$, which has difference vector $(1, 2, 3)$. □

For particular values of m and k, there can be several distinct difference vectors; these are denoted by $\mathbf{v}_j$. In Example 5.1.1, we saw that there were three distinct difference vectors when $m = 3$ and $k = 3$. We now define four scalars which are needed subsequently.

1. We define $c_{\mathbf{v}_j}$ to be the number of choice sets containing the treatment $00\ldots0$ with the difference vector $\mathbf{v}_j$.

2. We define $x_{\mathbf{v}_j;i}$ to be the number of times the difference i appears in the difference vector $\mathbf{v}_j$. Then $\sum_i x_{\mathbf{v}_j;i} = \binom{m}{2}$, since this is the total number of entries in $\mathbf{v}_j$.

3. We define $i_{\mathbf{v}_j}$ to be an indicator variable, where

$$i_{\mathbf{v}_j} = \begin{cases} 0 & \text{if no choice sets have the difference vector } \mathbf{v}_j, \\ 1 & \text{if all the choice sets with the difference vector } \mathbf{v}_j \text{ appear} \\ & \text{in the choice experiment.} \end{cases}$$

   At least one of the $i_{\mathbf{v}_j}$ values must be non-zero; otherwise the experiment contains no choice sets.

4. We define $a_{\mathbf{v}_j} = i_{\mathbf{v}_j}/N$. These are similar to the a_v defined in Section 4.1.1.

Using these definitions, we see that the total number of choice sets, N, is given by

$$N = \frac{2^k}{m} \sum_j c_{\mathbf{v}_j} i_{\mathbf{v}_j}.$$

■ EXAMPLE 5.1.3.
Let $m = 3$, $k = 3$. We have seen that there are three different difference vectors and we let $\mathbf{v}_1 = (1, 1, 2)$, $\mathbf{v}_2 = (1, 2, 3)$ and $\mathbf{v}_3 = (2, 2, 2)$. Table 5.4 is Table 5.3 reordered to show all the triples associated with each of the difference vectors. Consider just the

triples containing the treatment 000. Then we see that $c_{\mathbf{v}_1} = c_{\mathbf{v}_2} = 9$ and $c_{\mathbf{v}_3} = 3$, since $(000, 011, 101)$, $(000, 011, 110)$, and $(000, 101, 110)$ are the three choice sets with $m = 3$ which contain 000 and which have difference vector $(2, 2, 2)$. We also see that

$$\begin{array}{lll} x_{v_1;1} = 2 & x_{v_2;1} = 1 & x_{v_3;1} = 0 \\ x_{v_1;2} = 1 & x_{v_2;2} = 1 & x_{v_3;2} = 3 \\ x_{v_1;3} = 0 & x_{v_2;3} = 1 & x_{v_3;3} = 0 \\ \sum_i x_{\mathbf{v}_1;i} = 3 & \sum_i x_{\mathbf{v}_2;i} = 3 & \sum_i x_{\mathbf{v}_3;i} = 3. \end{array}$$

□

Table 5.4 All Possible Triples when $k = 3$ Sorted by Difference Vector

$\mathbf{v}_1 = (1, 1, 2)$	$\mathbf{v}_2 = (1, 2, 3)$	$\mathbf{v}_3 = (2, 2, 2)$
(000,001,010)	(000,001,110)	(000,011,101)
(000,001,011)	(000,001,111)	(000,011,110)
(000,001,100)	(000,010,101)	(000,101,110)
(000,001,101)	(000,010,111)	(001,010,100)
(000,010,011)	(000,011,100)	(001,010,111)
(000,010,100)	(000,011,111)	(001,100,111)
(000,010,110)	(000,100,111)	(010,100,111)
(000,100,101)	(000,101,111)	(011,101,110)
(000,100,110)	(000,110,111)	
(001,010,011)	(001,010,101)	
(001,011,101)	(001,010,110)	
(001,011,111)	(001,011,100)	
(001,100,101)	(001,011,110)	
(001,101,111)	(001,100,110)	
(010,011,110)	(001,101,110)	
(010,011,111)	(001,110,111)	
(010,100,110)	(010,011,100)	
(010,110,111)	(010,011,101)	
(011,101,111)	(010,100,101)	
(011,110,111)	(011,100,101)	
(100,101,110)	(010,101,110)	
(100,101,111)	(010,101,111)	
(100,110,111)	(011,100,110)	
(101,110,111)	(011,100,111)	

5.1.2 The Derivation of the Λ Matrix

In this section, we derive the Λ matrix when we assume that the class of competing designs consists of all designs in which all choice sets with a given difference vector are either all included in the choice experiment or none of them are. So, in Example 5.1.3, the competing designs are the choice sets from one or more of the columns of Table 5.4. Thus there are 7 competing designs to be considered in that case.

The matrix of contrasts is unaffected by the choice set size. However, we do need to derive the appropriate Λ matrix.

From the results in Section 3.3.1, we know that, if we assume that

$$\pi_1 = \pi_2 = \cdots = \pi_{2^k} = 1$$

(that is, all treatment combinations are equally attractive, the usual null hypothesis), then

$$\Lambda_{i_1,i_1} = \frac{m-1}{m^2} \sum_{i_2<i_3<\cdots<i_m} \lambda_{i_1,i_2,\ldots,i_m} \tag{5.1}$$

$$\Lambda_{i_1,i_2} = -\frac{1}{m^2} \sum_{i_3<i_4<\cdots<i_m} \lambda_{i_1,i_2,\ldots,i_m} \tag{5.2}$$

where

$$\lambda_{i_1,i_2,\ldots,i_m} = \frac{n_{i_1,i_2,\ldots,i_m}}{N}$$

and $n_{i_1,i_2,\ldots,i_m}$ indicates whether or not the choice experiment contains $(T_{i_1}, T_{i_2}, \ldots, T_{i_m})$ as a choice set. Thus the diagonal entries of Λ are $\frac{m-1}{m^2}$ times the proportion of choice sets containing the treatment combination T_{i_1}, and the off-diagonal entries are $\frac{1}{m^2}$ times the negative of the proportion of choice sets containing both T_{i_1} and T_{i_2}.

In Section 4.1.1, the general form of the Λ matrix for $m = 2$ was given as a linear combination of the identity matrix of order 2^k and the $D_{k,i}$ matrices. We now derive a similar result for any value of m.

■ LEMMA 5.1.2.
Under the usual null hypothesis,

$$\Lambda = \frac{m-1}{m^2} z I_{2^k} - \frac{1}{m^2} \sum_{i=1}^{k} y_i D_{k,i},$$

where

$$z = \sum_j c_{\mathbf{v}_j} a_{\mathbf{v}_j} = \frac{1}{m-1} \sum_{i=1}^{k} \binom{k}{i} y_i$$

and

$$y_i = \frac{2}{m} \binom{k}{i}^{-1} \sum_j c_{\mathbf{v}_j} a_{\mathbf{v}_j} x_{\mathbf{v}_j;i}.$$

The row and column sums of Λ are equal to 0.

Proof. We begin by counting the number of times each treatment combination appears in the design.

The number of times that the treatment combination $00\ldots0$ appears in the choice experiment is $\sum_j c_{\mathbf{v}_j} i_{\mathbf{v}_j}$, and this is the number of times that any treatment combination appears in the choice experiment because of the assumption that all choice sets with a given difference vector appear (or do not appear) in the choice experiment. (Throughout this section the summations over j are over all possible difference vectors $\mathbf{v}_j$ for the appropriate values of k and m.) So the proportion of choice sets in which any treatment combination appears is

$$z = \frac{\sum_j c_{\mathbf{v}_j} i_{\mathbf{v}_j}}{N} = \sum_j c_{\mathbf{v}_j} a_{\mathbf{v}_j}$$

as required.

Now we count the number of pairs in the choice experiment that have i attributes different. If choice sets with difference vector $\mathbf{v}_j$ are in the design, then each treatment combination appears in $c_{\mathbf{v}_j}$ choice sets with difference vector $\mathbf{v}_j$, and $x_{\mathbf{v}_j;i}$ of the differences are equal to i. Altogether, there are 2^k treatment combinations and each treatment combination can appear in any of the m positions in the choice set. Thus the total number of pairs in the choice experiment with i attributes different is

$$\frac{2^k}{m}\sum_j c_{\mathbf{v}_j} i_{\mathbf{v}_j} x_{\mathbf{v}_j;i}. \tag{5.3}$$

Considering the 2^k treatment combinations, the number of pairs with i attributes different is

$$\binom{k}{i}2^k\frac{1}{2}.$$

Each of these pairs appears in the choice experiment the same number of times, say r_i. Then the total number of pairs in the choice experiment with i attributes different can also be expressed as

$$\binom{k}{i}2^k\frac{1}{2}r_i. \tag{5.4}$$

Equating (5.3) and (5.4), we get

$$\binom{k}{i}2^k\frac{1}{2}r_i = \frac{2^k}{m}\sum_j c_{\mathbf{v}_j} i_{\mathbf{v}_j} x_{\mathbf{v}_j;i}.$$

So, if the proportion of choice sets in which each pair with i attributes different appears is $y_i = r_i/N$, then

$$\binom{k}{i}2^k\frac{1}{2}y_i = \frac{2^k}{mN}\sum_j c_{\mathbf{v}_j} i_{\mathbf{v}_j} x_{\mathbf{v}_j;i}.$$

Thus

$$\binom{k}{i}y_i = \frac{2}{m}\sum_j c_{\mathbf{v}_j} a_{\mathbf{v}_j} x_{\mathbf{v}_j;i},$$

from which we see that

$$y_i = \frac{2}{m}\binom{k}{i}^{-1}\sum_j c_{\mathbf{v}_j} a_{\mathbf{v}_j} x_{\mathbf{v}_j;i}.$$

Hence we have established that

$$\Lambda = \frac{m-1}{m^2}zI_{2^k} - \frac{1}{m^2}\sum_{i=1}^{k} y_i D_{k,i}.$$

Finally, we establish that the row and column sums of Λ are equal to 0. Summing over i, we get

$$\begin{aligned}
\frac{1}{m^2}\sum_{i=1}^{k}\binom{k}{i}y_i &= \frac{1}{m^2}\sum_{i=1}^{k}\frac{2}{m}\sum_j c_{\mathbf{v}_j}a_{\mathbf{v}_j}x_{\mathbf{v}_j;i} \\
&= \frac{1}{m^2}\frac{2}{m}\sum_j c_{\mathbf{v}_j}a_{\mathbf{v}_j}\sum_{i=1}^{k}x_{\mathbf{v}_j;i} \\
&= \frac{1}{m^2}\frac{2}{m}\sum_j c_{\mathbf{v}_j}a_{\mathbf{v}_j}\binom{m}{2} \\
&= \frac{m-1}{m^2}\sum_j c_{\mathbf{v}_j}a_{\mathbf{v}_j} \\
&= \frac{m-1}{m^2}z,
\end{aligned}$$

as required. □

■ EXAMPLE 5.1.4.

Let $m = 3$, $k = 3$, with $\mathbf{v}_j$, $c_{\mathbf{v}_j}$ and $x_{\mathbf{v}_j;i}$ as in Example 5.1.3. Assume that choice sets with difference vectors $\mathbf{v}_2$ and $\mathbf{v}_3$ only are in the choice experiment. Then $i_{\mathbf{v}_1} = 0$, $i_{\mathbf{v}_2} = i_{\mathbf{v}_3} = 1$, $N = 24 + 8 = 32$, $a_{\mathbf{v}_1} = 0$, and $a_{\mathbf{v}_2} = a_{\mathbf{v}_3} = 1/32$. Thus we have

$$\begin{aligned}
z &= \sum_j c_{\mathbf{v}_j}a_{\mathbf{v}_j} \\
&= (9 \times 0) + (9 \times \tfrac{1}{32}) + (3 \times \tfrac{1}{32}) \\
&= \frac{12}{32}
\end{aligned}$$

and

$$\begin{aligned}
y_1 &= \frac{2}{3}\binom{3}{1}^{-1}(c_{\mathbf{v}_1}a_{\mathbf{v}_1}x_{\mathbf{v}_1;1} + c_{\mathbf{v}_2}a_{\mathbf{v}_2}x_{\mathbf{v}_2;1} + c_{\mathbf{v}_3}a_{\mathbf{v}_3}x_{\mathbf{v}_3;1}) \\
&= \frac{2}{3} \times \frac{1}{3}(0 + 9 \times \frac{1}{32} \times 1 + 3 \times \frac{1}{32} \times 0) \\
&= \frac{2}{32},
\end{aligned}$$

$$\begin{aligned}
y_2 &= \frac{2}{3}\binom{3}{2}^{-1}(c_{\mathbf{v}_1}a_{\mathbf{v}_1}x_{\mathbf{v}_1;2} + c_{\mathbf{v}_2}a_{\mathbf{v}_2}x_{\mathbf{v}_2;2} + c_{\mathbf{v}_3}a_{\mathbf{v}_3}x_{\mathbf{v}_3;2}) \\
&= \frac{2}{3} \times \frac{1}{3}(0 + 9 \times \frac{1}{32} \times 1 + 3 \times \frac{1}{32} \times 3) \\
&= \frac{4}{32},
\end{aligned}$$

$$\begin{aligned} y_3 &= \frac{2}{3}\binom{3}{3}^{-1}(c_{\mathbf{v}_1}a_{\mathbf{v}_1}x_{\mathbf{v}_1;3} + c_{\mathbf{v}_2}a_{\mathbf{v}_2}x_{\mathbf{v}_2;3} + c_{\mathbf{v}_3}a_{\mathbf{v}_3}x_{\mathbf{v}_3;3}) \\ &= \frac{2}{3}(0 + 9 \times \frac{1}{32} \times 1 + 3 \times \frac{1}{32} \times 0) \\ &= \frac{6}{32}. \end{aligned}$$

Note that y_i is the proportion of choice sets in which each pair with i attributes different appears in the choice experiment. For example, each pair with one attribute different appears in 2 of the 32 choice sets in the choice experiment: the pair 000 and 001 has one attribute different and appears in two choice sets, (000,001,110) and (000,001,111), out of the 32 in the experiment.

Now that we have calculated the z and y_i values, we can calculate the Λ matrix. For the Λ matrix we need the $D_{k,i}$ matrices. $D_{3,1}$ and $D_{3,2}$ are given in Example 4.1.2 and the $D_{3,3}$ matrix consists of 1s down the back diagonal and 0s in all the other positions. We get

$$\begin{aligned} \Lambda &= \tfrac{2}{9}zI_8 - \tfrac{1}{9}(y_1 D_{3,1} + y_2 D_{3,2} + y_3 D_{3,3}) \\ &= \tfrac{1}{9\times 32}\left[24I_8 - 2D_{3,1} - 4D_{3,2} - 6D_{3,3}\right] \\ &= \frac{1}{288}\begin{bmatrix} 24 & -2 & -2 & -4 & -2 & -4 & -4 & -6 \\ -2 & 24 & -4 & -2 & -4 & -2 & -6 & -4 \\ -2 & -4 & 24 & -2 & -4 & -6 & -2 & -4 \\ -4 & -2 & -2 & 24 & -6 & -4 & -4 & -2 \\ -2 & -4 & -4 & -6 & 24 & -2 & -2 & -4 \\ -4 & -2 & -6 & -4 & -2 & 24 & -4 & -2 \\ -4 & -6 & -2 & -4 & -2 & -4 & 24 & -2 \\ -6 & -4 & -4 & -2 & -4 & -2 & -2 & 24 \end{bmatrix}. \end{aligned}$$

□

5.1.3 The Model for Main Effects Only

In this section, we evaluate the information matrix for estimating main effects only and use this to determine the optimal choice experiment for k binary attributes using choice sets of size m.

As in Section 4.1.3 we let $B_{2^k,M}$ be the contrast matrix for main effects and we let C_M be the $k \times k$ principal minor of $C = B_{2^k}\Lambda B'_{2^k}$ associated with the main effects. Thus we are evaluating the information matrix when $B_h = B_{2^k,M}$.

In Lemma 4.1.3 we have shown that

$$B_{2^k,M}D_{k,i} = \left[\binom{k-1}{i} - \binom{k-1}{i-1}\right] B_{2^k,M}$$

for all allowable i.

We now show that at this stage we do not need to make any assumptions about the contrasts in B_r since the class of competing designs that we have chosen ensures that $C_{hr} = \mathbf{0}$ for any choice of B_r. By definition,

$$\begin{aligned} C_{hr} &= \begin{bmatrix} B_h \\ B_r \end{bmatrix} \Lambda\, [B'_h\ B'_r] \\ &= \begin{bmatrix} B_h\Lambda B'_h & B_h\Lambda B'_r \\ B_r\Lambda B'_h & B_r\Lambda B'_r \end{bmatrix} \\ &= \begin{bmatrix} C_{hh} & C_{hr} \\ C_{rh} & C_{rr} \end{bmatrix}. \end{aligned}$$

Since $B_h = B_{2^k,M}$ we can see that

$$\begin{aligned} C_{hr} &= \left[\frac{1}{m^2}\sum_{i=1}^{k}\binom{k}{i}y_i\right] B_h I_{2^k} B_r' - \left[\frac{1}{m^2}\sum_{i=1}^{k} y_i\right] B_h D_{k,i} B_r' \\ &= \frac{1}{m^2}\left[\sum_{i=1}^{k}\binom{k}{i}y_i\right] B_{2^k,M}B_r' - \frac{1}{m^2}\sum_{i=1}^{k} y_i \left[\binom{k-1}{i} - \binom{k-1}{i-1}\right] B_{2^k,M}B_r' \\ &= \mathbf{0}_{k,2^k-k-a}. \end{aligned}$$

This is true for any choice of B_r as long as the set of competing designs remains the same so that the form of Λ stays as a linear combination of the $D_{k,i}$.

Using the result in Lemma 4.1.3, we now can get an explicit expression for the information matrix, C_M, for the k main effects.

■ LEMMA 5.1.3.
The information matrix for main effects under the null hypothesis is given by

$$C_M = \frac{2}{m^2}\left[\sum_{i=1}^{k} y_i\binom{k-1}{i-1}\right] I_k.$$

Proof. The information matrix for main effects under the null hypothesis is

$$\begin{aligned} C_M &= B_{2^k,M}\Lambda B_{2^k,M}' \\ &= B_{2^k,M}\left[\frac{m-1}{m^2} z I_{2^k} - \frac{1}{m^2}\sum_{i=1}^{k} y_i D_{k,i}\right] B_{2^k,M}' \\ &= \frac{1}{m^2}\sum_{i=1}^{k}\binom{k}{i} y_i I_k - \frac{1}{m^2}\sum_{i=1}^{k} y_i B_{2^k,M} D_{k,i} B_{2^k,M}' \\ &= \frac{1}{m^2}\left[\sum_{i=1}^{k}\binom{k}{i} y_i I_k - \sum_{i=1}^{k} y_i\left[\binom{k-1}{i} - \binom{k-1}{i-1}\right] I_k\right] \\ &= \frac{1}{m^2}\left[\sum_{i=1}^{k}\left[\binom{k}{i} - \binom{k-1}{i} + \binom{k-1}{i-1}\right] y_i\right] I_k. \end{aligned}$$

Since

$$\binom{k}{i} = \binom{k-1}{i} + \binom{k-1}{i-1},$$

$$C_M = \frac{2}{m^2}\left[\sum_{i=1}^{k} y_i\binom{k-1}{i-1}\right] I_k,$$

as required. □

Thus the determinant of C_M is

$$\det(C_M) = \left[\frac{2}{m^2}\sum_{i=1}^{k} y_i\binom{k-1}{i-1}\right]^k.$$

■ EXAMPLE 5.1.5.

Let $m = 3$ and $k = 3$. There are two ways that we can calculate C_M: by evaluating $C_M = B_{2^3,M} \Lambda B'_{2^3,M}$ or by using the expression for C_M given in Lemma 5.1.3. In this example we will do both.

The matrix $B_{2^3,M}$ is the contrast matrix for main effects and is given by

$$B_{2^3,M} = \frac{1}{2\sqrt{2}} \begin{bmatrix} -1 & -1 & -1 & -1 & 1 & 1 & 1 & 1 \\ -1 & -1 & 1 & 1 & -1 & -1 & 1 & 1 \\ -1 & 1 & -1 & 1 & -1 & 1 & -1 & 1 \end{bmatrix}.$$

In this example, we assume that all choice sets with difference vectors $\mathbf{v}_2$ or $\mathbf{v}_3$ are in the choice experiment. Then we can use the Λ matrix from Example 5.1.4 to calculate

$$C_M = B_{2^3,M} \Lambda B'_{2^3,M} = \frac{1}{9} I_3; \quad \text{thus} \quad \det(C_M) = \left(\frac{1}{9}\right)^3.$$

Alternatively, we can use Lemma 5.1.3 to calculate C_M. Then, using the y_i values given in Example 5.1.4,

$$\begin{aligned} C_M &= \frac{2}{3^2} \left[\sum_{i=1}^{3} y_i \binom{3-1}{i-1} \right] I_3 \\ &= \frac{2}{9} \left[\frac{2}{32} \binom{2}{0} + \frac{4}{32} \binom{2}{1} + \frac{6}{32} \binom{2}{2} \right] I_3 \\ &= \frac{2}{9} \left[\frac{2}{32} + \frac{8}{32} + \frac{6}{32} \right] I_3 \\ &= \frac{1}{9} I_3. \end{aligned}$$

□

To find the D-optimal design, we must maximize $\det(C_M)$ subject to the constraint $2^k z/m = 1$ (since the total number of choice sets must equal N).

Substituting $y_i = \frac{2}{m} \binom{k}{i}^{-1} \sum_j c_{\mathbf{v}_j} a_{\mathbf{v}_j} x_{\mathbf{v}_j;i}$ and using the fact that $\binom{k-1}{i-1}/\binom{k}{i} = i/k$ gives

$$\begin{aligned} \det(C_M) &= \left[\frac{2}{m^2} \sum_{i=1}^{k} \binom{k-1}{i-1} \frac{2}{m} \binom{k}{i}^{-1} \left(\sum_j c_{\mathbf{v}_j} x_{\mathbf{v}_j;i} a_{\mathbf{v}_j} \right) \right]^k \\ &= \left[\frac{4}{m^3 k} \sum_{i=1}^{k} i \left(\sum_j c_{\mathbf{v}_j} x_{\mathbf{v}_j;i} a_{\mathbf{v}_j} \right) \right]^k. \end{aligned}$$

The following theorem establishes that the D-optimal design, for estimating main effects only, is one which consists of choice sets in which the sum of the differences attains the maximum value given in Lemma 5.1.1.

■ THEOREM 5.1.1.

The D-optimal design for testing main effects only, when all other effects are assumed to be zero, is given by choice sets in which, for each $\mathbf{v}_j$ present,

$$\sum_{i=1}^{m(m-1)/2} d_{ij} = \begin{cases} (m^2-1)k/4, & m \text{ odd}, \\ m^2 k/4, & m \text{ even}, \end{cases}$$

and there is at least one $\mathbf{v}_j$ with a non-zero $a_{\mathbf{v}_j}$; that is, the choice experiment is non-empty. For the optimal designs

$$\det(C_{\text{opt},M}) = \begin{cases} \left(\frac{m^2-1}{m^2 2^k}\right)^k, & m \text{ odd}, \\ \left(\frac{1}{2^k}\right)^k, & m \text{ even}. \end{cases}$$

Proof. Recall that

$$a_{\mathbf{v}_j} = \frac{i_{\mathbf{v}_j}}{N} = \frac{i_{\mathbf{v}_j} m}{2^k \sum_h c_{\mathbf{v}_h} i_{\mathbf{v}_h}}$$

and that $i_{\mathbf{v}_j}$ is an indicator variable indicating the difference vectors corresponding to the choice sets that are included in the choice experiment. Thus at least one of these $i_{\mathbf{v}_j}$, and therefore $a_{\mathbf{v}_j}$, values must be non-zero.

Substituting for $a_{\mathbf{v}_j}$ in $\det(C_M)$, we have

$$\begin{aligned} \det(C_M) &= \left[\sum_{i=1}^{k} i\left(\frac{\sum_j c_{\mathbf{v}_j} x_{\mathbf{v}_j;i} i_{\mathbf{v}_j}}{m^2 k 2^{k-2} \sum_h c_{\mathbf{v}_h} i_{\mathbf{v}_h}}\right)\right]^k \\ &= \left[\frac{\sum_j c_{\mathbf{v}_j} i_{\mathbf{v}_j} \left(\sum_{i=1}^{k} i x_{\mathbf{v}_j;i}\right)}{m^2 k 2^{k-2} \sum_h c_{\mathbf{v}_h} i_{\mathbf{v}_h}}\right]^k. \end{aligned}$$

Since $x_{\mathbf{v}_j;i}$ denotes the number of times the difference i appears in the difference vector $\mathbf{v}_j$, multiplying by i and summing these is equivalent to summing the $\binom{m}{2}$ entries in $\mathbf{v}_j$; thus,

$$\sum_{i=1}^{k} i x_{\mathbf{v}_j;i} = \sum_{i=1}^{m(m-1)/2} d_{ij}.$$

Therefore

$$\det(C_M) = \left[\frac{\sum_j c_{\mathbf{v}_j} i_{\mathbf{v}_j} \left(\sum_{i=1}^{m(m-1)/2} d_{ij}\right)}{m^2 k 2^{k-2} \sum_h c_{\mathbf{v}_h} i_{\mathbf{v}_h}}\right]^k.$$

When m is even, Lemma 5.1.1 states that the maximum value that $\sum_{i=1}^{m(m-1)/2} d_{ij}$ can attain is $m^2k/4$. Then it follows that $\sum_{i=1}^{m(m-1)/2} d_{ij} = m^2k/4 - p_j$ for some $p_j \geq 0$. Thus

$$\begin{aligned} \det(C_M) &= \left[\frac{\sum_j c_{\mathbf{v}_j} i_{\mathbf{v}_j} (m^2k/4 - p_j)}{m^2 k 2^{k-2} \sum_h c_{\mathbf{v}_h} i_{\mathbf{v}_h}}\right]^k \\ &= \left[\frac{1}{2^k} - \frac{\sum_j p_j c_{\mathbf{v}_j} i_{\mathbf{v}_j}}{m^2 k 2^{k-2} \sum_h c_{\mathbf{v}_h} i_{\mathbf{v}_h}}\right]^k. \end{aligned}$$

The $c_{\mathbf{v}_j}$ values are all positive and the $i_{\mathbf{v}_j}$ are either 0 or 1; so, for m even, $\det(C_M)$ has a maximum of $(1/2^k)^k$ when $p_j = 0$ for all j. Thus we obtain the maximum $\det(C_M)$ when

$$\sum_{i=1}^{m(m-1)/2} d_{ij} = m^2k/4.$$

Similarly when m is odd, Lemma 5.1.1 states that the maximum value of $\sum_{i=1}^{m(m-1)/2} d_{ij}$ is $(m^2-1)k/4$. Then $\sum_{i=1}^{m(m-1)/2} d_{ij} = (m^2-1)k/4 - p_j$ for some $p_j \geq 0$. Substituting, we have

$$\begin{aligned}\det(C_M) &= \left[\frac{\sum_j c_{\mathbf{v}_j} i_{\mathbf{v}_j} \left((m^2-1)k/4 - p_j\right)}{m^2 k 2^{k-2} \sum_h c_{\mathbf{v}_h} i_{\mathbf{v}_h}}\right]^k \\ &= \left[\frac{m^2-1}{m^2 2^k} - \frac{\sum_j p_j c_{\mathbf{v}_j} i_{\mathbf{v}_j}}{m^2 k 2^{k-2} \sum_h c_{\mathbf{v}_h} i_{\mathbf{v}_h}}\right]^k.\end{aligned}$$

Since the $c_{\mathbf{v}_j}$ values are all positive and the $i_{\mathbf{v}_j}$ are either 0 or 1, $\det(C_M)$ has a maximum of $((m^2-1)/(m^2 2^k))^k$ when $p_j = 0$ for all j. For m odd, we obtain the maximum $\det(C_M)$ when

$$\sum_{i=1}^{m(m-1)/2} d_{ij} = (m^2-1)k/4.$$

Therefore, the maximum value of $\det(C_M)$ is

$$\det(C_{\text{opt},M}) = \begin{cases} \left(\frac{m^2-1}{m^2 2^k}\right)^k, & m \text{ odd}, \\ \left(\frac{1}{2^k}\right)^k, & m \text{ even}. \end{cases}$$

This occurs when

$$\sum_{i=1}^{m(m-1)/2} d_{ij} = \begin{cases} (m^2-1)k/4, & m \text{ odd}, \\ m^2 k/4, & m \text{ even}. \end{cases}$$ □

The D-efficiency relative to the optimal design is calculated using the expression

$$\text{Eff}_D = \left(\frac{\det(C_M)}{\det(C_{\text{opt},M})}\right)^{1/p},$$

where $p = k$, the number of main effects we estimate.

We now look at all possible designs for a small example.

■ EXAMPLE 5.1.6.

Recall that, for $m = 3$ and $k = 3$, there are three difference vectors: $\mathbf{v}_1 = (1,1,2)$, $\mathbf{v}_2 = (1,2,3)$ and $\mathbf{v}_3 = (2,2,2)$. All the possible designs for choice experiments in this situation are given in Table 5.5. Since m is odd, Theorem 5.1.1 states that the D-optimal designs have choice sets in which the entries in the difference vectors sum to $(m^2-1)k/4 = 2k = 6$ and a maximum $\det(C_{\text{opt},M}) = (1/9)^3$. The difference vectors $\mathbf{v}_2$ and $\mathbf{v}_3$ have entries which sum to 6; so there are three D-optimal designs:

1. All 24 triples with difference vector $\mathbf{v}_2$. So $a_{\mathbf{v}_2} = 1/24$ and $a_{\mathbf{v}_1} = a_{\mathbf{v}_3} = 0$.
2. All 8 triples with difference vector $\mathbf{v}_3$. So $a_{\mathbf{v}_3} = 1/8$ and $a_{\mathbf{v}_1} = a_{\mathbf{v}_2} = 0$.
3. All 32 triples with difference vectors $\mathbf{v}_2$ and $\mathbf{v}_3$. So $a_{\mathbf{v}_2} = a_{\mathbf{v}_3} = 1/32$ and $a_{\mathbf{v}_1} = 0$.

The smallest of these D-optimal designs is the second one, consisting of the following eight triples, each with difference vector $\mathbf{v}_3$:

(000, 011, 101), (000, 011, 110), (000, 101, 110), (001, 010, 100),
(001, 010, 111), (001, 100, 111), (010, 100, 111), (011, 101, 110). □

Table 5.5 All Possible Choice Experiment Designs for Binary Attributes when $k = 3$ and $m = 3$

$\mathbf{v}_j$ in Design	$\sum_i d_{ij}$	$i_{\mathbf{v}_1}$	$i_{\mathbf{v}_2}$	$i_{\mathbf{v}_3}$	N	$a_{\mathbf{v}_1}$	$a_{\mathbf{v}_2}$	$a_{\mathbf{v}_3}$	Efficiency (%)
$\mathbf{v}_1$	4	1	0	0	24	$\frac{1}{24}$	0	0	66.67
$\mathbf{v}_2$	6	0	1	0	24	0	$\frac{1}{24}$	0	100
$\mathbf{v}_3$	6	0	0	1	8	0	0	$\frac{1}{8}$	100
$\mathbf{v}_1$ & $\mathbf{v}_2$	4, 6	1	1	0	48	$\frac{1}{48}$	$\frac{1}{48}$	0	83.33
$\mathbf{v}_1$ & $\mathbf{v}_3$	4, 6	1	0	1	32	$\frac{1}{32}$	0	$\frac{1}{32}$	75
$\mathbf{v}_2$ & $\mathbf{v}_3$	6	0	1	1	32	0	$\frac{1}{32}$	$\frac{1}{32}$	100
$\mathbf{v}_1$, $\mathbf{v}_2$ & $\mathbf{v}_3$	4, 6	1	1	1	56	$\frac{1}{56}$	$\frac{1}{56}$	$\frac{1}{56}$	85.71

5.1.4 The Model for Main Effects and Two-Factor Interactions

To find the D-optimal designs for estimating main effects and two-factor interactions for any choice set size m, we generalize the results of Section 4.1.4.

As in Section 4.1.4, we let $B_{2^k,M}$ be the rows of B_{2^k} that correspond to main effects and we let $B_{2^k,T}$ be the rows of B_{2^k} that correspond to the two-factor interactions. The matrix associated with main effects and two-factor interactions is denoted by $B_{2^k,MT}$ and is the concatenation of $B_{2^k,M}$ and $B_{2^k,T}$. For the D-optimal design we evaluate the $\left[k + \binom{k}{2}\right] \times \left[k + \binom{k}{2}\right]$ principal minor of $C = B_{2^k}\Lambda B_{2^k}$. As we are working with the complete factorial, it is not important how the contrasts for interactions of more than two factors are divided between B_a and B_h.

In Lemma 4.1.5, we established that

$$B_{2^k,T}D_{k,i} = \left[\binom{k-2}{i} - 2\binom{k-2}{i-1} + \binom{k-2}{i-2}\right] B_{2^k,T}$$

for all allowable i.

■ **LEMMA 5.1.4.**
Under the null hypothesis, the information matrix for main effects plus two-factor interactions is given by

$$C_{MT} = \begin{bmatrix} \frac{2}{m^2}\left[\sum_{i=1}^k y_i\binom{k-1}{i-1}\right] I_k & \mathbf{0} \\ \mathbf{0} & \frac{4}{m^2}\left[\sum_{i=1}^k y_i\binom{k-2}{i-1}\right] I_{k(k-1)/2} \end{bmatrix}.$$

Proof. We let $B_h = B_{2^k,MT}$ and then

$$\begin{aligned} C_{MT} &= B_{2^k,MT}\Lambda B'_{2^k,MT} \\ &= \begin{bmatrix} B_{2^k,M}\Lambda B'_{2^k,M} & \mathbf{0} \\ \mathbf{0} & B_{2^k,T}\Lambda B'_{2^k,T} \end{bmatrix} \\ &= \begin{bmatrix} \frac{2}{m^2}\sum_{i=1}^k y_i\binom{k-1}{i-1} I_k & \mathbf{0} \\ \mathbf{0} & B_{2^k,T}\Lambda B'_{2^k,T} \end{bmatrix}. \end{aligned}$$

Now

$$B_{2^k,T}\Lambda B'_{2^k,T} = B_{2^k,T}\left[\frac{m-1}{m^2}zI_{2^k} - \frac{1}{m^2}\sum_{i=1}^{k} y_i D_{k,i}\right] B'_{2^k,T}$$
$$= \frac{1}{m^2}B_{2^k,T}\left[\sum_{i=1}^{k}\binom{k}{i}y_i I_{2^k} - \sum_{i=1}^{k} y_i D_{k,i}\right] B'_{2^k,T}$$
$$= \frac{1}{m^2}\left[\sum_{i=1}^{k} y_i\left[\binom{k}{i}B_{2^k,T} - B_{2^k,T}D_{k,i}\right]\right] B'_{2^k,T}$$
$$= \frac{1}{m^2}\left[\sum_{i=1}^{k} y_i\left[\binom{k}{i} - \binom{k-2}{i} + 2\binom{k-2}{i-1} - \binom{k-2}{i-2}\right]\right] I_{k(k-1)/2}.$$

As we noted previously,

$$\binom{k}{i} = \binom{k-1}{i} + \binom{k-1}{i-1};$$

so $\binom{k-1}{i-1} = \binom{k-2}{i-1} + \binom{k-2}{i-2}$ and $\binom{k-1}{i} = \binom{k-2}{i} + \binom{k-2}{i-1}$.

It follows that

$$\binom{k}{i} = \binom{k-2}{i} + 2\binom{k-2}{i-1} + \binom{k-2}{i-2},$$

and hence

$$\binom{k}{i} - \binom{k-2}{i} + 2\binom{k-2}{i-1} - \binom{k-2}{i-2} = 4\binom{k-2}{i-1}.$$

$$\text{Therefore}\quad B_{2^k,T}\Lambda B'_{2^k,T} = \frac{4}{m^2}\left[\sum_{i=1}^{k} y_i\binom{k-2}{i-1}\right] I_{k(k-1)/2},$$

and we have

$$C_{MT} = \begin{bmatrix} \frac{2}{m^2}\left[\sum_{i=1}^{k} y_i\binom{k-1}{i-1}\right] I_k & \mathbf{0} \\ \mathbf{0} & \frac{4}{m^2}\left[\sum_{i=1}^{k} y_i\binom{k-2}{i-1}\right] I_{k(k-1)/2} \end{bmatrix},$$

as required. □

Hence the determinant of C_{MT} is

$$\det(C_{MT}) = \left[\frac{2}{m^2}\sum_{i=1}^{k} y_i\binom{k-1}{i-1}\right]^k \times \left[\frac{4}{m^2}\sum_{i=1}^{k} y_i\binom{k-2}{i-1}\right]^{k(k-1)/2}.$$

■ **EXAMPLE 5.1.7.**

Let $m = 3$ and $k = 3$. We can calculate C_{MT} either by evaluating

$$C_{MT} = B_{2^3,MT}\Lambda B'_{2^3,MT},$$

or by using Lemma 5.1.4. In this example, we will do both.

The matrix $B_{2^3,MT}$ is the contrast matrix for main effects plus all two-factor interactions and is given by

$$B_{2^3,MT} = \frac{1}{2\sqrt{2}} \begin{bmatrix} -1 & -1 & -1 & -1 & 1 & 1 & 1 & 1 \\ -1 & -1 & 1 & 1 & -1 & -1 & 1 & 1 \\ -1 & 1 & -1 & 1 & -1 & 1 & -1 & 1 \\ 1 & 1 & -1 & -1 & -1 & -1 & 1 & 1 \\ 1 & -1 & 1 & -1 & -1 & 1 & -1 & 1 \\ 1 & -1 & -1 & 1 & 1 & -1 & -1 & 1 \end{bmatrix}.$$

In this example, we assume that all choice sets with difference vectors $\mathbf{v}_2$ or $\mathbf{v}_3$ are in the choice experiment. Then we can use the Λ matrix from Example 5.1.4 to calculate

$$C_{MT} = B_{2^3,MT}\Lambda B'_{2^3,MT} = \begin{bmatrix} \frac{1}{9}I_3 & \mathbf{0} \\ \mathbf{0} & \frac{1}{12}I_3 \end{bmatrix} \text{ and } \det(C_{MT}) = \left(\frac{1}{9}\right)^3 \left(\frac{1}{12}\right)^3.$$

We can also use Lemma 5.1.4 to calculate C_{MT}. From Example 5.1.5,

$$\left[\frac{2}{m^2}\sum_{i=1}^{k} y_i \binom{k-1}{i-1}\right]^k = \frac{1}{9}I_3.$$

Then, using the y_i values given in Example 5.1.4,

$$\begin{aligned} \frac{4}{m^2}\sum_{i=1}^{k} y_i \binom{k-2}{i-1} I_{k(k-1)/2} &= \frac{4}{3^2}\left[y_1\binom{1}{0} + y_2\binom{1}{1} + y_3\binom{1}{2}\right] I_3 \\ &= \frac{4}{9}\left[\frac{2}{32} + \frac{4}{32}\right] I_3 \\ &= \frac{1}{12}I_3. \end{aligned}$$

Hence

$$C_{MT} = \begin{bmatrix} \frac{1}{9}I_3 & \mathbf{0} \\ \mathbf{0} & \frac{1}{12}I_3 \end{bmatrix}.$$

□

To find the D-optimal design, we need to maximize $\det(C_{MT})$, subject to the constraint that $2^k z/m = 1$.

■ **THEOREM 5.1.2.**
The D-optimal design for testing main effects and two-factor interactions, when all other effects are assumed to be zero, is given by designs where

$$y_i = \begin{cases} \frac{m(m-1)}{2^k}\binom{k+1}{k/2}^{-1}, & k \text{ even and } i = k/2,\ k/2+1, \\ \frac{m(m-1)}{2^k}\binom{k}{(k+1)/2}^{-1}, & k \text{ odd and } i = (k+1)/2, \\ 0, & \text{otherwise}, \end{cases}$$

when this results in non-zero y_i values that correspond to difference vectors that actually exist. The maximum value of the determinant is then

$$\det(C_{MT}) = \begin{cases} \left(\frac{(m-1)(k+2)}{m(k+1)2^k}\right)^{k+k(k-1)/2} & k \text{ even,} \\ \left(\frac{(m-1)(k+1)}{mk2^k}\right)^{k+k(k-1)/2} & k \text{ odd.} \end{cases}$$

Proof. In Theorem 4.1.3, we proved that for $m = 2$ the D-optimal design for testing main effects and two-factor interactions is given by

$$a_v = \begin{cases} \left\{2^{k-1}\binom{k+1}{k/2}\right\}^{-1} & k \text{ even and } v = k/2,\ k/2+1, \\ \left\{2^{k-1}\binom{k}{(k+1)/2}\right\}^{-1} & k \text{ odd and } v = (k+1)/2, \\ 0 & \text{otherwise.} \end{cases}$$

In the proof of this theorem, the function $f = WZ^{(k-1)/2}$, and therefore $\det(C_{MT})$, is maximized subject to the constraint $\sum_{v=1}^{k}\binom{k}{v}x_v = 1$, where

$$W = \sum_{v=1}^{k}\binom{k-1}{v-1}x_v$$

and

$$Z = \sum_{v=1}^{k}\binom{k-2}{v-1}x_v.$$

For choice sets of size m the constraint is

$$\frac{2^k}{m}z = \frac{2^k}{m(m-1)}\binom{k}{i}y_i = 1.$$

In order to use the results in the proof of Theorem 4.1.3, we let

$$x_i = 2^k y_i/(m(m-1))$$

(suggested by Moore (2001) when $m = 3$). Then we have the same form of the constraint as we had for $m = 2$.

Now we let

$$W = \frac{2^k}{m(m-1)}\sum_{i=1}^{k}\binom{k-1}{i-1}y_i = \sum_{i=1}^{k}\binom{k-1}{i-1}x_i$$

and

$$Z = \frac{2^k}{m(m-1)}\sum_{i=1}^{k}\binom{k-2}{i-1}y_i = \sum_{i=1}^{k}\binom{k-2}{i-1}x_i.$$

Using the results in Theorem 4.1.3, the function $f = WZ^{(k-1)/2}$, and therefore

$$\det(C_{MT}) = \left[\frac{2}{m^2}\sum_{i=1}^{k}y_i\binom{k-1}{i-1}\right]^k \times \left[\frac{4}{m^2}\sum_{i=1}^{k}y_i\binom{k-2}{i-1}\right]^{k(k-1)/2}$$

is maximized subject to the constraint $\sum_{i=1}^{k}\binom{k}{i}x_i = 1$ for the same x_i values given above for the $m = 2$ case. Using $x_i = 2^k y_i/(m(m-1))$, we obtain the optimal designs in terms of the y_i values, as required.

We now determine the maximum value of the determinant at these y_i values. For k even, only two values of i will give the maximum determinant. These are $i = k/2$ and $i = k/2 + 1$, where

$$y_{k/2} = y_{k/2+1} = \frac{m(m-1)}{2^k}\binom{k+1}{k/2}^{-1}$$

and all other $y_i = 0$. Then

$$\begin{aligned}\det(C_{MT}) &= \left[\frac{2}{m^2}\sum_{i=k/2,k/2+1} y_i\binom{k-1}{i-1}\right]^k \times \left[\frac{4}{m^2}\sum_{i=k/2,k/2+1}^{k} y_i\binom{k-2}{i-1}\right]^{k(k-1)/2} \\ &= \left[\frac{2}{m^2}\frac{m(m-1)}{2^k}\binom{k+1}{k/2}^{-1}\left[\binom{k-1}{k/2-1}+\binom{k-1}{k/2}\right]\right]^k \\ &\quad \times \left[\frac{4}{m^2}\frac{m(m-1)}{2^k}\binom{k+1}{k/2}^{-1}\left[\binom{k-2}{k/2-1}+\binom{k-2}{k/2}\right]\right]^{k(k-1)/2}.\end{aligned}$$

Now $\binom{k-1}{k/2-1}+\binom{k-1}{k/2} = \binom{k}{k/2}$, $\binom{k-2}{k/2-1}+\binom{k-2}{k/2} = \binom{k-1}{k/2}$,

$$\frac{\binom{k}{k/2}}{\binom{k+1}{k/2}} = \frac{k+2}{2(k+1)} \quad\text{and}\quad \frac{\binom{k-1}{k/2}}{\binom{k+1}{k/2}} = \frac{k+2}{4(k+1)}.$$

Therefore, for k even,

$$\det(C_{MT}) = \left(\frac{(m-1)(k+2)}{m(k+1)2^k}\right)^{k+k(k-1)/2}.$$

For k odd, the only value of i that will give the maximum determinant is $i = (k+1)/2$, where

$$y_{(k+1)/2} = \frac{m(m-1)}{2^k}\binom{k}{(k+1)/2}^{-1}$$

and all other $y_i = 0$. Then

$$\begin{aligned}\det(C_{MT}) &= \left[\frac{2}{m^2}\frac{m(m-1)}{2^k}\binom{k}{(k+1)/2}^{-1}\binom{k-1}{(k+1)/2-1}\right]^k \\ &\quad \times \left[\frac{4}{m^2}\frac{m(m-1)}{2^k}\binom{k}{(k+1)/2}^{-1}\binom{k-2}{(k+1)/2-1}\right]^{k(k-1)/2}.\end{aligned}$$

Now $\dfrac{\binom{k-1}{(k+1)/2-1}}{\binom{k}{(k+1)/2}} = \dfrac{\binom{k-1}{(k-1)/2}}{\binom{k}{(k+1)/2}} = \dfrac{k+1}{2k}$ and $\dfrac{\binom{k-2}{(k+1)/2-1}}{\binom{k}{(k+1)/2}} = \dfrac{\binom{k-2}{(k-1)/2}}{\binom{k}{(k+1)/2}} = \dfrac{k+1}{4k}$.

Therefore, for k odd,

$$\det(C_{MT}) = \left(\frac{(m-1)(k+1)}{mk2^k}\right)^{k+k(k-1)/2}.$$

Hence the maximum value of the determinant is

$$\det(C_{MT}) = \begin{cases} \left(\frac{(m-1)(k+2)}{m(k+1)2^k}\right)^{k+k(k-1)/2}, & k \text{ even}, \\ \left(\frac{(m-1)(k+1)}{mk2^k}\right)^{k+k(k-1)/2}, & k \text{ odd}. \end{cases}$$

□

Note that for any design we can calculate the D-efficiency relative to the optimal design using the expression

$$\text{Eff}_D = \left(\frac{\det(C_{MT})}{\det(C_{\text{opt},MT})}\right)^{1/p},$$

where $p = k + k(k-1)/2$, the total number of main effects and two-factor interactions that we are estimating.

The following two examples illustrate the use of Theorem 5.1.2 to obtain choice sets with the maximum value of the determinant of C_{MT}.

■ EXAMPLE 5.1.8.
For $m = 3$ and $k = 3$, Theorem 5.1.2 states that the D-optimal design is given by

$$y_i = \begin{cases} \frac{3\times 2}{2^3}\binom{3}{2}^{-1} = \frac{1}{4}, & i = 2, \\ 0, & \text{otherwise}. \end{cases}$$

The maximum value of $\det(C_{MT})$ is

$$\det(C_{\text{opt},MT}) = \left(\frac{2\times 4}{3\times 3\times 2^3}\right)^{3+3} = \left(\frac{1}{9}\right)^6.$$

Since

$$\begin{aligned} y_1 &= 4a_{\mathbf{v}_1} + 2a_{\mathbf{v}_2} = 0, \\ y_2 &= 2a_{\mathbf{v}_1} + 2a_{\mathbf{v}_2} + 2a_{\mathbf{v}_3} = \frac{1}{4}, \\ y_3 &= 6a_{\mathbf{v}_2} = 0, \end{aligned}$$

we see that $a_{\mathbf{v}_1} = a_{\mathbf{v}_2} = 0$ and $a_{\mathbf{v}_3} = \frac{1}{8}$. Thus the D-optimal design consists of the 8 triples with difference vector $\mathbf{v}_3 = (2,2,2)$. Information about the 7 competing designs is shown in Table 5.6. □

■ EXAMPLE 5.1.9.
For $m = 3$ and $k = 4$ the possible difference vectors are $\mathbf{v}_1 = (1,1,2)$, $\mathbf{v}_2 = (1,2,3)$, $\mathbf{v}_3 = (1,3,4)$, $\mathbf{v}_4 = (2,2,2)$, $\mathbf{v}_5 = (2,2,4)$, and $\mathbf{v}_6 = (2,3,3)$. Using Theorem 5.1.2, the D-optimal design is given by

$$y_i = \begin{cases} \frac{3\times 2}{2^4}\binom{5}{2}^{-1} = \frac{6}{160}, & i = 2,3, \\ 0, & \text{otherwise}. \end{cases}$$

Table 5.6 The Efficiency, for the Estimation of Main Effects plus Two-Factor Interactions, of the 7 Competing Designs when $k = 3$ and $m = 3$

$\mathbf{v}_j$ in Design	$\det(C_{MT})$	Efficiency (%)
$\mathbf{v}_1$	$(\frac{2}{27})^3(\frac{1}{9})^3$	81.65
$\mathbf{v}_2$	$(\frac{1}{9})^3(\frac{2}{27})^3$	81.65
$\mathbf{v}_3$	$(\frac{1}{9})^6$	100
$\mathbf{v}_1$ & $\mathbf{v}_2$	$(\frac{5}{54})^6$	83.33
$\mathbf{v}_1$ & $\mathbf{v}_3$	$(\frac{1}{12})^3(\frac{1}{9})^3$	86.60
$\mathbf{v}_2$ & $\mathbf{v}_3$	$(\frac{1}{9})^3(\frac{1}{12})^3$	86.60
$\mathbf{v}_1$, $\mathbf{v}_2$ & $\mathbf{v}_3$	$(\frac{2}{21})^6$	85.71

The maximum value of $\det(C_{MT})$ is

$$\det(C_{\text{opt},MT}) = \left(\frac{2 \times 6}{3 \times 5 \times 2^4}\right)^{4+6} = \left(\frac{1}{20}\right)^{10}.$$

Since y_2 and y_3 are the only y_i values that are non-zero and $\mathbf{v}_4$ and $\mathbf{v}_6$ are the only difference vectors containing all 2s, all 3s, or a combination of 2s and 3s, we see that $a_{\mathbf{v}_j} = 0$ unless $j = 4, 6$. Thus we have

$$\begin{aligned}
y_1 &= 6a_{\mathbf{v}_1} + 6a_{\mathbf{v}_2} + 2a_{\mathbf{v}_3} = 0, \\
y_2 &= 2a_{\mathbf{v}_1} + 4a_{\mathbf{v}_2} + 4a_{\mathbf{v}_4} + 2a_{\mathbf{v}_5} + 2a_{\mathbf{v}_6} = 6/160, \\
y_3 &= 6a_{\mathbf{v}_2} + 2a_{\mathbf{v}_3} + 6a_{\mathbf{v}_6} = 6/160, \\
y_4 &= 8a_{\mathbf{v}_3} + 6a_{\mathbf{v}_5} = 0.
\end{aligned}$$

The solution is $a_{\mathbf{v}_1} = a_{\mathbf{v}_2} = a_{\mathbf{v}_3} = a_{\mathbf{v}_5} = 0$ and $a_{\mathbf{v}_4} = a_{\mathbf{v}_6} = 1/160$. Thus the D-optimal design consists of the 64 triples with difference vector $\mathbf{v}_4 = (2, 2, 2)$ and the 96 triples with difference vector $\mathbf{v}_6 = (2, 3, 3)$. If we use just the 64 triples with difference vector $\mathbf{v}_4$, then this design is 99.03% efficient, and for the 96 triples with difference vector $\mathbf{v}_6$ only, the design is 99.60% efficient. □

However, for some values of m and k, solutions to the y_i equations do not exist. For example, when $m = 3$ and $k \equiv 1 \pmod 4$, no solution exists; the following example illustrates this case.

■ **EXAMPLE 5.1.10.**
If we let $m = 3$ and $k = 5$, then the D-optimal design given by Theorem 5.1.2 is

$$y_3 = \frac{3 \times 2}{2^5}\binom{5}{3}^{-1}, \quad y_1 = y_2 = y_4 = y_5 = 0.$$

This means that triples with difference vector $(3, 3, 3)$ are required since any other triple would result in one of the other y_i values being non-zero. However, the only possible difference vectors are

$$\mathbf{v}_1 = (1, 1, 2),\ \ \mathbf{v}_2 = (1, 2, 3),\ \ \mathbf{v}_3 = (1, 3, 4),\ \ \mathbf{v}_4 = (1, 4, 5),\ \ \mathbf{v}_5 = (2, 2, 2),$$

$$\mathbf{v}_6 = (2, 2, 4), \quad \mathbf{v}_7 = (2, 3, 3), \quad \mathbf{v}_8 = (2, 3, 5), \quad \mathbf{v}_9 = (2, 4, 4), \quad \mathbf{v}_{10} = (3, 3, 4);$$

so no triple with the difference vector $(3, 3, 3)$ exists. By checking all possible designs, it can easily be shown that the optimal design consists of the 960 triples with difference vector $\mathbf{v}_7 = (2, 3, 3)$ and the 480 triples with difference vector $\mathbf{v}_{10} = (3, 3, 4)$, where $y_2 = 1/1440$, $y_3 = 9/1440$, $y_4 = 3/1440$ and $y_1 = y_5 = 0$. Thus $a_{\mathbf{v}_7} = a_{\mathbf{v}_{10}} = 1/1440$ and $a_{\mathbf{v}_j} = 0$ for $j = 1, 2, 3, 4, 5, 6, 8, 9$. The maximum obtainable determinant is

$$\det(C_{\text{opt},MT}) = \left(\frac{13}{540}\right)^{15}.$$

As Example 5.1.10 shows, the optimal designs derived in this and the preceding sections can become very large as the number of attributes increases. The question of how many choice sets can be included in a stated choice experiment has been considered by various authors. Brazell and Louviere (1995) show that choice experiments with up to 128 choice sets can be effective in parameter estimation. In the next section, we investigate the D-efficiency of small designs obtained from a generalization of the constant difference construction in Section 4.2.3.

5.1.5 Exercises

1. Let $k = 3$ and $m = 4$. Give the 35 sets of 4 treatment combinations with 000. Calculate the difference vector for each. Hence calculate the $c_{\mathbf{v}_j}$ and, for each $\mathbf{v}_j$, calculate the $x_{\mathbf{v}_j;i}$. Verify that the bound of Lemma 5.1.1 is correct.

2. Use Lemma 5.1.1 when $k = 3$ and $m = 4$. Can you get more than one set of rows in this case?

3. Let $k = 3$ and $m = 4$. Evaluate z, y_1, y_2 and y_3. Hence give an expression for Λ.

4. If $m = 2$ show that $y_i = a_i$.

5. Let $k = 3$ and $m = 4$. Using Theorem 5.1.1 give all the possible D-optimal designs for estimating main effects. Can you use a smaller subset of any of these to get a smaller design which is still D-optimal?

6. Let $k = 3$ and $m = 4$. Use Theorem 5.1.2 to give the y_i that would correspond to the D-optimal designs for estimating main effects plus two-factor interactions. Do the y_i correspond to an actual design? If so, give it. If not, try 2 or 3 designs "close to" the optimal y_i and compare them. Which would you recommend?

5.2 SMALL OPTIMAL AND NEAR-OPTIMAL DESIGNS FOR LARGER CHOICE SET SIZES

In this section, we give constructions for small optimal and near-optimal designs for choice sets of size m. The results are an extension of results in Section 4.2, where we gave constructions for optimal and near-optimal designs for estimating main effects only, and for estimating main effects plus two-factor interactions. Recall that the constructions there started with a fraction of resolution 3 (for estimating main effects only) or resolution 5 (for estimating main effects plus two-factor interactions). Pairs were formed by adding one or more generators to the treatment combinations in the fraction, where the addition is performed component-wise modulo 2. Each generator gave rise to a set of pairs. We had to assume that all contrasts other than the ones we were estimating were 0 however. So B_a contains all the contrasts that not in B_h and B_r is empty.

5.2.1 The Model for Main Effects Only

When estimating main effects only, for each attribute there must be at least one generator with a 1 in the corresponding position. Using one generator consisting of all 1s, which is equivalent to using the resolution 3 fraction and its foldover, results in a D-optimal design in a minimum number of pairs, in which all main effects can be estimated.

The next result gives generators for small optimal designs for estimating main effects when $m \geq 2$.

■ THEOREM 5.2.1.
Let F be a fractional factorial design of resolution at least 3 and with k factors. Let G be a set of generators $G = (\mathbf{g}_1, \mathbf{g}_2, \ldots, \mathbf{g}_m)$, where each $\mathbf{g}_i$ is a binary k-tuple, and $\mathbf{g}_1 = \mathbf{0}$. Let $\mathbf{v} = (d_1, d_2, \ldots, d_{m(m-1)/2})$ be the difference vector consisting of all the pairwise differences between the generators in G, where

$$\sum_{i=1}^{m(m-1)/2} d_i = \begin{cases} (m^2-1)k/4, & m \text{ odd}, \\ m^2k/4, & m \text{ even}. \end{cases}$$

Then the choice sets given by $(F, F+\mathbf{g}_2, \ldots, F+\mathbf{g}_m)$, where the addition is done component-wise modulo 2, are optimal for estimating main effects only.

Proof. Let $|F| = 2^{k-p}$ and let B_F be the columns of B_M corresponding to the treatment combinations in F. Thus $B_F B_F' = \frac{2^{k-p}}{2^k} I_k$. Let $D_{\mathbf{g}_i}$ be the diagonal matrix with an entry of 1 in position j if the jth entry of $\mathbf{g}_i$ is 0 and an entry of -1 in position j if the jth entry of $\mathbf{g}_i$ is 1. Then $D_{\mathbf{g}_i} B_F$ has the same columns as the columns of B_M for the treatment combinations in $F + \mathbf{g}_i$.

We construct the choice sets from F by adding the generators to F, using modulo 2 arithmetic. Then the choice sets are $(F, F+\mathbf{g}_2, \ldots, F+\mathbf{g}_m)$, where each row represents a choice set. Let N be the number of choice sets of size m (hence $N = 2^{k-p}$ if there are no repeated choice sets) and let n_{uv} be the number of choice sets that contain the pair of treatment combinations u, v. This number may be 0 if at least one of u and v does not appear in the choice sets. Let $n_{uv,cd}$ be the number of choice sets which contain u and v in columns c and d (unordered). Then $n_{uv} = \sum_{c,d} n_{uv,cd}$, and we define $n_{uu} = -\sum_{v \neq u} n_{uv}$. Then the values n_{uv} and n_{uu} are the entries of $m^2 N \Lambda$.

Consider two columns c and d. Then we know that column c contains the treatment combinations in $F + \mathbf{g}_c$ (in some order) and column d contains the treatment combinations in $F + \mathbf{g}_d$ in some order. If $F + \mathbf{g}_c = F + \mathbf{g}_d$, then $\mathbf{g}_c + \mathbf{g}_d \in F$ and we can write F as $F_1 \cup (F_1 + \mathbf{g}_c + \mathbf{g}_d)$. Then, using this order for the elements of F, we have that the submatrix of Λ corresponding to the elements in F can be written as (reordering rows and columns if necessary)

$$\Lambda_{cd} = \frac{1}{m^2 2^{k-p}} \begin{bmatrix} I_{2^{k-p-1}} & -I_{2^{k-p-1}} & \mathbf{0}_{2^{k-p-1},2^k-2^{k-p}} \\ -I_{2^{k-p-1}} & I_{2^{k-p-1}} & \mathbf{0}_{2^{k-p-1},2^k-2^{k-p}} \\ \mathbf{0}_{2^k-2^{k-p},2^{k-p-1}} & \mathbf{0}_{2^k-2^{k-p},2^{k-p-1}} & \mathbf{0}_{2^k-2^{k-p},2^k-2^{k-p}} \end{bmatrix}$$

since we know that each treatment combination in F gives rise to a distinct choice set. We can write B_M as $B_M = [B_{F_1} \quad D_{\mathbf{g}_c+\mathbf{g}_d}B_{F_1} \quad B_{\bar{F}}]$. Now

$$B_{F_1}B'_{F_1} = \frac{2^{k-p-1}}{2^k}I_k;$$

so

$$\begin{aligned}(m^2 2^{k-p})B_M\Lambda_{cd}B'_M &= B_{F_1}B'_{F_1} - D_{\mathbf{g}_c+\mathbf{g}_d}B_{F_1}B'_{F_1} \\ &\quad + D_{\mathbf{g}_c+\mathbf{g}_d}B_{F_1}B'_{F_1}D_{\mathbf{g}_c+\mathbf{g}_d} - B_{F_1}B'_{F_1}D_{\mathbf{g}_c+\mathbf{g}_d} \\ &= \frac{2^{k-p}}{2^k}(I_k - D_{\mathbf{g}_c+\mathbf{g}_d}).\end{aligned}$$

If $\mathbf{g}_c + \mathbf{g}_d \notin F$, then the same argument establishes that

$$B_M\Lambda_{cd}B'_M = \frac{2^{k-p+1}}{2^k} \times \frac{1}{m^2 2^{k-p}}(I_k - D_{\mathbf{g}_c+\mathbf{g}_d}).$$

Now we know that

$$\sum_{i=1}^{m(m-1)/2} d_i = \begin{cases} (m^2-1)k/4, & m \text{ odd}, \\ m^2k/4, & m \text{ even}. \end{cases}$$

Consider the contribution to the $\sum_i d_i$ in a choice set from just one attribute. The attribute will have x 0s and $(m-x)$ 1s to give a total of $x(m-x)$. This contribution is maximized by having $m/2$ 1s and $m/2$ 0s for m even, and $(m-1)/2$ 0s and $(m+1)/2$ 1s (or vice versa) for m odd. Thus the maximum contribution to $\sum_i d_i$ from any one attribute is $m^2/4$ for m even and $(m^2-1)/4$ for m odd. Thus we see that each attribute must contribute exactly this amount to $\sum_i d_i$ for the optimal designs.

Suppose $m = 2m_1 + 1$. Now $\sum_{c,d} D_{\mathbf{g}_c+\mathbf{g}_d}$ is summing $\binom{m_1}{2} + \binom{m_1+1}{2}$ entries of 1 in each diagonal position and $m_1(m_1+1)$ entries of -1 in each diagonal position. Thus, if m is odd,

$$\sum_{c,d} D_{\mathbf{g}_c+\mathbf{g}_d} = \left(\binom{m_1}{2} + \binom{m_1+1}{2} - m_1(m_1+1)\right) I_k = -\frac{m-1}{2}I_k.$$

Similarly, if m is even,

$$\sum_{c,d} D_{\mathbf{g}_c+\mathbf{g}_d} = -\frac{m}{2}I_k.$$

Suppose that none of the $\mathbf{g}_i$ is in F and that $\mathbf{g}_c + \mathbf{g}_d \notin F$ for any pair c and d. Let A be the set of treatment combinations that appear in the choice experiment. Then

$$(m^2 2^{k-p})\left[\, B_F \quad D_{\mathbf{g}_2}B_F \quad \ldots \quad D_{\mathbf{g}_c+\mathbf{g}_d}B_F \quad B_{\bar{A}} \,\right]\sum_{c,d}\Lambda_{cd}\begin{bmatrix} B'_F \\ B'_F D_{\mathbf{g}_2} \\ \vdots \\ B'_F D_{\mathbf{g}_c+\mathbf{g}_d} \\ B'_{\bar{A}} \end{bmatrix}$$

$$= \sum_{c,d} \left(\frac{2^{k-p+1}}{2^k} (I_k - D_{\mathbf{g}_c+\mathbf{g}_d}) \right)$$

$$= \begin{cases} \frac{2^{k-p+1}}{2^k} \left(\binom{m}{2} + \frac{m-1}{2} \right) I_k, & m \text{ odd}, \\ \frac{2^{k-p+1}}{2^k} \left(\binom{m}{2} + \frac{m}{2} \right) I_k, & m \text{ even} \end{cases}$$

$$= \begin{cases} \frac{2^{k-p+1}}{2^k} \times \frac{m^2-1}{2} I_k, & m \text{ odd}, \\ \frac{2^{k-p+1}}{2^k} \times \frac{m^2}{2} I_k, & m \text{ even}. \end{cases}$$

Thus we have

$$C_M = \begin{cases} \frac{m^2-1}{m^2 2^k} I_k, & m \text{ odd}, \\ \frac{1}{2^k} I_k, & m \text{ even}, \end{cases}$$

as required.

Now suppose that, for at least one pair of columns, the number of pairs is 2^{k-p-1} (that is, $F + \mathbf{g}_c = F + \mathbf{g}_d$ for some c, d). Then we see that

$$N B_M \Lambda_{cd} B'_M = \frac{2^{k-p}}{2^k} (I_k - D_{\mathbf{g}_c+\mathbf{g}_d}).$$

Although this works out for the pairs, once the pairs are considered as part of the larger choice sets, then the pairs will in fact appear twice and so we need to use

$$2N B_M \Lambda_{cd} B'_M = \frac{2^{k-p+1}}{2^k} (I_k - D_{\mathbf{g}_c+\mathbf{g}_d})$$

when evaluating C_M. Thus the proof from before can be used.

The only situation that is not covered by the above proof is when m is a power of 2. In that case, the optimal set of generators must form a subgroup and the choice sets are this subgroup and its distinct cosets formed by adding elements of F. Making this observation, a straightforward modification of the proof above establishes the result. □

■ EXAMPLE 5.2.1.

Let $m = 5$ and $k = 9$. To obtain an optimal design for estimating main effects, we require a fraction F of the 2^9 factorial which has resolution at least 3. The 16 treatment combinations given in the first column of Table 5.7 are a 1/32 fraction of resolution 3 with defining contrast $I \equiv BCE \equiv CDF \equiv ACG \equiv ABH \equiv ADJ$. To obtain the choice sets, we need $m = 5$ generators $G = (\mathbf{g}_1, \mathbf{g}_2, \mathbf{g}_3, \mathbf{g}_4, \mathbf{g}_5)$, where $\mathbf{g}_1 = \mathbf{0}$, so that the differences in the difference vector sum to $(m^2 - 1)k/4 = 6k = 54$. One set of generators that satisfies this condition is

$$G = (000000000,\ 000000111,\ 111111000,\ 000111111,\ 111111111),$$

which has the difference vector $(3, 3, 3, 3, 6, 6, 6, 6, 9, 9)$. The 16 choice sets are given by $(F,\ F+\mathbf{g}_2,\ F+\mathbf{g}_3,\ F+\mathbf{g}_4,\ F+\mathbf{g}_5)$, where the addition is done component-wise modulo 2. The choice sets are given in Table 5.7, where the choice sets are represented by the rows. The B matrix has one row for each of the 9 main effects and 80 columns for the treatment combinations. This matrix is normalized by dividing the entries by $\sqrt{2^9}$. The C_M matrix for these choice sets is $\frac{3}{1600} I_9$, and the design is therefore 100% efficient. □

Table 5.7 Optimal Choice Sets for Estimating Main Effects Only for $m = 5$ and $k = 9$.

F	$F + \mathbf{g}_2$	$F + \mathbf{g}_3$	$F + \mathbf{g}_4$	$F + \mathbf{g}_5$
000000000	000000111	111111000	000111111	111111111
000101001	000101110	111010001	000010110	111010110
001011100	001011011	110100100	001100011	110100011
001110101	001110010	110001101	001001010	110001010
010010010	010010101	101101010	010101101	101101101
010111011	010111100	101000011	010000100	101000100
011001110	011001001	100110110	011110001	100110001
011100111	011100000	100011111	011011000	100011000
100000111	100000000	011111111	100111000	011111000
100101110	100101001	011010110	100010001	011010001
101011011	101011100	010100011	101100100	010100100
101110010	101110101	010001010	101001101	010001101
110010101	110010010	001101101	110101010	001101010
110111100	110111011	001000100	110000011	001000011
111001001	111001110	000110001	111110110	000110110
111100000	111100111	000011000	111011111	000011111

5.2.2 The Model for Main Effects and Two-Factor Interactions

To estimate main effects plus two-factor interactions in paired comparisons, a construction is given in Lemma 4.2.1. It starts with a resolution 5 (or greater) fraction of the complete 2^k factorial and a set of generators. The set of generators needs to satisfy two conditions:

1. For each attribute, there must be at least one generator with a 1 in the corresponding position (to estimate main effects);

2. For any two attributes there must be at least one generator in which the corresponding positions have a 0 and a 1 (to estimate the two-factor interactions).

These sets of generators are added to the fraction to obtain near-optimal pairs. This method can easily be extended to obtain near-optimal choice sets of size m.

Let $G_j = (\mathbf{g}_{1j}, \mathbf{g}_{2j}, \ldots, \mathbf{g}_{mj})$, where $\mathbf{g}_{1j} = \mathbf{0}$, be binary k-tuples which we will call generators. Let $\mathbf{v}_j = (d_{1j}, d_{2j}, \ldots, d_{m(m-1)/2,j})$ be the difference vector consisting of all the pairwise differences between the generators in G_j. The possible $\mathbf{v}_j$ vectors are those difference vectors determined in Theorem 5.1.2 for an optimal design for the particular values of m and k. Thus the G_j are not unique, and several different near-optimal designs are possible. The construction of the choice sets proceeds as follows:

1. Start with a resolution 5 (or greater) fraction F of the complete 2^k factorial design. Let F have 2^{k-p} treatment combinations.

2. Add the the elements of the set of generators, G_1, to F, where the addition is done component-wise modulo 2, to form 2^{k-p} choice sets of size m.

3. Repeat step 2 to form another 2^{k-p} choice sets of size m, until all main effects and two-factor interactions can be estimated. Thus for each attribute there must be at least one generator with a 1 in the corresponding position (to estimate main effects) and,

for any two attributes, there must be at least one generator in which the corresponding positions have a 0 and a 1 (to estimate the two-factor interactions).

The number of times that step 2 is repeated will depend on the number of attributes and the choice set size. In some cases step 2 is only repeated once; so the number of choice sets required in this instance for a near-optimal design is 2^{k-p+1}.

■ **EXAMPLE 5.2.2.**
Let $m = 3$ and $k = 4$. There are no fractions of the 2^4 factorial which are resolution 5; so we must use the 16 treatment combinations from the complete factorial. These treatment combinations are given in the first column of Table 5.8. In Example 5.1.9, the difference vectors for the optimal designs are $(2,2,2)$ and $(2,3,3)$; so $\mathbf{g}_{2j}$ and $\mathbf{g}_{3j}$ must be chosen so that the difference vector for G_j is either $(2,2,2)$ or $(2,3,3)$. We choose

$$G_1 = (\mathbf{0}, \mathbf{g}_{21}, \mathbf{g}_{31}) = (0000, 1100, 0110) \text{ with } \mathbf{v}_1 = (2,2,2)$$

and form $2^\ell = 16$ triples by adding the elements of G_1 to F. From these we can estimate all the main effects and two-factor interactions except the main effect of the fourth attribute. So we repeat step 2 adding

$$G_2 = (\mathbf{0}, \mathbf{g}_{22}, \mathbf{g}_{32}) = (0000, 1100, 0111), \text{ where } \mathbf{v}_2 = (2,3,3),$$

to F to get an additional 16 triples. If we use G_2 only then the two-factor interaction between attributes 3 and 4 cannot be estimated. With both G_1 and G_2, all main effects and two-factor interactions can now be estimated; so we have no need to generate any more triples. The 32 triples shown in Table 5.8 form a design with

$$\det(C_{MT}) = \left(\frac{1}{18}\right)^8 \left(\frac{1}{36}\right)^2.$$

From Example 5.1.9, the maximum value of the determinant is

$$\det(C_{\text{opt},MT}) = \left(\frac{1}{20}\right)^{10}.$$

Thus the design in Table 5.8 is 96.73% efficient and is therefore near-optimal. This design is much smaller than the optimal design in Example 5.1.9 which consists of 160 triples. □

5.2.3 Dominating Options

If all the binary attributes involved in an experiment have ordered levels, so that 1 is preferred 0, say, by all respondents, then we see that 000…0 will never be chosen in any choice set in which it appears and 111…1 will be chosen in every choice set in which it appears. For the estimation of main effects it is not possible to have choice sets in which, effectively, only a subset of the k attributes are involved and the remaining attributes have the same levels for all options in the choice set. So provided we work with a fractional factorial which does not contain 000…0 or 111…1 and do not add any generator which results in either of these two treatment combinations then dominating alternatives are not an issue in this situation.

If we want to estimate main effects and two-factor interactions optimally, then we will need to consider fractions and sets of generators on a case-by-case basis. For example, for the triples in Table 5.4 we see that only 2 of the 8 triples with difference vector $\mathbf{v}_3$ do not have a dominated or dominating option and that 12 of the 24 triples with difference vector $\mathbf{v}_2$ do not have such an option.

Table 5.8 Near-Optimal Choice Sets for Estimating Main Effects and Two-Factor Interactions for $m = 3$ and $k = 4$.

F	$F + \mathbf{g}_{21}$	$F + \mathbf{g}_{31}$	F	$F + \mathbf{g}_{22}$	$F + \mathbf{g}_{32}$
0000	1100	0110	0000	1100	0111
0001	1101	0111	0001	1101	0110
0010	1110	0100	0010	1110	0101
0011	1111	0101	0011	1111	0100
0100	1000	0010	0100	1000	0011
0101	1001	0011	0101	1001	0010
0110	1010	0000	0110	1010	0001
0111	1011	0001	0111	1011	0000
1000	0100	1110	1000	0100	1111
1001	0101	1111	1001	0101	1110
1010	0110	1100	1010	0110	1101
1011	0111	1101	1011	0111	1100
1100	0000	1010	1100	0000	1011
1101	0001	1011	1101	0001	1010
1110	0010	1000	1110	0010	1001
1111	0011	1001	1111	0011	1000

5.2.4 Exercises

1. Let $k = 6$ and $m = 3$. Using Theorem 5.2.1, find a set of triples that is optimal for estimating main effects.

2. Let $k = 5$ and $m = 5$. Find a choice experiment in which all main effects and two-factor interactions can be estimated.

5.3 REFERENCES AND COMMENTS

Most of the results in this chapter originally appeared in Burgess and Street (2003). Readers can find software to construct choice sets from an initial factorial design and sets of generators, as well as calculate the corresponding information matrix and variance-covariance matrix, at the following website: http://maths.science.uts.edu.au/maths/wiki/SPExpts.

The need for designs that perform well but have a smaller number of choice sets is partly a cognitive issue (see Iyengar and Lepper (2000), Schwartz et al. (2002), and Iyengar et al. (2004)), and partly a cost issue.

CHAPTER 6

DESIGNS FOR ASYMMETRIC ATTRIBUTES

In this chapter we extend the results of the previous two chapters from binary attributes to attributes with any number of levels. There is no requirement that all the attributes in a particular situation have the same number of levels.

We begin by considering an example of a choice experiment with asymmetric attributes from Maddala et al. (2002).

■ **EXAMPLE 6.0.1.**

There are 6 attributes with 3, 4, 5, 3, 5, and 2 levels, respectively, in a choice experiment examining preferences for HIV testing methods. The attributes, together with the attribute levels, are given in Table 6.1, and one choice set from the study is given in Table 6.2. Each respondent was presented with 11 choice sets and, for each of these, was asked to choose one of two options. The respondents were all surveyed at HIV testing locations and so a forced choice experiment was appropriate. □

Thus we are considering the design of experiments in which options are described by k attributes and the qth attribute has $\ell_q \geq 2$ levels. The choice sets may have any number of options, denoted by $m \geq 2$, although all choice sets in a particular experiment have all choice sets of the same size; choice experiments in which choice sets may have different sizes, and the optimal choice set size, are considered in Section 7.2. Once again we are only considering forced choice experiments; choice experiments in which "none" is an option are considered in Section 7.1.1. Throughout this chapter, we assume there are k attributes, that the qth attribute has ℓ_q levels which are $0, 1, \ldots, \ell_q - 1$ for $q = 1, \ldots, k$ and that $L = \prod_{q=1}^{k} \ell_q$ is the number of possible treatment combinations.

Table 6.1 Attributes and Levels for the Study Examining Preferences for HIV Testing Methods

Attribute	Attribute Levels
Location	Doctor's office Public clinic Home
Price	$0 $10 $50 $100
Sample collection	Draw blood Swab mouth Urine sample Prick finger
Timeliness/accuracy	Results in 1–2 weeks; almost always accurate immediate results; almost always accurate immediate results; less accurate
Privacy/anonymity	Only you know; not linked Phones; not linked In person; not linked Phone; linked In person; linked
Counseling	Talk to a counselor Read brochure and then talk to counselor

Table 6.2 One Choice Set from the Study Examining Preferences for HIV Testing Methods

Attribute	Option A	Option B
Location	Doctor's office	Public clinic
Price	$100	$10
Sample collection	Swab mouth	Urine sample
Timeliness/accuracy	Results in 1–2 weeks; almost always accurate	immediate results; less accurate
Privacy/anonymity	In person; not linked	Only you know; not linked
Counseling	Talk to a counselor to	Read brochure and then talk counselor

Suppose that you have already narrowed down your choice of HIV testing methods to the two alternatives above.
Which of these two would you choose? *(tick one only)*
Option A ○ Option B ○

In Chapter 5 we defined a difference vector which did not differentiate between the attributes since all attributes had two levels. For asymmetric attributes, we define a difference vector which records the differences by attributes. We then derive a general expression for the information matrix Λ and use that to determine the optimal designs for estimating main effects only. We do not determine the optimal designs for main effects and two-factor interactions, but we do give an expression for the determinant (to allow comparison of designs), and some tables giving the optimal designs for small values of k, m, and ℓ_q, $q = 1, \ldots, k$.

6.1 DIFFERENCE VECTORS

In Section 5.1.1, we defined a difference vector for choice sets in which all the attributes had two levels. We now define the difference vector for a choice set with options described by k asymmetric attributes with levels $\ell_1, \ell_2, \ldots, \ell_k$. We are interested in the number of attributes with equal levels and the number with different levels in the choice set, as this is linked to how efficiently main effects and interaction effects can be estimated. In a choice set of size m there are $\binom{m}{2}$ pairs of options (or treatment combinations) in the choice set and we record the pairwise differences between the attributes in a difference vector. For example, for treatment combinations 2401 and 1403, the levels of the first and fourth attributes are different while the levels of the second and third attributes are the same. We write this difference as 1001. Thus each entry in the difference vector is a binary k-tuple which indicates whether the levels of the attributes are the same or different. (When all attributes are binary we merely record the sum of the entries in each k-tuple and not the k-tuple itself; see Section 5.1.1.) We assume that all choice sets with a particular difference vector are either all in the experiment or none of them is included.

In general, let

$$\mathbf{v} = (\mathbf{d}_1, \mathbf{d}_2, \ldots, \mathbf{d}_{m(m-1)/2})$$

be a difference vector, where

$$\mathbf{d}_r = i_1 i_2 \ldots i_k \quad \text{for } r = 1, 2, \ldots, \binom{m}{2}$$

and

$$i_q = \begin{cases} 1 & \text{if the levels of attribute } q \text{ are different in the } r\text{th pairwise} \\ & \text{comparison of two treatment combinations in the choice set,} \\ 0 & \text{otherwise.} \end{cases}$$

Since we can write the treatment combinations in the choice set in any order, the order of the comparison of pairs of treatment combinations is not important; so we assume that any difference vector has $\mathbf{d}_1 \leq \mathbf{d}_2 \leq \cdots \leq \mathbf{d}_{m(m-1)/2}$.

■ EXAMPLE 6.1.1.

Suppose $m = 3$ and that there are $k = 2$ attributes with levels $\ell_1 = 2$ and $\ell_2 = 3$. We consider the attribute differences in all the pairwise comparisons of treatment combinations in the choice set (00,10,12). The first and second treatment combinations have the first attribute different and the second attribute the same; so that entry in the difference vector is $\mathbf{d}_1$=10. The first and third treatment combinations have both attributes different; so that entry in the difference vector is $\mathbf{d}_2$=11. When comparing the second and third treatment combinations, only the second attribute is different; so the final entry in the difference vector is $\mathbf{d}_3$=01. Thus the difference vector for choice set (00,10,12) is $\mathbf{v} = (\mathbf{d}_1, \mathbf{d}_2, \mathbf{d}_3)$=(10,11,01) which we write as $\mathbf{v} = (01, 10, 11)$.

$$\underbrace{0\,\overbrace{0,\;1}^{d_1=10,}\;\overbrace{0,\;1}^{d_3=01}\,2}_{d_2=11}$$

Choice sets (00,10,12) and (01,02,11) have difference vectors (10,11,01) and (01,10,11), respectively. These difference vectors are considered to be the same, since we can reorder the options in a choice set, and both are written as (01,10,11). However, the entries 01 and 10 in the difference vector denote which factor is different and are not considered to be the same. In Table 6.3 we list all the $\binom{6}{3} = 20$ choice sets of size 3 (or triples) together with the corresponding difference vector. □

Table 6.3 All Possible Choice Sets when $m = 3$, $k = 2$, $\ell_1 = 2$, and $\ell_2 = 3$

Choice Set	Difference Vector	Choice Set	Difference Vector
(00,01,02)	(01,01,01)	(01,02,10)	(01,11,11)
(00,01,10)	(01,10,11)	(01,02,11)	(01,10,11)
(00,01,11)	(01,10,11)	(01,02,12)	(01,10,11)
(00,01,12)	(01,11,11)	(01,10,11)	(01,10,11)
(00,02,10)	(01,10,11)	(01,10,12)	(01,11,11)
(00,02,11)	(01,11,11)	(01,11,12)	(01,10,11)
(00,02,12)	(01,10,11)	(02,10,11)	(01,11,11)
(00,10,11)	(01,10,11)	(02,10,12)	(01,10,11)
(00,10,12)	(01,10,11)	(02,11,12)	(01,10,11)
(00,11,12)	(01,11,11)	(10,11,12)	(01,01,01)

We now establish the upper bound for the sum of the differences in a difference vector for attribute q.

■ **THEOREM 6.1.1.**

For a particular difference vector $\mathbf{v}$, *for a given* m, *the least upper bound for the sum of the differences for a particular attribute* q *is*

$$S_q = \begin{cases} (m^2-1)/4 & \ell_q = 2,\ m \text{ odd}, \\ m^2/4 & \ell_q = 2,\ m \text{ even}, \\ (m^2 - (\ell_q x^2 + 2xy + y))/2 & 2 < \ell_q < m, \\ m(m-1)/2 & \ell_q \geq m, \end{cases}$$

where positive integers x *and* y *satisfy the equation* $m = \ell_q x + y$ *for* $0 \leq y < \ell_q$.

Proof. The upper bound for the sum of the differences for two-level attributes for m odd and even has been established in Lemma 5.1.1. When $\ell_q \geq m$, there are enough levels of the factor so that the level can change in each treatment in the choice set. There are $\binom{m}{2}$ entries in the difference vector and, in this case, each entry will be 1. Therefore the maximum sum of the differences is $\binom{m}{2}$.

When $2 < \ell_q < m$, we write the treatment combinations in the choice set as the rows of an $m \times k$ array. Suppose that, in column q of this array, p_1 of the entries are 0, p_2 of the entries are 1, and so on until p_{ℓ_q} of the entries are $\ell_q - 1$, where $\sum_{i=1}^{\ell_q} p_i = m$. By looking at the pairwise differences between the entries in column q, the contribution to the sum of the differences S_q is

$$\sum_{i=1}^{\ell_q-1} \sum_{j=i+1}^{\ell_q} p_i p_j = \tfrac{1}{2} \sum_{i=1}^{\ell_q} p_i(m - p_i) = \tfrac{1}{2}m^2 - \tfrac{1}{2} \sum_{i=1}^{\ell_q} p_i^2.$$

We wish to maximize $(m^2 - \sum_{i=1}^{\ell_q} p_i^2)/2$ subject to the constraint $\sum_{i=1}^{\ell_q} p_i = m$, and we do this by minimizing $\sum_{i=1}^{\ell_q} p_i^2$.

Now the global minimum has $p_i = m/\ell_q$. Suppose $m = \ell_q x + y$. Then

$$b_1 \ p_i\text{s are equal to } \lfloor m/\ell_q \rfloor = x$$

and

$$b_2 \ p_i\text{s are equal to } \lfloor m/\ell_q \rfloor + 1,$$

where $b_1 + b_2 = \ell_q$. Thus

$$\ell_q x + y = m = b_1 x + b_2(x + 1).$$

Hence $b_2 = y$ and $b_1 = \ell_q - y$, and the minimum value of $\sum_{i=1}^{\ell_q} p_i^2$ is

$$\sum_{i=1}^{\ell_q} p_i^2 = b_1 x^2 + b_2(x + 1)^2 = \ell_q x^2 + 2xy + y.$$

Hence, for column (or attribute) q, the maximum contribution to S_q is

$$(m^2 - (\ell_q x^2 + 2xy + y))/2 \text{ where } m = \ell_q x + y.$$ □

■ EXAMPLE 6.1.2.

Let $m = 3$ and $k = 2$ with $\ell_1 = 2$ and $\ell_2 = 3$. Considering Table 6.3, we see that the possible difference vectors are $(01, 01, 01)$ with $S_1 = 0+0+0 = 0$ and $S_2 = 1+1+1 = 3$; $(01, 10, 11)$ with $S_1 = 0 + 1 + 1 = 2$ and $S_2 = 1 + 0 + 1 = 2$; and $(01, 11, 11)$ with $S_1 = 0 + 1 + 1 = 2$ and $S_2 = 1 + 1 + 1 = 3$. Using Theorem 6.1.1, the upper bound for the attribute with two levels is $S_1 = (m^2 - 1)/4 = 2$ and for the attribute with three levels $S_2 = m(m - 1)/2 = 3$. Only the difference vector (01,11,11) achieves the upper bound for both attributes. □

■ EXAMPLE 6.1.3.

Let $m = 4$ and $k = 2$ with $\ell_1 = 2$ and $\ell_2 = 3$. All possible choice sets and corresponding difference vectors are shown in Table 6.4. The possible difference vectors are $(01, 01, 01, 10, 11, 11)$ with $S_1 = 0+0+0+1+1+1 = 3$ and $S_2 = 1+1+1+0+1+1 = 5$; $(01, 01, 10, 10, 11, 11)$ with $S_1 = 0+0+1+1+1+1 = 4$ and $S_2 = 1+1+0+0+1+1 = 4$; $(01, 01, 10, 11, 11, 11)$ with $S_1 = 0+0+1+1+1+1 = 4$ and $S_2 = 1+1+0+1+1+1 = 5$. Using Theorem 6.1.1, the upper bound for the attribute with two levels is $S_1 = m^2/4 = 4$. Now $2 < \ell_2 < m$ so we solve $4 = 3x + y$ for x and y. Thus $x = y = 1$ and the upper bound for the attribute with 3 levels is $S_2 = (m^2 - (\ell_q x^2 + 2xy + y))/2 = 5$. Only the final difference vector achieves the upper bound for both attributes. □

Table 6.4 All Possible Choice Sets when $m = 4$, $k = 2$, $\ell_1 = 2$, and $\ell_2 = 3$

Choice Set	Difference Vector	Choice Set	Difference Vector
(00,01,02,10)	(01,01,01,10,11,11)	(01,02,10,11)	(01,01,10,11,11,11)
(00,01,02,11)	(01,01,01,10,11,11)	(01,02,10,12)	(01,01,10,11,11,11)
(00,01,02,12)	(01,01,01,10,11,11)	(01,02,11,12)	(01,01,10,10,11,11)
(00,01,10,11)	(01,01,10,10,11,11)	(01,10,11,12)	(01,01,01,10,11,11)
(00,01,10,12)	(01,01,10,11,11,11)	(02,10,11,12)	(01,01,01,10,11,11)
(00,01,11,12)	(01,01,10,11,11,11)		
(00,02,10,11)	(01,01,10,11,11,11)		
(00,02,10,12)	(01,01,10,10,11,11)		
(00,02,11,12)	(01,01,10,11,11,11)		
(00,10,11,12)	(01,01,01,10,11,11)		

For particular values of m and k, there are usually several difference vectors. We denote these by $\mathbf{v}_j$ and add a subscript j to the previous definitions of the difference vector entries. Thus

$$\mathbf{v}_j = (\mathbf{d}_{1j}, \mathbf{d}_{2j}, \ldots, \mathbf{d}_{m(m-1)/2,j}),$$

where $\mathbf{d}_{uj} = i_1 i_2 \ldots i_k$ and $i_q = 1$ or 0 as before. As we said above, in this chapter we restrict the set of competing designs so that all the choice sets with a particular difference vector appear equally often.

As in Section 5.1.1, we define four scalars that are needed subsequently.

1. $c_{\mathbf{v}_j}$ is the number of choice sets containing the treatment $00\ldots0$ with the difference vector $\mathbf{v}_j$.

2. $x_{\mathbf{v}_j;\mathbf{d}}$ is the number of times the difference $\mathbf{d}$ appears in the difference vector $\mathbf{v}_j$. Then $\sum_{\mathbf{d}} x_{\mathbf{v}_j;\mathbf{d}} = \binom{m}{2}$, since this is the total number of entries in $\mathbf{v}_j$.

3. $i_{\mathbf{v}_j}$ is an indicator variable, where

$$i_{\mathbf{v}_j} = \begin{cases} 0 & \text{if no choice sets have the difference vector } \mathbf{v}_j, \\ 1 & \text{if all the choice sets with the difference vector } \mathbf{v}_j \text{ appear} \\ & \text{in the choice experiment.} \end{cases}$$

   At least one of the $i_{\mathbf{v}_j}$ values must be non-zero, otherwise the experiment contains no choice sets.

4. $a_{\mathbf{v}_j} = i_{\mathbf{v}_j}/N$.

Using these definitions we see that the total number of choice sets, N, is given by

$$N = \frac{L}{m} \sum_j c_{\mathbf{v}_j} i_{\mathbf{v}_j}.$$

■ **EXAMPLE 6.1.4.**

Let $m = 3$, $k = 2$ with $\ell_1 = 2$ and $\ell_2 = 3$. Using all treatment combinations from the complete 2×3 factorial, there are $\binom{6}{3} = 20$ distinct choice sets of size 3, two with difference vector $\mathbf{v}_1 = (01, 01, 01)$, 12 with difference vector $\mathbf{v}_2 = (01, 10, 11)$ and 6 with difference vector $\mathbf{v}_3 = (01, 11, 11)$. Table 6.5 is Table 6.3 reordered to show all the choice sets associated with each of the difference vectors. Consider the choice sets containing the treatment combination 00: (00,01,02) with difference vector $\mathbf{v}_1$; (00,01,10), (00,01,11), (00,02,10), (00,02,12), (00,10,11) and (00,10,12) with difference vector $\mathbf{v}_2$; and (00,01,12), (00,02,11) and (00,11,12) with difference vector $\mathbf{v}_3$. Thus we have

$$c_{\mathbf{v}_1} = 1, \ c_{\mathbf{v}_2} = 6, \text{ and } c_{\mathbf{v}_3} = 3.$$

Now we consider the $x_{\mathbf{v}_j;\mathbf{d}}$ values. For $\mathbf{v}_1$ the only possible value of $\mathbf{d}$ is 01, which appears 3 times; so $x_{\mathbf{v}_1;01} = 3$. The $x_{\mathbf{v}_j;\mathbf{d}}$ values are given below.

$$\begin{array}{lll} x_{v_1;01} = 3 & x_{v_2;01} = 1 & x_{v_3;01} = 1 \\ x_{v_1;10} = 0 & x_{v_2;10} = 1 & x_{v_3;10} = 0 \\ x_{v_1;11} = 0 & x_{v_2;11} = 1 & x_{v_3;11} = 2 \\ \sum_{\mathbf{d}} x_{\mathbf{v}_1;\mathbf{d}} = 3 & \sum_{\mathbf{d}} x_{\mathbf{v}_2;\mathbf{d}} = 3 & \sum_{\mathbf{d}} x_{\mathbf{v}_3;\mathbf{d}} = 3. \end{array}$$

Table 6.5 All Possible Triples when $k = 2$, $\ell_1 = 2$, and $\ell_2 = 3$ Sorted by Difference Vector

$\mathbf{v}_1 = (01, 01, 01)$	$\mathbf{v}_2 = (01, 10, 11)$	$\mathbf{v}_3 = (01, 11, 11)$
(00,01,02)	(00,01,10)	(00,01,12)
(10,11,12)	(00,01,11)	(00,02,11)
	(00,02,10)	(00,11,12)
	(00,02,12)	(01,02,10)
	(00,10,11)	(01,10,12)
	(00,10,12)	(02,10,11)
	(01,02,11)	
	(01,02,12)	
	(01,10,11)	
	(01,11,12)	
	(02,10,12)	
	(02,11,12)	

6.1.1 Exercises

1. Give all the choice sets of size 3 when $k = 2$, $\ell_1 = 2$, and $\ell_2 = 4$ and give the values of $c_{\mathbf{v}_j}$ and $x_{\mathbf{v}_j;\mathbf{d}}$.

2. Let $m = 2$, $k = 3$, $\ell_1 = 2$, $\ell_2 = 3$, and $\ell_3 = 4$. Find the values of S_1, S_2, and S_3. Find all the possible values of $\mathbf{v}$ and of $c_{\mathbf{v}}$.

3. Let $m = 3$, $k = 3$, $\ell_1 = 2$, $\ell_2 = 3$, and $\ell_3 = 4$. Find the values of S_1, S_2, and S_3. Find all the possible values of $\mathbf{v}$ and of $c_{\mathbf{v}}$. (Hint: You do not need to find all $\binom{23}{2}$ triples with 000 to do this. Instead try counting differences.)

6.2 THE DERIVATION OF THE INFORMATION MATRIX Λ

The class of competing designs consists of all designs in which all choice sets with a given difference vector are either all included in the choice experiment or none of them is included.

The matrix of contrasts is unaffected by the choice set size. However, we need to derive the appropriate Λ matrix.

From the results in Equations (3.4) and (3.5), we know that if we assume that

$$\pi_1 = \pi_2 = \cdots = \pi_L = 1$$

(that is, all items are equally attractive, the usual null hypothesis), then

$$\Lambda_{i_1,i_1} = \frac{m-1}{m^2} \sum_{i_2<i_3<\cdots<i_m} \lambda_{i_1,i_2,\ldots,i_m}$$

and

$$\Lambda_{i_1,i_2} = -\frac{1}{m^2} \sum_{i_3<i_4<\cdots<i_m} \lambda_{i_1,i_2,\ldots,i_m},$$

where

$$\lambda_{i_1,i_2,\ldots,i_m} = \frac{n_{i_1,i_2,\ldots,i_m}}{N}$$

and $n_{i_1,i_2,\ldots,i_m}$ indicates whether or not $(T_{i_1}, T_{i_2}, \ldots, T_{i_m})$ is a choice set in the choice experiment. Thus the diagonal entries of Λ are $(m-1)/m^2$ times the proportion of choice sets containing the treatment combination T_{i_1} and the off-diagonal entries are $1/m^2$ times the negative of the proportion of choice sets containing both T_{i_1} and T_{i_2}.

In Section 4.1.1, the general form of the Λ matrix for $m = 2$ was given as a linear combination of the identity matrix of order 2^k and the $D_{k,v}$ matrices. We want to do something similar here. The general form of Λ is a linear combination of the identity matrix and some D matrices. Now, however, the D matrices need to indicate which attributes are different, not just that v attributes are different.

We define $D_{\mathbf{d}}$ to be a $(0,1)$ matrix of order L with rows and columns labeled by the treatment combinations, and with a 1 in position (x, y) if the attribute differences between treatment combinations x and y are represented by the binary k-tuple $\mathbf{d}$.

In general, $D_{\mathbf{d}} = M_{i_1} \otimes M_{i_2} \otimes \cdots \otimes M_{i_k}$, where

$$M_{i_q} = \begin{cases} I_{\ell_q} & i_q = 0, \\ J_{\ell_q} - I_{\ell_q} & i_q = 1 \end{cases}$$

and J_{ℓ_q} is a matrix of 1s of order ℓ_q.

■ EXAMPLE 6.2.1.

Let $m = 3$, $k = 2$, $\ell_1 = 2$, and $\ell_2 = 3$. The treatment combinations are $T_1 = 00, T_2 = 01$, $T_3 = 02$, $T_4 = 10$, $T_5 = 11$, $T_6 = 12$. From Example 6.1.4, we see that the distinct d_u

entries are 01, 10, and 11; so we need D_{01}, D_{10}, and D_{11}.

$$\begin{aligned}
D_{01} &= M_0 \otimes M_1 \\
&= I_{\ell_1} \otimes (J_{\ell_2} - I_{\ell_2}) \\
&= I_2 \otimes (J_3 - I_3) \\
&= \begin{bmatrix} J_3 - I_3 & \mathbf{0} \\ \mathbf{0} & J_3 - I_3 \end{bmatrix} \\
&= \begin{bmatrix} 0 & 1 & 1 & 0 & 0 & 0 \\ 1 & 0 & 1 & 0 & 0 & 0 \\ 1 & 1 & 0 & 0 & 0 & 0 \\ 0 & 0 & 0 & 0 & 1 & 1 \\ 0 & 0 & 0 & 1 & 0 & 1 \\ 0 & 0 & 0 & 1 & 1 & 0 \end{bmatrix}.
\end{aligned}$$

This matrix tells us which pairs of treatment combinations have difference 01. Thus we are considering all the pairs where the first attribute is the same and the second attribute is different. For example, the (2,3) position is 1, which means that the treatment combinations $T_2 = 01$ and $T_3 = 02$ have a difference of 01. Similarly,

$$\begin{aligned}
D_{10} &= M_1 \otimes M_0 \\
&= (J_{\ell_1} - I_{\ell_1}) \otimes I_{\ell_2} \\
&= (J_2 - I_2) \otimes I_3 \\
&= \begin{bmatrix} \mathbf{0} & I_3 \\ I_3 & \mathbf{0} \end{bmatrix} \\
&= \begin{bmatrix} 0 & 0 & 0 & 1 & 0 & 0 \\ 0 & 0 & 0 & 0 & 1 & 0 \\ 0 & 0 & 0 & 0 & 0 & 1 \\ 1 & 0 & 0 & 0 & 0 & 0 \\ 0 & 1 & 0 & 0 & 0 & 0 \\ 0 & 0 & 1 & 0 & 0 & 0 \end{bmatrix}
\end{aligned}$$

and

$$\begin{aligned}
D_{11} &= M_1 \otimes M_1 \\
&= (J_{\ell_1} - I_{\ell_1}) \otimes (J_{\ell_2} - I_{\ell_2}) \\
&= (J_2 - I_2) \otimes (J_3 - I_3) \\
&= \begin{bmatrix} \mathbf{0} & J_3 - I_3 \\ J_3 - I_3 & \mathbf{0} \end{bmatrix} \\
&= \begin{bmatrix} 0 & 0 & 0 & 0 & 1 & 1 \\ 0 & 0 & 0 & 1 & 0 & 1 \\ 0 & 0 & 0 & 1 & 1 & 0 \\ 0 & 1 & 1 & 0 & 0 & 0 \\ 1 & 0 & 1 & 0 & 0 & 0 \\ 1 & 1 & 0 & 0 & 0 & 0 \end{bmatrix}.
\end{aligned}$$

□

In the following lemma we give the general form of the Λ matrix for any choice set size for asymmetric attributes.

■ LEMMA 6.2.1.
Under the usual null hypothesis

$$\Lambda = \frac{m-1}{m^2} z I_L - \frac{1}{m^2} \sum_{\mathbf{d}} y_{\mathbf{d}} D_{\mathbf{d}}$$

where

$$y_{\mathbf{d}} = \frac{2}{m \prod_{q=1}^{k} (\ell_q - 1)^{i_q}} \sum_j c_{\mathbf{v}_j} a_{\mathbf{v}_j} x_{\mathbf{v}_j;\mathbf{d}}$$

and

$$z = \sum_j c_{\mathbf{v}_j} a_{\mathbf{v}_j} = \frac{1}{m-1} \sum_{\mathbf{d}} (\prod_{q=1}^{k} (\ell_q - 1)^{i_q}) y_{\mathbf{d}}.$$

The summations over j and $\mathbf{d}$ are over all possible difference vectors $\mathbf{v}_j$ and all distinct difference vector entries $\mathbf{d}$, respectively.

Proof. The diagonal elements of Λ are

$$\Lambda_{i_1,i_1} = \frac{m-1}{m^2} \sum_{i_2 < i_3 < \cdots < i_m} \lambda_{i_1,i_2,\ldots,i_m}$$

where

$$\begin{aligned} \lambda_{i_1,i_2,\ldots,i_m} &= \frac{n_{i_1,i_2,\ldots,i_m}}{N} \\ &= \begin{cases} \frac{1}{N} & \text{if } (T_{i_1}, T_{i_2}, \ldots, T_{i_m}) \text{ is a choice set} \\ & \text{in the choice experiment,} \\ 0 & \text{otherwise.} \end{cases} \end{aligned}$$

Summing the $\lambda_{i_1,i_2,\ldots,i_m}$ over $i_2, \ldots, i_m$, we get

$$\begin{aligned} \sum_{i_2 < i_3 < \cdots < i_m} \lambda_{i_1,i_2,\ldots,i_m} &= \frac{\text{the number of choice sets containing } T_{i_1}}{N} \\ &= \frac{\sum_j c_{\mathbf{v}_j} i_{\mathbf{v}_j}}{N} \\ &= \sum_j c_{\mathbf{v}_j} a_{\mathbf{v}_j} \\ &= z. \end{aligned}$$

Then the diagonal elements of Λ are

$$\frac{m-1}{m^2} z.$$

Now the off-diagonal elements of Λ are

$$\begin{aligned} \Lambda_{i_1,i_2} &= -\frac{1}{m^2} \sum_{i_3 < i_4 < \cdots < i_m} \lambda_{i_1,i_2,\ldots,i_m} \\ &= -\frac{1}{m^2} \times \frac{\text{number of choice sets in which } T_{i_1} \text{ \& } T_{i_2} \text{ appear together}}{N}. \end{aligned}$$

In order to calculate the number of choice sets in which T_{i_1} and T_{i_2} appear together, we first need to calculate the number of pairs in the experiment with difference $\mathbf{d}$. If choice sets with difference vector $\mathbf{v}_j$ are in the design, then each treatment combination appears in $c_{\mathbf{v}_j}$ choice sets with difference vector $\mathbf{v}_j$ and $x_{\mathbf{v}_j;\mathbf{d}}$ of the differences are equal to $\mathbf{d}$. Altogether there are L treatment combinations, and each treatment combination can appear in any of the m positions in the choice set. Thus the total number of pairs in the choice experiment with an attribute difference of $\mathbf{d}$ is

$$\frac{L}{m}\sum_j c_{\mathbf{v}_j} i_{\mathbf{v}_j} x_{\mathbf{v}_j;\mathbf{d}}. \tag{6.1}$$

Consider any treatment combination $\mathbf{t}$. To construct another treatment combination with difference $\mathbf{d}$ from $\mathbf{t}$, we must include the same level for all attributes corresponding to $i_q = 0$ in $\mathbf{d}$, while if $i_q = 1$, then any of the other $(\ell_q - 1)$ distinct levels can be used. So, in total, there are

$$\prod_{q=1}^{k}(\ell_q - 1)^{i_q}$$

treatment combinations with difference $\mathbf{d}$ from $\mathbf{t}$. If we consider the L treatment combinations, the number of pairs with an attribute difference of $\mathbf{d}$ is

$$\frac{1}{2}L \times (\text{the number of ways difference } \mathbf{d} \text{ can occur}) = \frac{1}{2}L\prod_{q=1}^{k}(\ell_q - 1)^{i_q}.$$

Each of these pairs appears in the choice experiment the same number of times, say $r_{\mathbf{d}}$. Then the total number of pairs in the choice experiment with attribute difference $\mathbf{d}$ can also be expressed as

$$\frac{1}{2}L\prod_{q=1}^{k}(\ell_q - 1)^{i_q} r_{\mathbf{d}}. \tag{6.2}$$

Equating (6.1) and (6.2), we get

$$\frac{1}{2}L\prod_{q=1}^{k}(\ell_q - 1)^{i_q} r_{\mathbf{d}} = \frac{L}{m}\sum_j c_{\mathbf{v}_j} i_{\mathbf{v}_j} x_{\mathbf{v}_j;\mathbf{d}}.$$

So, if the proportion of choice sets in which each pair with attribute difference $\mathbf{d}$ appears is $y_{\mathbf{d}} = r_{\mathbf{d}}/N$, then

$$\frac{1}{2}L\prod_{q=1}^{k}(\ell_q - 1)^{i_q} y_{\mathbf{d}} = \frac{L}{mN}\sum_j c_{\mathbf{v}_j} i_{\mathbf{v}_j} x_{\mathbf{v}_j;\mathbf{d}}.$$

Thus

$$\prod_{q=1}^{k}(\ell_q - 1)^{i_q} y_{\mathbf{d}} = \frac{2}{m}\sum_j c_{\mathbf{v}_j} a_{\mathbf{v}_j} x_{\mathbf{v}_j;\mathbf{d}},$$

from which we see that

$$y_{\mathbf{d}} = \frac{2}{m\prod_{q=1}^{k}(\ell_q - 1)^{i_q}}\sum_j c_{\mathbf{v}_j} a_{\mathbf{v}_j} x_{\mathbf{v}_j;\mathbf{d}}.$$

Then the off-diagonal elements of Λ are

$$-\frac{1}{m^2}\sum_{\mathbf{d}} y_{\mathbf{d}} D_{\mathbf{d}}.$$

Hence we have established that

$$\Lambda = \frac{m-1}{m^2} z I_L - \frac{1}{m^2}\sum_{\mathbf{d}} y_{\mathbf{d}} D_{\mathbf{d}}$$

as required.

Finally we establish that the row and column sums of Λ are equal to 0. Summing over $\mathbf{d}$ we get

$$\begin{aligned}
\frac{1}{m^2}\sum_{\mathbf{d}}\prod_{q=1}^{k}(\ell_q-1)^{i_q} y_{\mathbf{d}} &= \frac{1}{m^2}\sum_{\mathbf{d}}\frac{2}{m}\sum_{j} c_{\mathbf{v}_j} a_{\mathbf{v}_j} x_{\mathbf{v}_j;\mathbf{d}} \\
&= \frac{2}{m^3}\sum_{j} c_{\mathbf{v}_j} a_{\mathbf{v}_j}\sum_{\mathbf{d}} x_{\mathbf{v}_j;\mathbf{d}} \\
&= \frac{2}{m^3}\sum_{j} c_{\mathbf{v}_j} a_{\mathbf{v}_j}\binom{m}{2} \\
&= \frac{m-1}{m^2}\sum_{j} c_{\mathbf{v}_j} a_{\mathbf{v}_j} \\
&= \frac{m-1}{m^2} z,
\end{aligned}$$

as required. □

■ EXAMPLE 6.2.2.

Let $m = 3$, $k = 2$ with $\ell_1 = 2$ and $\ell_2 = 3$. Assume that the choice sets in the choice experiment are the ones with difference vectors $\mathbf{v}_1$ and $\mathbf{v}_3$ only. Then the total number of choice sets is $N = 2 + 6 = 8$ and we have

$$i_{\mathbf{v}_1} = i_{\mathbf{v}_3} = 1 \text{ and } i_{\mathbf{v}_2} = 0$$

and hence

$$a_{\mathbf{v}_1} = a_{\mathbf{v}_3} = \tfrac{1}{N} = \tfrac{1}{8} \text{ and } a_{\mathbf{v}_2} = 0.$$

From Example 6.1.4 we have the $c_{\mathbf{v}_j}$ and $x_{\mathbf{v}_j;\mathbf{d}}$ values; now we need to calculate the $y_{\mathbf{d}}$ values.

For $\mathbf{d} = i_1 i_2 = 01$,

$$\prod_{q=1}^{k}(\ell_q-1)^{i_q} = (\ell_1-1)^{i_1}(\ell_2-1)^{i_2} = (2-1)^0(3-1)^1 = 2.$$

For $\mathbf{d} = i_1 i_2 = 10$,

$$\prod_{q=1}^{k}(\ell_q-1)^{i_q} = (\ell_1-1)^{i_1}(\ell_2-1)^{i_2} = (2-1)^1(3-1)^0 = 1.$$

For $\mathbf{d} = i_1 i_2 = 11$,

$$\prod_{q=1}^{k} (\ell_q - 1)^{i_q} = (\ell_1 - 1)^{i_1} (\ell_2 - 1)^{i_2} = (2-1)^1 (3-1)^1 = 2.$$

Then we have

$$\begin{aligned}
y_{01} &= \frac{2}{3(2-1)^0(3-1)^1} (c_{\mathbf{v}_1} a_{\mathbf{v}_1} x_{\mathbf{v}_1;01} + c_{\mathbf{v}_2} a_{\mathbf{v}_2} x_{\mathbf{v}_2;01} + c_{\mathbf{v}_3} a_{\mathbf{v}_3} x_{\mathbf{v}_3;01}) \\
&= \frac{2}{3 \times 2} (1 \times \frac{1}{8} \times 3 + 6 \times 0 \times 1 + 3 \times \frac{1}{8} \times 1) \\
&= \frac{2}{8}, \\
y_{10} &= \frac{2}{3(2-1)^1(3-1)^0} (c_{\mathbf{v}_1} a_{\mathbf{v}_1} x_{\mathbf{v}_1;10} + c_{\mathbf{v}_2} a_{\mathbf{v}_2} x_{\mathbf{v}_2;10} + c_{\mathbf{v}_3} a_{\mathbf{v}_3} x_{\mathbf{v}_3;10}) \\
&= \frac{2}{3 \times 1} (1 \times \frac{1}{8} \times 0 + 6 \times 0 \times 1 + 3 \times \frac{1}{8} \times 0) \\
&= 0, \\
y_{11} &= \frac{2}{3(2-1)^1(3-1)^1} (c_{\mathbf{v}_1} a_{\mathbf{v}_1} x_{\mathbf{v}_1;11} + c_{\mathbf{v}_2} a_{\mathbf{v}_2} x_{\mathbf{v}_2;11} + c_{\mathbf{v}_3} a_{\mathbf{v}_3} x_{\mathbf{v}_3;11}) \\
&= \frac{2}{3 \times 2} (1 \times \frac{1}{8} \times 0 + 6 \times 0 \times 1 + 3 \times \frac{1}{8} \times 2) \\
&= \frac{2}{8},
\end{aligned}$$

and

$$\begin{aligned}
z &= \sum_j c_{\mathbf{v}_j} a_{\mathbf{v}_j} \\
&= (1 \times \frac{1}{8}) + (6 \times 0) + (3 \times \frac{1}{8}) \\
&= \frac{4}{8}.
\end{aligned}$$

Note that $y_{\mathbf{d}}$ is the proportion of choice sets in which each pair of treatment combinations with attribute difference $\mathbf{d}$ appears in the choice experiment. For example, there are no pairs in the choice experiment with attribute difference $\mathbf{d} = 10$ (that is, $y_{10} = 0$). The pairs with attribute difference $\mathbf{d} = 01$ in the experiment are

00	01
00	02
11	12
01	02
10	12
10	11.

Each of these pairs appears in only two of the 8 choice sets in the experiment (that is, $y_{01} = 2/8$). Similarly, the pairs with attribute difference $\mathbf{d} = 11$ appear in two out of the

8 choice sets in the choice experiment (that is, $y_{11} = 2/8$). For example, the pair 00,12 appears in choice sets (00, 01, 12) and (00, 11, 12).

Now that we have calculated the $y_{\mathbf{d}}$ and z values, we can use the $D_{\mathbf{d}}$ matrices given in Example 6.2.1 to calculate the Λ matrix. In general we have

$$\Lambda = \frac{m-1}{m^2} z I_L - \frac{1}{m^2} \sum_{\mathbf{d}} y_{\mathbf{d}} D_{\mathbf{d}}.$$

Thus when $k = 2$ and $m = 3$ we have

$$\begin{aligned}
\Lambda &= \frac{2 \times 4}{3^2 \times 8} I_6 - \frac{1}{9}(y_{01} D_{01} + y_{10} D_{10} + y_{11} D_{11}) \\
&= \frac{1}{72}\left[8 I_6 - 2D_{01} - 2D_{11}\right] \\
&= \frac{1}{72} \begin{bmatrix} 8 & -2 & -2 & 0 & -2 & -2 \\ -2 & 8 & -2 & -2 & 0 & -2 \\ -2 & -2 & 8 & -2 & -2 & 0 \\ 0 & -2 & -2 & 8 & -2 & -2 \\ -2 & 0 & -2 & -2 & 8 & -2 \\ -2 & -2 & 0 & -2 & -2 & 8 \end{bmatrix}.
\end{aligned}$$

□

6.2.1 Exercises

1. Let $m = 2$, $k = 2$, $\ell_1 = 3$, and $\ell_2 = 4$. Find the Λ matrix for the pairs with difference vector (11).
2. Let $m = 3$, $k = 3$, $\ell_1 = 2$, $\ell_2 = 3$ and $\ell_3 = 4$. Give the Λ matrix for the triples with difference vector (011, 111, 111). Give the Λ matrix for the other possible difference vectors in this case.

6.3 THE MODEL FOR MAIN EFFECTS ONLY

In this section, we evaluate the information matrix for estimating main effects only and use this to determine the optimal choice experiment for asymmetric attributes using choice sets of size m.

We let B_M be the normalized contrast matrix for main effects and we let C_M be the information matrix for estimating main effects only. Thus $C_M = B_M \Lambda B_M'$ is the $\sum_{q=1}^{k}(\ell_q - 1) \times \sum_{q=1}^{k}(\ell_q - 1)$ information matrix.

We let B_{ℓ_q} be a normalized contrast matrix for main effects for a factor with ℓ_q levels. Then a normalized contrast matrix for main effects for a $\ell_1 \times \ell_2 \times \cdots \times \ell_k$ factorial is

$$B_M = \begin{bmatrix} B_{\ell_1} \otimes \frac{1}{\sqrt{\ell_2}} \mathbf{j}'_{\ell_2} \otimes \cdots \otimes \frac{1}{\sqrt{\ell_k}} \mathbf{j}'_{\ell_k} \\ \frac{1}{\sqrt{\ell_1}} \mathbf{j}'_{\ell_1} \otimes B_{\ell_2} \otimes \cdots \otimes \frac{1}{\sqrt{\ell_k}} \mathbf{j}'_{\ell_k} \\ \vdots \\ \frac{1}{\sqrt{\ell_1}} \mathbf{j}'_{\ell_1} \otimes \frac{1}{\sqrt{\ell_2}} \mathbf{j}'_{\ell_2} \otimes \cdots \otimes B_{\ell_k} \end{bmatrix}.$$

■ **EXAMPLE 6.3.1.**

Let $m = 3$, $k = 2$, $\ell_1 = 2$, and $\ell_2 = 3$. Then the contrast matrix for the main effect of the first attribute is

$$B_{\ell_1} = B_2 = \begin{bmatrix} \frac{-1}{\sqrt{2}} & \frac{1}{\sqrt{2}} \end{bmatrix}$$

and for the second attribute is

$$B_{\ell_2} = B_3 = \begin{bmatrix} -\frac{1}{\sqrt{2}} & 0 & \frac{1}{\sqrt{2}} \\ \frac{1}{\sqrt{6}} & \frac{-2}{\sqrt{6}} & \frac{1}{\sqrt{6}} \end{bmatrix}.$$

Then

$$\begin{aligned} B_M &= \begin{bmatrix} B_{\ell_1} \otimes \frac{1}{\sqrt{\ell_2}}\mathbf{j}'_{\ell_2} \\ \frac{1}{\sqrt{\ell_1}}\mathbf{j}'_{\ell_1} \otimes B_{\ell_2} \end{bmatrix} \\ &= \begin{bmatrix} \begin{bmatrix} \frac{-1}{\sqrt{2}} & \frac{1}{\sqrt{2}} \end{bmatrix} \otimes \frac{1}{\sqrt{3}} \begin{bmatrix} 1 & 1 & 1 \end{bmatrix} \\ \frac{1}{\sqrt{2}} \begin{bmatrix} 1 & 1 \end{bmatrix} \otimes \begin{bmatrix} -\frac{1}{\sqrt{2}} & 0 & \frac{1}{\sqrt{2}} \\ \frac{1}{\sqrt{6}} & \frac{-2}{\sqrt{6}} & \frac{1}{\sqrt{6}} \end{bmatrix} \end{bmatrix} \\ &= \\ &= \begin{bmatrix} \frac{-1}{\sqrt{6}} & \frac{-1}{\sqrt{6}} & \frac{-1}{\sqrt{6}} & \frac{1}{\sqrt{6}} & \frac{1}{\sqrt{6}} & \frac{1}{\sqrt{6}} \\ -\frac{1}{2} & 0 & \frac{1}{2} & -\frac{1}{2} & 0 & \frac{1}{2} \\ \frac{1}{2\sqrt{3}} & \frac{-1}{\sqrt{3}} & \frac{1}{2\sqrt{3}} & \frac{1}{2\sqrt{3}} & \frac{-1}{\sqrt{3}} & \frac{1}{2\sqrt{3}} \end{bmatrix}. \end{aligned}$$

□

Although B_2 is uniquely determined, up to a change in the order in which the rows and columns are written down, this is not true if $\ell_q \geq 3$, as the next example shows.

■ **EXAMPLE 6.3.2.**

Let $\ell_q = 4$. Then here are three quite distinct possibilities for B_4:

$$\begin{bmatrix} \frac{-3}{2\sqrt{5}} & \frac{-1}{2\sqrt{5}} & \frac{1}{2\sqrt{5}} & \frac{3}{2\sqrt{5}} \\ \frac{1}{2} & \frac{-1}{2} & \frac{-1}{2} & \frac{1}{2} \\ \frac{-1}{2\sqrt{5}} & \frac{3}{2\sqrt{5}} & \frac{-3}{2\sqrt{5}} & \frac{1}{2\sqrt{5}} \end{bmatrix}; \begin{bmatrix} \frac{-1}{2} & \frac{-1}{2} & \frac{1}{2} & \frac{1}{2} \\ \frac{-1}{\sqrt{2}} & \frac{1}{\sqrt{2}} & 0 & 0 \\ 0 & 0 & \frac{-1}{\sqrt{2}} & \frac{1}{\sqrt{2}} \end{bmatrix};$$

$$\text{and } \frac{1}{2} \begin{bmatrix} -1 & -1 & 1 & 1 \\ -1 & 1 & -1 & 1 \\ 1 & -1 & -1 & 1 \end{bmatrix}.$$

The first matrix consists of the linear, quadratic and cubic contrasts, which are given in Exercise 2.1.4.6. When there is interest in particular contrasts between the levels of an attribute, then contrasts such as those in the second matrix can be used. The contrasts in the third matrix represent the main effects and two-factor interaction of two 2-level pseudo-factors (see Section 2.2.3). □

We will see subsequently that the choice of B can have an effect on the form of the C matrix but not on the efficiency of the design.

If ℓ_q is the product of ℓ_1 and ℓ_2, say, then one easy way to get a contrast matrix B_{ℓ_q} is to take the Kronecker product of the matrices B_{ℓ_1} (with the constant row adjoined) and

B_{ℓ_2} (with the constant row adjoined) and then discard the constant row to get the required matrix. The following example illustrates this construction.

■ EXAMPLE 6.3.3.
Suppose that $\ell_1 = 2$, $\ell_2 = 3$ and $\ell_3 = 6$. Then we can get the matrix B_6 by discarding the first row of the matrix

$$\begin{bmatrix} \frac{1}{\sqrt{2}} & \frac{1}{\sqrt{2}} \\ \frac{-1}{\sqrt{2}} & \frac{1}{\sqrt{2}} \end{bmatrix} \otimes \begin{bmatrix} \frac{1}{\sqrt{3}} & \frac{1}{\sqrt{3}} & \frac{1}{\sqrt{3}} \\ \frac{-1}{\sqrt{2}} & 0 & \frac{1}{\sqrt{2}} \\ \frac{1}{\sqrt{6}} & \frac{-2}{\sqrt{6}} & \frac{1}{\sqrt{6}} \end{bmatrix} = \begin{bmatrix} \frac{1}{\sqrt{6}} & \frac{1}{\sqrt{6}} & \frac{1}{\sqrt{6}} & \frac{1}{\sqrt{6}} & \frac{1}{\sqrt{6}} & \frac{1}{\sqrt{6}} \\ -\frac{1}{2} & 0 & \frac{1}{2} & -\frac{1}{2} & 0 & \frac{1}{2} \\ \frac{1}{2\sqrt{3}} & \frac{-1}{\sqrt{3}} & \frac{1}{2\sqrt{3}} & \frac{1}{2\sqrt{3}} & \frac{-1}{\sqrt{3}} & \frac{1}{2\sqrt{3}} \\ \frac{-1}{\sqrt{6}} & \frac{-1}{\sqrt{6}} & \frac{-1}{\sqrt{6}} & \frac{1}{\sqrt{6}} & \frac{1}{\sqrt{6}} & \frac{1}{\sqrt{6}} \\ \frac{1}{2} & 0 & -\frac{1}{2} & -\frac{1}{2} & 0 & \frac{1}{2} \\ \frac{-1}{2\sqrt{3}} & \frac{1}{\sqrt{3}} & \frac{-1}{2\sqrt{3}} & \frac{1}{2\sqrt{3}} & \frac{-1}{\sqrt{3}} & \frac{1}{2\sqrt{3}} \end{bmatrix}.$$

□

In the following lemma, we derive an expression for $B_M D_{\mathbf{d}} B'_M$ for attribute q; this result is then used to obtain the block diagonal matrices of C_M for each attribute and an expression for $\det(C_M)$.

■ LEMMA 6.3.1.
The $(\ell_q - 1) \times (\ell_q - 1)$ block matrix of $B_M D_{\mathbf{d}} B'_M$ for attribute q is

$$\left[\prod_{j=1, j\neq q}^{k} (\ell_j - 1)^{i_j}\right] (-1)^{i_q} I_{\ell_q - 1}.$$

Proof. Recall that $D_{\mathbf{d}} = M_{i_1} \otimes M_{i_2} \otimes \cdots \otimes M_{i_k}$, where

$$M_{i_q} = \begin{cases} I_{\ell_q}, & i_q = 0, \\ J_{\ell_q} - I_{\ell_q}, & i_q = 1 \end{cases},$$

and J_{ℓ_q} is a matrix of ones of order ℓ_q.

For a particular attribute difference $\mathbf{d}$,

$$B_M D_{\mathbf{d}} B_M' = B_M[M_{i_1} \otimes M_{i_2} \otimes \cdots \otimes M_{i_k}]B_M'$$

$$= \begin{bmatrix} B_{\ell_1} \otimes \frac{1}{\sqrt{\ell_2}}\mathbf{j}_{\ell_2}' \otimes \cdots \otimes \frac{1}{\sqrt{\ell_k}}\mathbf{j}_{\ell_k}' \\ \frac{1}{\sqrt{\ell_1}}\mathbf{j}_{\ell_1}' \otimes B_{\ell_2} \otimes \cdots \otimes \frac{1}{\sqrt{\ell_k}}\mathbf{j}_{\ell_k}' \\ \vdots \\ \frac{1}{\sqrt{\ell_1}}\mathbf{j}_{\ell_1}' \otimes \frac{1}{\sqrt{\ell_2}}\mathbf{j}_{\ell_2}' \otimes \cdots \otimes B_{\ell_k} \end{bmatrix} [M_{i_1} \otimes M_{i_2} \otimes \cdots \otimes M_{i_k}]B_M'$$

$$= \begin{bmatrix} B_{\ell_1}M_{i_1} \otimes \frac{1}{\sqrt{\ell_2}}\mathbf{j}_{\ell_2}'M_{i_2} \otimes \cdots \otimes \frac{1}{\sqrt{\ell_k}}\mathbf{j}_{\ell_k}'M_{i_k} \\ \frac{1}{\sqrt{\ell_1}}\mathbf{j}_{\ell_1}'M_{i_1} \otimes B_{\ell_2}M_{i_2} \otimes \cdots \otimes \frac{1}{\sqrt{\ell_k}}\mathbf{j}_{\ell_k}'M_{i_k} \\ \vdots \\ \frac{1}{\sqrt{\ell_1}}\mathbf{j}_{\ell_1}'M_{i_1} \otimes \frac{1}{\sqrt{\ell_2}}\mathbf{j}_{\ell_2}'M_{i_2} \otimes \cdots \otimes B_{\ell_k}M_{i_k} \end{bmatrix} B_M'$$

$$= \begin{bmatrix} B_{\ell_1}M_{i_1}B_{\ell_1}' \otimes \frac{1}{\sqrt{\ell_2}}\mathbf{j}_{\ell_2}'M_{i_2}\frac{1}{\sqrt{\ell_2}}\mathbf{j}_{\ell_2} \otimes \cdots \otimes \frac{1}{\sqrt{\ell_k}}\mathbf{j}_{\ell_k}'M_{i_k}\frac{1}{\sqrt{\ell_k}}\mathbf{j}_{\ell_k} \\ \frac{1}{\sqrt{\ell_1}}\mathbf{j}_{\ell_1}'M_{i_1}\frac{1}{\sqrt{\ell_1}}\mathbf{j}_{\ell_1} \otimes B_{\ell_2}M_{i_2}B_{\ell_2}' \otimes \cdots \otimes \frac{1}{\sqrt{\ell_k}}\mathbf{j}_{\ell_k}'M_{i_k}\frac{1}{\sqrt{\ell_k}}\mathbf{j}_{\ell_k} \\ \vdots \\ \frac{1}{\sqrt{\ell_1}}\mathbf{j}_{\ell_1}'M_{i_1}\frac{1}{\sqrt{\ell_1}}\mathbf{j}_{\ell_1} \otimes \frac{1}{\sqrt{\ell_2}}\mathbf{j}_{\ell_2}'M_{i_2}\frac{1}{\sqrt{\ell_2}}\mathbf{j}_{\ell_2} \otimes \cdots \otimes B_{\ell_k}M_{i_k}B_{\ell_k}' \end{bmatrix}.$$

If $M_{i_q} = I_{\ell_q}$, then

$$\frac{1}{\sqrt{\ell_q}}\mathbf{j}_{\ell_q}'M_{i_q}\frac{1}{\sqrt{\ell_q}}\mathbf{j}_{\ell_q} = \frac{1}{\ell_q}\mathbf{j}_{\ell_q}'\mathbf{j}_{\ell_q} = \frac{1}{\ell_q} \times \ell_q = 1$$

and

$$B_{\ell_q}M_{i_q}B_{\ell_q}' = B_{\ell_q}I_{\ell_q}B_{\ell_q}' = I_{\ell_q-1}.$$

If $M_{i_q} = J_{\ell_q} - I_{\ell_q}$, then

$$\begin{aligned} \frac{1}{\sqrt{\ell_q}}\mathbf{j}_{\ell_q}'M_{i_q}\frac{1}{\sqrt{\ell_q}}\mathbf{j}_{\ell_q} &= \frac{1}{\ell_q}\mathbf{j}_{\ell_q}'(J_{\ell_q} - I_{\ell_q})\mathbf{j}_{\ell_q} \\ &= \frac{1}{\ell_q}(\mathbf{j}_{\ell_q}'J_{\ell_q}\mathbf{j}_{\ell_q}) - \frac{1}{\ell_q}(\mathbf{j}_{\ell_q}'\mathbf{j}_{\ell_q}) \\ &= \frac{1}{\ell_q}(\ell_q^2 - \ell_q) \\ &= \ell_q - 1 \end{aligned}$$

and

$$\begin{aligned} B_{\ell_q}M_{i_q}B_{\ell_q}' &= B_{\ell_q}(J_{\ell_q} - I_{\ell_q})B_{\ell_q}' \\ &= B_{\ell_q}J_{\ell_q}B_{\ell_q}' - B_{\ell_q}B_{\ell_q}' \\ &= \mathbf{0} - I_{\ell_q-1} \\ &= -I_{\ell_q-1}. \end{aligned}$$

Hence

$$B_{\ell_q} M_{i_q} B'_{\ell_q} = \begin{cases} I_{\ell_q} & \text{if } i_q = 0, \\ -I_{\ell_q} & \text{if } i_q = 1 \end{cases} = (-1)^{i_q} I_{\ell_q}$$

and

$$\frac{1}{\sqrt{\ell_q}} \mathbf{j}'_{\ell_q} M_{i_q} \frac{1}{\sqrt{\ell_q}} \mathbf{j}_{\ell_q} = \begin{cases} 1 & \text{if } i_q = 0, \\ \ell_q - 1 & \text{if } i_q = 1 \end{cases} = (\ell_q - 1)^{i_q}.$$

Now the $(\ell_q - 1) \times (\ell_q - 1)$ block matrix of $B_M D_{\mathbf{d}} B'_M$ for attribute q is

$$\frac{1}{\sqrt{\ell_1}} \mathbf{j}'_{\ell_1} M_{i_1} \frac{1}{\sqrt{\ell_1}} \mathbf{j}_{\ell_1} \otimes \mathbf{j}'_{\ell_2} M_{i_2} \frac{1}{\sqrt{\ell_2}} \mathbf{j}_{\ell_2} \otimes \cdots \otimes B_{\ell_q} M_{i_q} B'_{\ell_q} \otimes \cdots \otimes \frac{1}{\sqrt{\ell_k}} \mathbf{j}'_{\ell_k} M_{i_k} \frac{1}{\sqrt{\ell_k}} \mathbf{j}_{\ell_k},$$

where there is one $B_{\ell_q} M_{i_q} B'_{\ell_q}$ term and $(k-1)$ terms of the form $\frac{1}{\sqrt{\ell_j}} \mathbf{j}'_{\ell_j} M_{i_j} \frac{1}{\sqrt{\ell_j}} \mathbf{j}_{\ell_j}$ for $j = 1, \ldots, k; j \neq q$. Then for attribute q,

$$B_M D_{\mathbf{d}} B'_M = \left[\prod_{j=1, j \neq q}^{k} (\ell_j - 1)^{i_j} \right] (-1)^{i_q} I_{\ell_q - 1},$$

as required. □

Note that the proof did not depend on the specific form of the B matrix; see Exercise 6.3.1.2.

Also observe that the same argument used in the proof can be used to show that C_{hr} is a zero matrix for any choice of B_h and B_r, as long as no contrast appears in both.

Using Lemma 6.3.1, for attribute q we can calculate the block matrix of the information matrix $B_{\ell_q} \Lambda B'_{\ell_q}$ that is relevant to the main effect for that attribute.

■ **LEMMA 6.3.2.**

Under the null hypothesis, the block matrix of the information matrix for the main effect of attribute q is given by

$$B_{\ell_q} \Lambda B'_{\ell_q} = \frac{1}{m^2} \sum_{\mathbf{d}} \left\{ y_{\mathbf{d}} \left[\prod_{j=1, j \neq q}^{k} (\ell_j - 1)^{i_j} \right] \left[(\ell_q - 1)^{i_q} - (-1)^{i_q} \right] \right\} I_{\ell_q - 1},$$

and the determinant of the information matrix C_M for all the attributes is then

$$\det(C_M) = \prod_{q=1}^{k} \left[\frac{1}{m^2} \sum_{\mathbf{d}} \left\{ y_{\mathbf{d}} \left(\prod_{j=1, j \neq q}^{k} (\ell_j - 1)^{i_j} \right) \left[(\ell_q - 1)^{i_q} - (-1)^{i_q} \right] \right\} \right]^{\ell_q - 1}.$$

Proof.

$$B_{\ell_q}\Lambda B'_{\ell_q} = B_{\ell_q}\left[\frac{m-1}{m^2}zI_L - \frac{1}{m^2}\sum_{\mathbf{d}} y_{\mathbf{d}}D_{\mathbf{d}}\right]B'_{\ell_q}$$

$$= \frac{m-1}{m^2}\frac{1}{m-1}\sum_{\mathbf{d}}\left[y_{\mathbf{d}}\left(\prod_{j=1}^{k}(\ell_j-1)^{i_j}\right)\right]B_{\ell_q}B'_{\ell_q} - \frac{1}{m^2}\sum_{\mathbf{d}} y_{\mathbf{d}}B_{\ell_q}D_{\mathbf{d}}B'_{\ell_q}$$

$$= \frac{1}{m^2}\left[\sum_{\mathbf{d}}\left[y_{\mathbf{d}}\left(\prod_{j=1}^{k}(\ell_j-1)^{i_j}\right)\right]I_{\ell_q-1}\right.$$

$$\left. - \sum_{\mathbf{d}}\left[y_{\mathbf{d}}\left(\prod_{j=1,j\neq q}^{k}(\ell_j-1)^{i_j}\right)(-1)^{i_q}\right]I_{\ell_q-1}\right]$$

$$= \frac{1}{m^2}\sum_{\mathbf{d}}\left\{y_{\mathbf{d}}\left[\left(\prod_{j=1}^{k}(\ell_j-1)^{i_j}\right) - (-1)^{i_q}\prod_{j=1,j\neq q}^{k}(\ell_j-1)^{i_j}\right]\right\}I_{\ell_q-1}$$

$$= \frac{1}{m^2}\sum_{\mathbf{d}}\left\{y_{\mathbf{d}}\left[\prod_{j=1,j\neq q}^{k}(\ell_j-1)^{i_j}\right]\left[(\ell_q-1)^{i_q} - (-1)^{i_q}\right]\right\}I_{\ell_q-1}.$$

For attribute q, the determinant of $B_{\ell_q}\Lambda_k B'_{\ell_q}$ is

$$\left[\frac{1}{m^2}\sum_{\mathbf{d}}\left[y_{\mathbf{d}}\left(\prod_{j=1,j\neq q}^{k}(\ell_j-1)^{i_j}\right)\left[(\ell_q-1)^{i_q} - (-1)^{i_q}\right]\right]\right]^{\ell_q-1}.$$

Then the determinant of C_M is

$$\det(C_M) = \prod_{q=1}^{k}\left[\frac{1}{m^2}\sum_{\mathbf{d}}\left\{y_{\mathbf{d}}\left(\prod_{j=1,j\neq q}^{k}(\ell_j-1)^{i_j}\right)\left[(\ell_q-1)^{i_q} - (-1)^{i_q}\right]\right\}\right]^{\ell_q-1}.$$

□

■ EXAMPLE 6.3.4.

Let $m = 3$, $k = 2$, $\ell_1 = 2$, and $\ell_2 = 3$. Assume that the choice sets in the choice experiment are the ones with difference vectors $\mathbf{v}_1$ and $\mathbf{v}_3$ only. First we calculate C_M by evaluating

$$C_M = B_M\Lambda B'_M.$$

We also calculate C_M by using Lemma 6.3.2. The matrices Λ and B_M are given in Examples 6.2.2 and 6.3.1, respectively. Then

$$C_M = B_M\Lambda B'_M = \begin{bmatrix} \frac{1}{9} & \vdots & 0 & 0 \\ \cdots & \cdot & \cdots & \cdots \\ 0 & \vdots & \frac{1}{6} & 0 \\ 0 & \vdots & 0 & \frac{1}{6} \end{bmatrix}.$$

Alternatively, using Lemma 6.3.2 and the $y_{\mathbf{d}}$ from Example 6.2.2, the $(\ell_1-1)\times(\ell_1-1)$ block matrix of C_M for the first attribute (that is, $q = 1$) is

$$
\begin{aligned}
B_2\Lambda B_2' &= \frac{1}{m^2}\sum_{\mathbf{d}}\left\{y_{\mathbf{d}}\left(\prod_{j=1,j\neq q}^{k}(\ell_j-1)^{i_j}\right)\left[(\ell_q-1)^{i_q} \;-\; (-1)^{i_q}\right]\right\}I_{\ell_q-1}\\
&= \frac{1}{3^2}\sum_{\mathbf{d}}\left\{y_{\mathbf{d}}\left(\prod_{j=2}^{2}(\ell_j-1)^{i_j}\right)\left[(\ell_1-1)^{i_1} \;-\; (-1)^{i_1}\right]\right\}I_{\ell_1-1}\\
&= \frac{1}{9}\sum_{\mathbf{d}}\left\{y_{\mathbf{d}}(\ell_2-1)^{i_2}\left[(\ell_1-1)^{i_1} \;-\; (-1)^{i_1}\right]\right\}I_1\\
&= \frac{1}{9}\Big\{y_{01}(3-1)^1[(2-1)^0-(-1)^0]\\
&\qquad + y_{10}(3-1)^0[(2-1)^1-(-1)^1]\\
&\qquad + y_{11}(3-1)^1[(2-1)^1-(-1)^1]\Big\}\\
&= \frac{1}{9}[2y_{10}+4y_{11}]\\
&= \frac{1}{9}[(2\times 0)+(4\times\frac{2}{8})]\\
&= \frac{1}{9}.
\end{aligned}
$$

Similarly, the $(\ell_2-1)\times(\ell_2-1)$ block matrix of C_M for the second attribute (that is, $q=2$) is

$$
\begin{aligned}
B_3\Lambda B_3' &= \frac{1}{3^2}\sum_{\mathbf{d}}\left\{y_{\mathbf{d}}\left(\prod_{j=1,j\neq 2}^{2}(\ell_j-1)^{i_j}\right)\left[(\ell_2-1)^{i_2} \;-\; (-1)^{i_2}\right]\right\}I_{\ell_2-1}\\
&= \frac{1}{9}\sum_{\mathbf{d}}\left\{y_{\mathbf{d}}(\ell_1-1)^{i_1}\left[(\ell_2-1)^{i_2} \;-\; (-1)^{i_2}\right]\right\}I_2\\
&= \frac{1}{9}\Big\{y_{01}(2-1)^0[(3-1)^1-(-1)^1]\\
&\qquad + y_{10}(2-1)^1[(3-1)^0-(-1)^0]\\
&\qquad + y_{11}(2-1)^1[(3-1)^1-(-1)^1]\Big\}I_2\\
&= \frac{1}{9}[3y_{01}+3y_{11}]I_2\\
&= \frac{1}{9}[(3\times\frac{2}{8}))+(3\times\frac{2}{8}))]I_2\\
&= \frac{1}{6}I_2.
\end{aligned}
$$

Then, as before,

$$
C_M = \begin{bmatrix} \frac{1}{9} & \vdots & 0 & 0 \\ \cdots & \cdots & \cdots & \cdots \\ 0 & \vdots & \frac{1}{6} & 0 \\ 0 & \vdots & 0 & \frac{1}{6} \end{bmatrix}. \qquad \square
$$

To find the D-optimal design, we must maximize $\det(C_M)$, subject to the constraint $\prod_{q=1}^{k} \ell_q z/m = 1$.

The following theorem establishes that the D-optimal design, for estimating main effects only, is one which consists of choice sets in which the sum of the differences attains the maximum value given in Theorem 6.1.1.

■ THEOREM 6.3.1.

The D-optimal design, for estimating main effects only, is given by choice sets in which there is at least one $\mathbf{v}_j$ with a non-zero $a_{\mathbf{v}_j}$ (that is, the choice set is non-empty) and, for each $\mathbf{v}_j$ present, for each attribute q,

$$S_q = \begin{cases} (m^2-1)/4, & \ell_q = 2,\ m \text{ odd}, \\ m^2/4, & \ell_q = 2,\ m \text{ even}, \\ (m^2 - (\ell_q x^2 + 2xy + y))/2, & 2 < \ell_q < m, \\ m(m-1)/2, & \ell_q \geq m, \end{cases}$$

where positive integers x and y satisfy the equation $m = \ell_q x + y$ for $0 \leq y < \ell_q$. The maximum value of the determinant of C_M is given by

$$\det(C_{\text{opt},M}) = \prod_{q=1}^{k} \left[\frac{2\ell_q S_q}{m^2 L(\ell_q - 1)} \right]^{\ell_q - 1}.$$

Proof. From Lemma 6.3.2,

$$\det(C_M) = \prod_{q=1}^{k} \left[\frac{1}{m^2} \sum_{\mathbf{d}} \left\{ y_{\mathbf{d}} \left(\prod_{j=1, j\neq q}^{k} (\ell_j - 1)^{i_j} \right) \left[(\ell_q - 1)^{i_q} - (-1)^{i_q} \right] \right\} \right]^{\ell_q - 1}.$$

Substituting

$$y_{\mathbf{d}} = \frac{2}{m \prod_{q=1}^{k} (\ell_q - 1)^{i_q}} \sum_j c_{\mathbf{v}_j} a_{\mathbf{v}_j} x_{\mathbf{v}_j;\mathbf{d}}$$

gives

$$\begin{aligned} \det(C_M) &= \prod_{q=1}^{k} \left[\frac{1}{m^2} \sum_{\mathbf{d}} \left\{ \frac{2 \prod_{j=1, j\neq q}^{k} (\ell_j - 1)^{i_j}}{m \prod_{j=1}^{k} (\ell_j - 1)^{i_j}} \right.\right. \\ &\qquad \times \left.\left. \left[(\ell_q - 1)^{i_q} - (-1)^{i_q} \right] \sum_j c_{\mathbf{v}_j} a_{\mathbf{v}_j} x_{\mathbf{v}_j;\mathbf{d}} \right\} \right]^{\ell_q - 1} \\ &= \prod_{q=1}^{k} \left[\frac{2}{m^3} \sum_{\mathbf{d}} \left[\frac{(\ell_q - 1)^{i_q} - (-1)^{i_q}}{(\ell_q - 1)^{i_q}} \sum_j c_{\mathbf{v}_j} a_{\mathbf{v}_j} x_{\mathbf{v}_j;\mathbf{d}} \right] \right]^{\ell_q - 1} \\ &= \prod_{q=1}^{k} \left[\frac{2}{m^3} \sum_{\mathbf{d}} \left[\left(1 - \left(\frac{1}{1 - \ell_q} \right)^{i_q} \right) \sum_j c_{\mathbf{v}_j} a_{\mathbf{v}_j} x_{\mathbf{v}_j;\mathbf{d}} \right] \right]^{\ell_q - 1}. \end{aligned}$$

Now

$$1 - \left(\frac{1}{1 - \ell_q} \right)^{i_q} = \begin{cases} \frac{\ell_q}{\ell_q - 1} & \text{if } i_q = 1, \\ 0 & \text{if } i_q = 0. \end{cases}$$

So

$$
\begin{aligned}
\det(C_M) &= \prod_{q=1}^{k}\left[\frac{2}{m^3}\sum_{\mathbf{d},i_q=1}\left[\frac{\ell_q}{\ell_q-1}\sum_j c_{\mathbf{v}_j}a_{\mathbf{v}_j}x_{\mathbf{v}_j;\mathbf{d}}\right]\right]^{\ell_q-1}\\
&= \prod_{q=1}^{k}\left[\frac{2\ell_q}{m^3(\ell_q-1)}\sum_j\left[c_{\mathbf{v}_j}a_{\mathbf{v}_j}\sum_{\mathbf{d},i_q=1}x_{\mathbf{v}_j;\mathbf{d}}\right]\right]^{\ell_q-1}.
\end{aligned}
$$

Since

$$\sum_{\mathbf{d},i_q=1} x_{\mathbf{v}_j;\mathbf{d}} = \text{the number of differences for attribute } q,$$

$\det(C_M)$ is maximized when $\sum_{\mathbf{d},i_q=1} x_{\mathbf{v}_j;\mathbf{d}}$ is maximized ; this is achieved when

$$\sum_{\mathbf{d},i_q=1} x_{\mathbf{v}_j;\mathbf{d}} = S_q.$$

Then the maximum $\det(C_M)$ is given by

$$
\begin{aligned}
\det(C_{\text{opt},M}) &= \prod_{q=1}^{k}\left[\frac{2\ell_q}{m^3(\ell_q-1)}\sum_j c_{\mathbf{v}_j}a_{\mathbf{v}_j}S_q\right]^{\ell_q-1}\\
&= \prod_{q=1}^{k}\left[\frac{2\ell_q S_q}{m^3(\ell_q-1)N}\sum_j c_{\mathbf{v}_j}i_{\mathbf{v}_j}\right]^{\ell_q-1}\\
&= \prod_{q=1}^{k}\left[\frac{2\ell_q S_q}{m^3(\ell_q-1)}\frac{m}{L\sum_j c_{\mathbf{v}_j}i_{\mathbf{v}_j}}\sum_j c_{\mathbf{v}_j}i_{\mathbf{v}_j}\right]^{\ell_q-1}\\
&= \prod_{q=1}^{k}\left[\frac{2\ell_q S_q}{m^2L(\ell_q-1)}\right]^{\ell_q-1}
\end{aligned}
$$

since $N = (L\sum_j c_{\mathbf{v}_j}i_{\mathbf{v}_j}/m)$. □

The D-efficiency relative to the optimal design is calculated using the expression

$$\text{Eff}_D = \left(\frac{\det(C_M)}{\det(C_{\text{opt},M})}\right)^{1/p}$$

where $p = \sum_q(\ell_q - 1)$, the number of main effects we estimate.

■ EXAMPLE 6.3.5.

Recall that for $m = 3$ and $k = 2$ with $\ell_1 = 2$ and $\ell_2 = 3$, there are three difference vectors $\mathbf{v}_1 = (01, 01, 01)$, $\mathbf{v}_2 = (01, 10, 11)$ and $\mathbf{v}_3 = (01, 11, 11)$. Theorem 6.3.1 states that the D-optimal design has choice sets in which the entries in the difference vectors sum to

$S_1 = (m^2 - 1)/4 = 2$ for attribute 1, and $S_2 = m(m-1)/2 = 3$ for attribute 2, and

$$\begin{aligned}\det(C_{\text{opt},M}) &= \left[\frac{2\ell_1 S_1}{m^2 L(\ell_1 - 1)}\right]^{\ell_1 - 1}\left[\frac{2\ell_2 S_2}{m^2 L(\ell_1 - 2)}\right]^{\ell_2 - 1}\\ &= \left[\frac{2 \times 2 \times 2}{3^2 \times 2 \times 3 \times (2-1)}\right]^{2-1}\left[\frac{2 \times 3 \times 3}{3^2 \times 2 \times 3 \times (3-1)}\right]^{3-1}\\ &= \left(\frac{4}{27}\right)\left(\frac{1}{6}\right)^2.\end{aligned}$$

Only difference vector $\mathbf{v}_3$ satisfies $S_1 = 2$ and $S_2 = 3$; so the D-optimal design consists of the six triples, each with difference vector $\mathbf{v}_3$. All possible choice experiment designs are given in Table 6.6. In this example the design consisting of choice sets from the difference vector $\mathbf{v}_3$ only has the maximum $\det(C_M)$ of $(4/27)(1/6)^2$. The efficiencies of the other designs are calculated relative to this maximum. Recall that the total number of parameters to be estimated is

$$p = \sum_{q=1}^{k}(\ell_q - 1) = (2-1) + (3-1) = 3.$$

Then, for example, the efficiency of the design consisting of choice sets with difference vector $\mathbf{v}_2$ only is

$$\text{Eff}_D = 100\left[\frac{\det(C_M)}{\det(C_{\text{opt},M})}\right]^{1/3} = 100\left[\frac{(\frac{4}{27})(\frac{1}{9})^2}{(\frac{4}{27})(\frac{1}{6})^2}\right]^{1/3} = 76.31\%.$$

Choice sets with difference vector $\mathbf{v}_1$ only, have an efficiency of 0 because the level of the first attribute does not change across the options in the choice sets and therefore the main effect cannot be estimated. □

6.3.1 Exercises

1. Let $m = 3$, $k = 3$, $\ell_1 = 2$, $\ell_2 = 3$, and $\ell_3 = 4$. Give the maximum possible determinant of the information matrix for estimating main effects for this situation. Find a set of choice sets that realize this maximum. Can you find fewer choice sets that are just as good?

2. Show that Lemma 6.3.1 holds for each of the B matrices given in Example 6.3.2.

6.4 CONSTRUCTING OPTIMAL DESIGNS FOR MAIN EFFECTS ONLY

In Theorem 5.2.1 a construction for small optimal designs for two-level attributes for choice sets of size m is given. A fraction of a factorial design is required as a starting design then the choice sets are formed by adding one or more sets of generators. For an optimal design for testing main effects only, the generators must have a difference vector in which the sum of the differences is the maximum.

Using the same method, we now give a construction for optimal choice sets of size m for the estimation of main effects only for asymmetric attributes with two or more levels, using

Table 6.6 All Possible Choice Experiments for $k = 2$, $\ell_1 = 2$, and $\ell_2 = 3$ when $m = 3$

$\mathbf{v}_j$ in Design	$i_{\mathbf{v}_1}$ $i_{\mathbf{v}_2}$ $i_{\mathbf{v}_3}$	N	S_1 S_2	$\det(C_M)$	Efficiency (%)
$\mathbf{v}_1$	1 0 0	2	0 3	$0(\frac{1}{6})^2$	0
$\mathbf{v}_2$	0 1 0	12	2 2	$(\frac{4}{27})(\frac{1}{9})^2$	76.31
$\mathbf{v}_3$	0 0 1	6	2 3	$(\frac{4}{27})(\frac{1}{6})^2$	100
$\mathbf{v}_1$ & $\mathbf{v}_2$	1 1 0	14	0 3 2 2	$(\frac{8}{63})(\frac{5}{42})^2$	75.90
$\mathbf{v}_1$ & $\mathbf{v}_3$	1 0 1	8	0 3 2 3	$(\frac{1}{9})(\frac{1}{6})^2$	90.86
$\mathbf{v}_2$ & $\mathbf{v}_3$	0 1 1	18	2 2 2 3	$(\frac{4}{27})(\frac{7}{54})^2$	84.57
$\mathbf{v}_1$, $\mathbf{v}_2$ & $\mathbf{v}_3$	1 1 1	20	0 3 2 2 2 3	$(\frac{2}{15})^3$	83.20

the complete factorial as the starting design. In order to do this, we need to consider the magnitude of the differences between the attributes in a choice set rather than just whether the attributes are different, as we did in the previous sections of this chapter.

In a choice set of size m there will be $m(m-1)$ differences in the levels of the attributes between pairs of treatment combinations. Let $\mathbf{e} = (e_1 e_2 \ldots e_k)$ represent the differences between one pair of treatment combinations. The differences are calculated component-wise modulo ℓ_q for attribute q.

EXAMPLE 6.4.1.

Suppose $m = 3$ and $k = 2$ with $\ell_1 = 2$ and $\ell_2 = 3$. For the choice set (00, 10, 12) the $m(m-1) = 6$ differences are

$$\begin{aligned} 00 - 10 &\equiv 10, \\ 10 - 00 &\equiv 10, \\ 00 - 12 &\equiv 11, \\ 12 - 00 &\equiv 12, \\ 10 - 12 &\equiv 01, \\ 12 - 10 &\equiv 02. \end{aligned}$$

For the first attribute, the difference 0 appears twice and 1 appears 4 times; for the second attribute the differences 0, 1, and 2 all appear twice. □

In the following theorem we give a construction for optimal choice sets for estimating main effects only.

THEOREM 6.4.1.

Let F be the complete factorial for k attributes where the qth attribute has ℓ_q levels. Suppose that we choose a set of m generators $G = \{\mathbf{g}_1 = \mathbf{0}, \mathbf{g}_2, \ldots, \mathbf{g}_m\}$ such that $\mathbf{g}_i \neq \mathbf{g}_j$ for $i \neq j$. Suppose that $\mathbf{g}_i = (g_{i1}, g_{i2}, \ldots, g_{ik})$ for $i = 1, \ldots, m$ and suppose that the multiset of differences for attribute q $\{\pm(g_{i_1 q} - g_{i_2 q}) \mid 1 \leq i_1, i_2 \leq m,\ i_1 \neq i_2\}$ contains each non-zero difference modulo ℓ_q equally often. Then the choice sets given by the rows of $F + \mathbf{g}_1, F + \mathbf{g}_2, \ldots, F + \mathbf{g}_m$, for one or more sets of generators G, are optimal for the estimation of main effects only, provided that there are as few zero differences as possible in each choice set.

Proof. Let P_{ℓ_q, e_q} be an $\ell_q \times \ell_q$ (0,1) matrix, where there is a 1 in position (t_1, t_2) if $t_2 - t_1 = e_q$. Thus

$$P_{\ell_q,0} = I_{\ell_q} \text{and} \sum_{e_q=1}^{\ell_q - 1} P_{\ell_q, e_q} = J_{\ell_q} - I_{\ell_q}.$$

Now

$$P_{\ell_1,e_1} \otimes P_{\ell_2,e_2} \otimes \cdots \otimes P_{\ell_k,e_k}$$

indicates those combinations $\mathbf{t}_1$ and $\mathbf{t}_2$ with $\mathbf{t}_2 - \mathbf{t}_1 = (e_1 e_2 \ldots e_k)$, where all differences are calculated component-wise modulo ℓ_q.

We let $\alpha_{\mathbf{e}}$ be the total number of times that $\mathbf{e} = (e_1 e_2 \ldots e_k)$ appears as a difference between elements in G. Thus we know that

$$\sum_{e_1} \cdots \sum_{e_{i-1}} \sum_{e_{i+1}} \cdots \sum_{e_k} \alpha_{\mathbf{e}} = \alpha_i,$$

independent of e_i for each $e_i \neq 0$. We also know that

$$\Lambda = \frac{1}{m^2N}\left[\left(\sum_{e_1}\cdots\sum_{e_k}\alpha_{\mathbf{e}}\right)(P_{\ell_1,0}\otimes P_{\ell_2,0}\otimes\cdots\otimes P_{\ell_k,0}) - \sum_{e_1}\cdots\sum_{e_k}\alpha_{\mathbf{e}}(P_{\ell_1,e_1}\otimes P_{\ell_2,e_2}\otimes\cdots\otimes P_{\ell_k,e_k})\right]$$

where N is the number of choice sets in the experiment.

Consider

$$\begin{aligned}\sum_{e_1}\cdots\sum_{e_k}\alpha_{\mathbf{e}} &= (\ell_i-1)\alpha_i + \sum_{e_1}\cdots\sum_{e_{i-1}}\sum_{e_i+1}\cdots\sum_{e_k}\alpha_{e_1\ldots e_{i-1}0e_{i+1}\ldots e_k}\\ &= (\ell_i-1)\alpha_i+\alpha_{i0}\\ &= \alpha\end{aligned}$$

for $i = 1,\ldots,k$. Note that $(\ell_1-1)\alpha_{10} = (\ell_2-1)\alpha_{20} = \cdots = (\ell_k-1)\alpha_{k0} = \alpha$.

Using the contrast matrix B_M from Section 6.3, it can easily be shown that the information matrix, $C = B_M\Lambda B_M'$, is a block diagonal matrix where the qth block diagonal for attribute q is given by $\ell_q\alpha_q I_{\ell_q-1}/(m^2N)$. Now we wish to have as few zero differences as possible. Recall that $m = \ell_q x + y$ where $0 \leq y < \ell_q$. Then we need to have y entries which are repeated $x+1$ times each and $\ell_q - y$ entries that are repeated x times each. This gives $(x+1)xy + x(x-1)(\ell_q-y)$ differences that are zero; so the number of non-zero differences is $m(m-1)-(x+1)xy-x(x-1)(\ell_q-y) = 2S_q$ where S_q is defined in Lemma 5.1.1. So considering all the choice sets, the total number of differences for attribute q is $2S_qN$. We also know that each non-zero level of attribute q appears α_q times as a difference between elements of the sets of generators. Hence the total number of non-zero differences for attribute q is $(\ell_q-1)\alpha_q L$.

Equating gives $2S_qN = (\ell_q-1)\alpha_q L$. Thus the coefficient of the qth block diagonal matrix for the designs constructed in this way is

$$\frac{\ell_q\alpha_q}{m^2N} = \frac{2S_q}{(\ell_q-1)(\prod_{i\neq q}\ell_i)m^2};$$

hence the designs are optimal by Theorem 5.1.1. □

While the optimal designs have equal replication of all of the possible Kronecker products of the P_{ℓ_q}, designs in which the Λ can be written as a linear combination of the Kronecker products of the P_{ℓ_q} still give rise to information matrices in which the main effects are independent of each other, as the next result shows.

■ COROLLARY 6.4.1.

$B_M(P_{\ell_1,e_1}\otimes P_{\ell_2,e_2}\otimes\cdots\otimes P_{\ell_k,e_k})B_M'$ is a block diagonal matrix.

Proof. Using the method of Lemma 6.3.1, we have

$$B_M(P_{\ell_1,e_1}\otimes P_{\ell_2,e_2}\otimes\cdots\otimes P_{\ell_k,e_k})B_M'$$

$$= \begin{bmatrix} B_{\ell_1}P_{\ell_1,e_1}B_{\ell_1}'\otimes\frac{1}{\sqrt{\ell_2}}\mathbf{j}_{\ell_2}'P_{\ell_2,e_2}\frac{1}{\sqrt{\ell_2}}\mathbf{j}_{\ell_2}\otimes\cdots\otimes\frac{1}{\sqrt{\ell_k}}\mathbf{j}_{\ell_k}'P_{\ell_k,e_k}\frac{1}{\sqrt{\ell_k}}\mathbf{j}_{\ell_k}\\ \frac{1}{\sqrt{\ell_1}}\mathbf{j}_{\ell_1}'P_{\ell_1,e_1}\frac{1}{\sqrt{\ell_1}}\mathbf{j}_{\ell_1}\otimes B_{\ell_2}P_{\ell_2,e_2}B_{\ell_2}'\otimes\cdots\otimes\frac{1}{\sqrt{\ell_k}}\mathbf{j}_{\ell_k}'P_{\ell_k,e_k}\frac{1}{\sqrt{\ell_k}}\mathbf{j}_{\ell_k}\\ \vdots\\ \frac{1}{\sqrt{\ell_1}}\mathbf{j}_{\ell_1}'P_{\ell_1,e_1}\frac{1}{\sqrt{\ell_1}}\mathbf{j}_{\ell_1}\otimes\frac{1}{\sqrt{\ell_2}}\mathbf{j}_{\ell_2}'P_{\ell_2,e_2}\frac{1}{\sqrt{\ell_2}}\mathbf{j}_{\ell_2}\otimes\cdots\otimes B_{\ell_k}P_{\ell_k,e_k}B_{\ell_k}'\end{bmatrix}.$$

Now $\mathbf{j}'_{\ell_q} P_{\ell_q,e_q} = \mathbf{j}'_{\ell_q}$ so $\frac{1}{\sqrt{\ell_q}}\mathbf{j}'_{\ell_q} P_{\ell_q,e_q} \frac{1}{\sqrt{\ell_q}}\mathbf{j}_{\ell_q} = 1$ and $B_{\ell_q} P_{\ell_q,e_q}$ gives a matrix with the columns of B_{ℓ_q} permuted. So we get the required result. □

How do we go about using the construction in Theorem 6.4.1? There are probably many ways of doing this but here is one technique which works. We begin by calculating the values of x and y (where $m = \ell_q x + y$) so we know that we have y values (between 0 and $\ell_q - 1$) that are repeated $x + 1$ times each and $(\ell_q - y)$ values which are repeated x times each. We then partition the values between 0 and $\ell_q - 1$ into two disjoint sets, one containing y entries and the other containing the remaining $(\ell_q - y)$ entries. There are $\binom{\ell_q}{y}$ ways to do this. For each partition we calculate the differences that arise from a vector with m entries in which the entries in the set with y entries are each repeated $x + 1$ times and the entries in the other set are each repeated x times. All such vectors have as few 0 differences as possible in the $m(m - 1)$ differences. Next we partition the vectors into sets based on the number of times each non-zero entry modulo ℓ_q appears as a difference. We then choose how many vectors to take from each partition so that we have all non-zero differences appearing equally often over the set of vectors chosen. If there are several attributes, perhaps with different numbers of levels, then we must choose the same number of vectors for each attribute. Once we have the vectors for each attribute then we can calculate the entries for the sets of generators by choosing one entry from each vector for each generator in such a way that no generator is repeated. A couple of examples should make this clearer.

■ EXAMPLE 6.4.2.

Suppose that $k = 2$, $\ell_1 = 2$ and $\ell_2 = 3$, and that $m = 3$.

Since $3 = 2 \times 1 + 1$, $x = y = 1$ and we have two partitions for the first attribute, where the entries in the first set are repeated twice and those in the second set are repeated once. Hence the vectors for the first attribute are (0,0,1) and (0,1,1). Each of these vectors has the difference 0 appearing twice and the difference 1 appearing 4 times. Based on the differences, there is only one partition of the vectors.

Since $3 = 3 \times 1$, $y = 0$ and we have one partition, {0,1,2}, and hence one vector (0,1,2) for the second attribute. Each non-zero difference appears three times in this vector.

To get equal replication of the non-zero differences, we need only choose one vector for each attribute. Hence we calculate our generators by choosing the first position for each of the three generators from (0,0,1) and the second position from (0,1,2). We have always assumed, without loss of generality, that the first generator is $\mathbf{g}_1 = \mathbf{0} = 00$. So the other two generators are 01 and 12, or 02 and 11. Thus the two possible sets of generators, G, are (00,01,12) and (00,02,11). The choice sets that arise from these sets of generators are given in Table 6.7. Notice that since the order of the options within each choice set and of the choice sets within the experiment is immaterial; in this case both sets of generators give rise to the same choice experiment. □

■ EXAMPLE 6.4.3.

Suppose that $k = 2$, $\ell_1 = \ell_2 = 4$ and $m = 6$.

Since $6 = 4 \times 1 + 2$, $y > 0$ and there are $\binom{4}{2} = 6$ partitions to consider where the entries in the first set are repeated twice and those in the second set are repeated once. Thus we get the following 6 vectors to consider: (0,0,1,1,2,3), (0,0,1,2,2,3), (0,0,1,2,3,3), (0,1,1,2,2,3), (0,1,1,2,3,3), (0,1,2,2,3,3). We can partition these vectors into two sets based on the differences between the elements of the vectors in each set. If we let

$$A = \{(0,0,1,1,2,3), (0,0,1,2,3,3), (0,1,1,2,2,3), (0,1,2,2,3,3)\}$$

Table 6.7 Choice Sets when $k = 2$, $\ell_1 = 2$, $\ell_2 = 3$, and $m = 3$

Option A	Option B	Option C	Option A	Option B	Option C
0 0	0 1	1 2	0 0	0 2	1 1
0 1	0 2	1 0	0 1	0 0	1 2
0 2	0 0	1 1	0 2	0 1	1 0
1 0	1 1	0 2	1 0	1 2	0 1
1 1	1 2	0 0	1 1	1 0	0 2
1 2	1 0	0 1	1 2	1 1	0 0
	Design 1			Design 2	

and

$$B = \{(0,0,1,2,2,3),(0,1,1,2,3,3)\},$$

then the differences from any vector in A are 0, which appears four times (as it should by construction), 1 and 3, which appear nine times each, and 2, which appears eight times. The differences from any vector in B are 0, which appears four times, 1 and 3, which appear eight times each, and 2, which appears ten times.

We want to choose a set of partitions such that each non-zero difference appears equally often. Suppose that we have x_1 vectors from A and x_2 vectors from B. Then the number of times 1 appears as a difference equals the number of times that 3 appears and we need only equate this to the number of times that 2 appears. Thus $9x_1 + 8x_2 = 8x_1 + 10x_2$. Solving we have $x_1 = 2x_2$ and so we let $x_1 = 2$ and $x_2 = 1$. So for each attribute we choose two vectors from A (possibly the same vector twice) and one vector from B.

Thus we could use (0,0,1,1,2,3) and (0,0,1,2,3,3) from A and (0,0,1,2,2,3) from B as the entries in the generators for the first attribute and (0,1,2,2,3,3) and (0,0,1,1,2,3) from A and (0,0,1,2,2,3) from B for the second attribute.

Now we need to pair up the entries from these sets to get the actual sets of generators and we must do so without getting any repeated generator ($\mathbf{g}_i$) within a set of generators (G), or any repeated set of generators. Recall that we prefer to have 00 as one generator in each set of generators. So we could pair the first set from A for the first attribute with the set from B for the second attribute to get the set of generators G_1=(00,01,10,12,22,33), pair the second set from A for the first attribute with the first set from A for the second attribute to get the set of generators G_2=(00,01,12,23,32,33), and finally pair the remaining sets to get the set of generators G_3=(00,01,10,21,22,33). These three sets of generators give 48 choice sets of size 6 which are 100% efficient for estimating main effects. Pairing in a different way gives another optimal design with three sets of generators;

$$G_1 = (00,02,10,11,23,32),$$

$$G_2 = (00,03,30,11,21,22),$$

$$G_3 = (00,01,13,22,32,33),$$

for example.

Suppose instead that our initial vectors for each of the two attributes has the vectors (0,0,1,1,2,3) and (0,0,1,2,3,3) from A, the vector (0,0,1,2,2,3) from B for the first attribute, and vectors (0,1,1,2,2,3) and (0,1,2,2,3,3) from A and (0,1,1,2,3,3) from B for the second attribute.

Pairing, we can get the sets of generators

$$G_1 = (00, 01, 11, 12, 22, 33),$$

$$G_2 = (00, 01, 11, 22, 23, 33),$$

$$G_3 = (00, 01, 12, 23, 32, 33),$$

which results in an optimal design. However, there are only 40 distinct choice sets and if these alone are used, then the design is 99.99% efficient. It is usual to remove repeated choice sets because their presence can confuse the respondents. The repeated choice sets arise because there are three pairs with difference 22 (00 and 22; 01 and 23; 11 and 33). Hence the differences 1 and 3 only appear four times and the difference 2 only appears five times from this pairing. Since 1 and 3 appear eight times and 2 appears nine times as differences from any vector in A this suggests that using either G_1 or G_3, together with G_2, will give 24 sets which are 100% efficient (since each non-zero difference will be represented 13 times over the two sets); this is indeed the case.

It is also worth noting that we can use different contrast matrices to calculate the C, and therefore C^{-1}, matrices. Although the structure of the C^{-1} matrices may be different, the determinants, and hence the efficiencies, will be equal. If we use as B_4 the contrast matrix based on the orthogonal polynomial contrasts for the 40 distinct choice sets discussed in the previous paragraph, we have

$$B_4 = \begin{bmatrix} \frac{-3}{2\sqrt{5}} & \frac{-1}{2\sqrt{5}} & \frac{1}{2\sqrt{5}} & \frac{3}{2\sqrt{5}} \\ \frac{1}{2} & \frac{-1}{2} & \frac{-1}{2} & \frac{1}{2} \\ \frac{-1}{2\sqrt{5}} & \frac{3}{2\sqrt{5}} & \frac{-3}{2\sqrt{5}} & \frac{1}{2\sqrt{5}} \end{bmatrix}$$

and

$$C_M^{-1} = \left[\begin{array}{ccc:ccc} \frac{7884}{473} & 0 & \frac{-72}{473} & 0 & 0 & 0 \\ 0 & \frac{720}{473} & 0 & 0 & 0 & 0 \\ \frac{-72}{473} & 0 & \frac{7776}{473} & 0 & 0 & 0 \\ \hdashline 0 & 0 & 0 & \frac{7884}{473} & 0 & \frac{-72}{473} \\ 0 & 0 & 0 & 0 & \frac{720}{473} & 0 \\ 0 & 0 & 0 & \frac{-72}{473} & 0 & \frac{7776}{473} \end{array}\right].$$

If, instead, we use the 2^2 factorial contrast for B_4 (see Example 6.3.2), then

$$B_4 = \frac{1}{2}\begin{bmatrix} -1 & -1 & 1 & 1 \\ -1 & 1 & -1 & 1 \\ 1 & -1 & -1 & 1 \end{bmatrix}$$

and

$$C_M^{-1} = \left[\begin{array}{ccc:ccc} \frac{720}{43} & 0 & 0 & 0 & 0 & 0 \\ 0 & \frac{180}{11} & 0 & 0 & 0 & 0 \\ 0 & 0 & \frac{720}{43} & 0 & 0 & 0 \\ \hdashline 0 & 0 & 0 & \frac{720}{43} & 0 & 0 \\ 0 & 0 & 0 & 0 & \frac{180}{11} & 0 \\ 0 & 0 & 0 & 0 & 0 & \frac{720}{43} \end{array}\right].$$

In both cases, the determinant of the C matrix is 4.75099×10^{-8} and the efficiency is 99.99%. □

■ EXAMPLE 6.4.4.
Let $k = 3$, $\ell_1 = 2$, $\ell_2 = 3$, and $\ell_3 = 6$.

Since we want each set of generators to contain 000 we only consider vectors which contain 0 for each attribute.

Suppose $m = 2$. For the first attribute we have one vector (0,1) with difference 1 twice. For the second attribute we have two vectors, (0,1) and (0,2), both giving each non-zero difference once. For the third attribute there are five vectors which form three sets in a partition based on differences: (0,1) and (0,5) with non-zero differences 1 and 5 once each; (0,2) and (0,4) with non-zero differences 2 and 4 once each; and (0,3) with non-zero difference 3 twice. So for the third attribute we need to have two vectors from each of the first two sets of the partition and one from the third set to ensure that all non-zero differences mod 6 appear equally often. Consequently we must have 5 sets of generators altogether. For the second attribute we can either repeat the same vector 5 times or have both vectors represented. One possible solution is (0,1), (0,1), (0,1), (0,1), and (0,1) for attribute 1; (0,1), (0,2), (0,1), (0,2), and (0,1) for attribute 2; and (0,1), (0,2), (0,3), (0,4), and (0,5) for attribute 3. Using these sets in order gives the five sets of generators

$$G_1 = (000, 111), \; G_2 = (000, 122), \; G_3 = (000, 113),$$

$$G_4 = (000, 124), \; G_5 = (000, 115).$$

There are 180 pairs and the design is 100% efficient.

Suppose $m = 3$. There is one partition of the vectors for the first attribute with entries (0,0,1) and (0,1,1). There is one vector for the second attribute: (0,1,2). There are $\binom{5}{2} = 10$ vectors for the third attribute (since we must include 0) and these are partitioned into three sets. The first set consists of (0,1,2), (0,1,5) and (0,4,5) with non-zero differences 1 and 5 twice each and 2 and 4 once each. The second set has (0,2,4) with non-zero differences 2 and 4 thrice each. The third set has the remaining 6 vectors and each of these has non-zero differences 1, 2, 4, and 5 once each and 3 twice. Suppose that we have x_1 vectors from the first partition for attribute 3, x_2 from the second and x_3 from the third. Because we want to have each non-zero difference appearing equally often, we get $2x_1 + x_3 = x_1 + 3x_2 + x_3 = 2x_3$. Suppose $x_2 = 0$. Then $x_1 = x_3 = 0$ which is a contradiction. Instead, try $x_2 = 1$. Then $2x_1 + x_3 = x_1 + x_3 + 3$; so $x_1 = 3$, and hence $x_3 = 6$. Thus for an optimal design we need to have 10 sets of generators and 360 choice sets. We could use (0,1,1) in all 10 sets of generators for attribute 1, (0,1,2) in all 10 sets of generators for attribute 2 and (0,1,2), (0,1,3), (0,1,4), (0,1,5), (0,2,3), (0,2,4), (0,2,5), (0,3,4), (0,3,5), and (0,4,5) for attribute 3. This gives the 10 sets of generators

$$G_1 = (000, 111, 122), \; G_2 = (000, 111, 123), \; G_3 = (000, 111, 124),$$

$$G_4 = (000, 111, 125), \; G_5 = (000, 112, 123), \; G_6 = (000, 112, 124),$$

$$G_7 = (000, 112, 125), \; G_8 = (000, 113, 124), \; G_9 = (000, 113, 125),$$

$$G_{10} = (000, 114, 125).$$

Although there is nothing in the results that we have presented here that would allow you to calculate this, it is true that using only G_1 gives a design with 36 choice sets which is 97.80% efficient and using only G_2 gives 36 choice sets which are 99.39% efficient.

In both cases the C_M and C_M^{-1} matrices are block diagonal. This improvement is not surprising; G_2 has a more equal representation of the non-zero differences for the third attribute than does G_2. On the other hand, using only G_6 with elements (0,2,4) for the third attribute results in $\det(C_M) = 0$, which is not surprising since only two of the differences are represented.

For $m = 4$, a similar argument shows that 5 sets of generators are required. For $m = 5$, the third attribute only requires one set of generators, and 36 choice sets are 100% efficient in this case. □

We have seen in the previous example that often it is only necessary to use one set of generators to get a few choice sets that are near-optimal. In some cases, one set of generators will give rise to an optimal design if there is a difference set for the appropriate values of ℓ_q and m. For example, there is an optimal set of 56 choice sets when $k = 4$, $\ell_1 = 2$, $\ell_2 = 4$, $\ell_3 = 7$, and $m = 4$ obtained by using the set of generators G=(000,011,122,134). This is because the differences from the set $\{0,1,2,4\}$ contain each non-zero difference modulo 7 exactly once. Other difference sets given in Section 2.4 can be used to construct small sets of optimal generators.

6.4.1 Exercises

1. Let $m = 3$, $k = 3$, $\ell_1 = 2$, $\ell_2 = 3$, and $\ell_3 = 4$. Use the ideas in this section to find some small but good designs for estimating main effects.

2. Green (1974) gave a construction for choice experiments designed to estimate main effects. His idea involved using the runs of an orthogonal array to correspond to the treatments of an incomplete block design and to let each block of the BIBD correspond to one choice set of size m in the choice experiment.

   The properties of these choice experiments are not readily determined in general, as they will depend on both the BIBD and the OA used.

   Suppose that we have $k = 7$ binary attributes. Use the OA[8,7,2,2] and a (8,14,7,4,3) BIBD and Green's construction to get the 14 choice sets. Show that these choice sets have $C_M = \frac{3}{448} I_7$ and are 85.7143% efficient for the estimation of main effects.

3. Use Green's construction, described in Exercise 6.4.1.2, with $k = 5$ attributes and $\ell_q = 4$ levels, $q = 1, 2, \ldots, 5$ to construct a choice experiment with 16 choice sets of size 4. What is the efficiency of the resulting design for the estimation of main effects?

6.5 THE MODEL FOR MAIN EFFECTS AND TWO-FACTOR INTERACTIONS

To find the D-optimal designs for estimating main effects and two-factor interactions for k attributes with any number of levels and for any choice set size m, we now evaluate the $p \times p$ information matrix $C_{MT} = B_{MT} \Lambda B'_{MT}$, where

$$p = \sum_{q=1}^{k} (\ell_q - 1) + \sum_{q_1=1}^{k-1} \sum_{q_2=q_1+1}^{k} (\ell_{q_1} - 1)(\ell_{q_2} - 1).$$

As in Section 6.3, we let B_M be the normalized rows of B that correspond to main effects, and we now let B_T be the normalized rows of B that correspond to the two-factor interactions. The contrast matrix associated with main effects and two-factor interactions is denoted by B_{MT} and is the concatenation of B_M and B_T

$$B_{MT} = \begin{bmatrix} B_M \\ B_T \end{bmatrix},$$

where

$$B_M = \begin{bmatrix} B_{\ell_1} \otimes \frac{1}{\sqrt{\ell_2}}\mathbf{j}'_{\ell_2} \otimes \cdots \otimes \frac{1}{\sqrt{\ell_k}}\mathbf{j}'_{\ell_k} \\ \frac{1}{\sqrt{\ell_1}}\mathbf{j}'_{\ell_1} \otimes B_{\ell_2} \otimes \cdots \otimes \frac{1}{\sqrt{\ell_k}}\mathbf{j}'_{\ell_k} \\ \vdots \\ \frac{1}{\sqrt{\ell_1}}\mathbf{j}'_{\ell_1} \otimes \frac{1}{\sqrt{\ell_2}}\mathbf{j}'_{\ell_2} \otimes \cdots \otimes B_{\ell_k} \end{bmatrix},$$

$$B_T = \begin{bmatrix} B_{\ell_1} \otimes B_{\ell_2} \otimes \frac{1}{\sqrt{\ell_3}}\mathbf{j}'_{\ell_3} \otimes \cdots \otimes \frac{1}{\sqrt{\ell_k}}\mathbf{j}'_{\ell_k} \\ B_{\ell_1} \otimes \frac{1}{\sqrt{\ell_2}}\mathbf{j}'_{\ell_2} \otimes B_{\ell_3} \otimes \cdots \otimes \frac{1}{\sqrt{\ell_k}}\mathbf{j}'_{\ell_k} \\ \vdots \\ \frac{1}{\sqrt{\ell_1}}\mathbf{j}'_{\ell_1} \otimes \cdots \otimes \frac{1}{\sqrt{\ell_{k-2}}}\mathbf{j}'_{\ell_{k-2}} \otimes B_{\ell_{k-1}} \otimes B_{\ell_k} \end{bmatrix}.$$

The following example illustrates the construction of the contrast matrix for main effects and two-factor interactions.

■ EXAMPLE 6.5.1.
Let $k = 2$ and $\ell_1 = 2$, $\ell_2 = 3$. From Example 6.3.1, the contrast matrix for main effects is given by

$$B_M = \begin{bmatrix} \frac{-1}{\sqrt{6}} & \frac{-1}{\sqrt{6}} & \frac{-1}{\sqrt{6}} & \frac{1}{\sqrt{6}} & \frac{1}{\sqrt{6}} & \frac{1}{\sqrt{6}} \\ -\frac{1}{2} & 0 & \frac{1}{2} & -\frac{1}{2} & 0 & \frac{1}{2} \\ \frac{1}{2\sqrt{3}} & \frac{-1}{\sqrt{3}} & \frac{1}{2\sqrt{3}} & \frac{1}{2\sqrt{3}} & \frac{-1}{\sqrt{3}} & \frac{1}{2\sqrt{3}} \end{bmatrix}.$$

The contrast matrix for the two-factor interactions using B_2 and B_3 from Example 6.3.1 is

$$\begin{aligned} B_T &= \begin{bmatrix} B_{\ell_1} \otimes B_{\ell_2} \end{bmatrix} \\ &= \begin{bmatrix} B_2 \otimes B_3 \end{bmatrix} \\ &= \begin{bmatrix} \begin{bmatrix} \frac{-1}{\sqrt{2}} & \frac{1}{\sqrt{2}} \end{bmatrix} \otimes \begin{bmatrix} -\frac{1}{\sqrt{2}} & 0 & \frac{1}{\sqrt{2}} \\ \frac{1}{\sqrt{6}} & \frac{-2}{\sqrt{6}} & \frac{1}{\sqrt{6}} \end{bmatrix} \end{bmatrix} \\ &= \begin{bmatrix} \frac{1}{2} & 0 & -\frac{1}{2} & -\frac{1}{2} & 0 & \frac{1}{2} \\ \frac{-1}{2\sqrt{3}} & \frac{1}{\sqrt{3}} & \frac{-1}{2\sqrt{3}} & \frac{1}{2\sqrt{3}} & \frac{-1}{\sqrt{3}} & \frac{1}{2\sqrt{3}} \end{bmatrix}. \end{aligned}$$

Then

$$B_{MT} = \begin{bmatrix} B_M \\ B_T \end{bmatrix}$$

$$= \begin{bmatrix} \frac{-1}{\sqrt{6}} & \frac{-1}{\sqrt{6}} & \frac{-1}{\sqrt{6}} & \frac{1}{\sqrt{6}} & \frac{1}{\sqrt{6}} & \frac{1}{\sqrt{6}} \\ -\frac{1}{2} & 0 & \frac{1}{2} & -\frac{1}{2} & 0 & \frac{1}{2} \\ \frac{1}{2\sqrt{3}} & \frac{-1}{\sqrt{3}} & \frac{1}{2\sqrt{3}} & \frac{1}{2\sqrt{3}} & \frac{-1}{\sqrt{3}} & \frac{1}{2\sqrt{3}} \\ \frac{1}{2} & 0 & -\frac{1}{2} & -\frac{1}{2} & 0 & \frac{1}{2} \\ \frac{-1}{2\sqrt{3}} & \frac{1}{\sqrt{3}} & \frac{-1}{2\sqrt{3}} & \frac{1}{2\sqrt{3}} & \frac{-1}{\sqrt{3}} & \frac{1}{2\sqrt{3}} \end{bmatrix}. \qquad \square$$

In the following lemma, we derive an expression for $B_{MT} D_{\mathbf{d}} B'_{MT}$ which is then used to obtain the determinant of C_{MT}.

■ **LEMMA 6.5.1.**

$$B_{MT} D_{\mathbf{d}} B'_{MT} = \begin{bmatrix} B_M D_{\mathbf{d}} B'_M & \mathbf{0} \\ \mathbf{0} & B_T D_{\mathbf{d}} B'_T \end{bmatrix}$$

where the $(\ell_q - 1) \times (\ell_q - 1)$ block matrix of $B_M D_{\mathbf{d}} B'_M$ for attribute q is

$$\left[\prod_{j=1, j \neq q}^{k} (\ell_j - 1)^{i_j} \right] (-1)^{i_q} I_{\ell_q - 1}$$

and the $(\ell_{q_1} - 1)(\ell_{q_2} - 1) \times (\ell_{q_1} - 1)(\ell_{q_2} - 1)$ block matrix of $B_T D_{\mathbf{d}} B'_T$ for attributes q_1 and q_2 is

$$\left[\prod_{j=1, j \neq q_1, q_2}^{k} (\ell_j - 1)^{i_j} \right] (-1)^{i_{q_1}} (-1)^{i_{q_2}} I_{(\ell_{q_1} - 1)(\ell_{q_2} - 1)}.$$

Proof.

$$\begin{aligned} B_{MT} D_{\mathbf{d}} B'_{MT} &= \begin{bmatrix} B_M \\ B_T \end{bmatrix} D_{\mathbf{d}} \begin{bmatrix} B'_M & B'_T \end{bmatrix} \\ &= \begin{bmatrix} B_M D_{\mathbf{d}} B'_M & B_M D_{\mathbf{d}} B'_T \\ B_T D_{\mathbf{d}} B'_M & B_T D_{\mathbf{d}} B'_T \end{bmatrix}. \end{aligned}$$

First, we consider $B_M D_{\mathbf{d}} B'_M$. From Lemma 6.3.1, we see that the $(\ell_q - 1) \times (\ell_q - 1)$ block matrix of $B_M D_{\mathbf{d}} B'_M$ for attribute q is

$$\left[\prod_{j=1, j \neq q}^{k} (\ell_j - 1)^{i_j} \right] (-1)^{i_q} I_{\ell_q - 1}.$$

We now show that the off-diagonal block matrices $B_M D_{\mathbf{d}} B_T'$ and $B_T D_{\mathbf{d}} B_M'$ are both equal to $\mathbf{0}$. For a particular attribute difference $\mathbf{d}$,

$$B_M D_{\mathbf{d}} B_T' = B_M[M_{i_1} \otimes M_{i_2} \otimes \cdots \otimes M_{i_k}]B_T'$$

$$= \begin{bmatrix} B_{\ell_1} \otimes \frac{1}{\sqrt{\ell_2}}\mathbf{j}_{\ell_2}' \otimes \cdots \otimes \frac{1}{\sqrt{\ell_k}}\mathbf{j}_{\ell_k}' \\ \frac{1}{\sqrt{\ell_1}}\mathbf{j}_{\ell_1}' \otimes B_{\ell_2} \otimes \cdots \otimes \frac{1}{\sqrt{\ell_k}}\mathbf{j}_{\ell_k}' \\ \vdots \\ \frac{1}{\sqrt{\ell_1}}\mathbf{j}_{\ell_1}' \otimes \frac{1}{\sqrt{\ell_2}}\mathbf{j}_{\ell_2}' \otimes \cdots \otimes B_{\ell_k} \end{bmatrix} [M_{i_1} \otimes M_{i_2} \otimes \cdots \otimes M_{i_k}]B_T'$$

$$= \begin{bmatrix} B_{\ell_1}M_{i_1} \otimes \frac{1}{\sqrt{\ell_2}}\mathbf{j}_{\ell_2}'M_{i_2} \otimes \cdots \otimes \frac{1}{\sqrt{\ell_k}}\mathbf{j}_{\ell_k}'M_{i_k} \\ \frac{1}{\sqrt{\ell_1}}\mathbf{j}_{\ell_1}'M_{i_1} \otimes B_{\ell_2}M_{i_2} \otimes \cdots \otimes \frac{1}{\sqrt{\ell_k}}\mathbf{j}_{\ell_k}'M_{i_k} \\ \vdots \\ \frac{1}{\sqrt{\ell_1}}\mathbf{j}_{\ell_1}'M_{i_1} \otimes \frac{1}{\sqrt{\ell_2}}\mathbf{j}_{\ell_2}'M_{i_2} \otimes \cdots \otimes B_{\ell_k}M_{i_k} \end{bmatrix} B_T'$$

$$= \begin{bmatrix} B_{\ell_1}M_{i_1}B_{\ell_1}' \otimes \frac{1}{\sqrt{\ell_2}}\mathbf{j}_{\ell_2}'M_{i_2}B_{\ell_2}' \otimes \cdots \otimes \frac{1}{\sqrt{\ell_k}}\mathbf{j}_{\ell_k}'M_{i_k}\frac{1}{\sqrt{\ell_k}}\mathbf{j}_{\ell_k} \\ \frac{1}{\sqrt{\ell_1}}\mathbf{j}_{\ell_1}'M_{i_1}B_{\ell_1}' \otimes B_{\ell_2}M_{i_2}\frac{1}{\sqrt{\ell_2}}\mathbf{j}_{\ell_2} \otimes \cdots \otimes \frac{1}{\sqrt{\ell_k}}\mathbf{j}_{\ell_k}'M_{i_k}\frac{1}{\sqrt{\ell_k}}\mathbf{j}_{\ell_k} \\ \vdots \\ \frac{1}{\sqrt{\ell_1}}\mathbf{j}_{\ell_1}'M_{i_1}\frac{1}{\sqrt{\ell_1}}\mathbf{j}_{\ell_1} \otimes \cdots \otimes \frac{1}{\sqrt{\ell_{k-1}}}\mathbf{j}_{\ell_{k-1}}'M_{i_{k-1}}B_{\ell_{k-1}}' \otimes B_{\ell_k}M_{i_k}B_{\ell_k}' \end{bmatrix}.$$

If $M_{i_q} = I_{\ell_q}$, then

$$\frac{1}{\sqrt{\ell_q}}\mathbf{j}_{\ell_q}'M_{i_q}B_{\ell_q}' = \frac{1}{\sqrt{\ell_q}}\mathbf{j}_{\ell_q}'B_{\ell_q}' = \mathbf{0}_{\ell_q-1}.$$

If $M_{i_q} = J_{\ell_q} - I_{\ell_q}$, then

$$\begin{aligned} \frac{1}{\sqrt{\ell_q}}\mathbf{j}_{\ell_q}'M_{i_q}B_{\ell_q}' &= \frac{1}{\sqrt{\ell_q}}\mathbf{j}_{\ell_q}'(J_{\ell_q} - I_{\ell_q})B_{\ell_q}' \\ &= \frac{1}{\sqrt{\ell_q}}\mathbf{j}_{\ell_q}'J_{\ell_q}B_{\ell_q}' - \frac{1}{\sqrt{\ell_q}}\mathbf{j}_{\ell_q}'B_{\ell_q}' \\ &= \mathbf{0}_{\ell_q-1}. \end{aligned}$$

Hence $B_M D_{\mathbf{d}} B_T' = \mathbf{0}$ and $B_{k,T} D_{\mathbf{d}} B_{k,M}' = (B_{k,M} D_{\mathbf{d}} B_{k,T}')' = \mathbf{0}$.

Now we consider $B_T D_{\mathbf{d}} B_T'$. For a particular attribute difference $\mathbf{d}$,

$$B_T D_{\mathbf{d}} B_T' = B_T[M_{i_1} \otimes M_{i_2} \otimes \cdots \otimes M_{i_k}]B_T'$$

$$= \begin{bmatrix} B_{\ell_1} \otimes B_{\ell_2} \otimes \frac{1}{\sqrt{\ell_3}}\mathbf{j}'_{\ell_3} \otimes \cdots \otimes \frac{1}{\sqrt{\ell_k}}\mathbf{j}'_{\ell_k} \\ B_{\ell_1} \otimes \frac{1}{\sqrt{\ell_2}}\mathbf{j}'_{\ell_2} \otimes B_{\ell_3} \otimes \cdots \otimes \frac{1}{\sqrt{\ell_k}}\mathbf{j}'_{\ell_k} \\ \vdots \\ \frac{1}{\sqrt{\ell_1}}\mathbf{j}'_{\ell_1} \otimes \cdots \otimes \frac{1}{\sqrt{\ell_{k-2}}}\mathbf{j}'_{\ell_{k-2}} \otimes B_{\ell_{k-1}} \otimes B_{\ell_k} \end{bmatrix} [M_{i_1} \otimes M_{i_2} \otimes \cdots \otimes M_{i_k}]B'_T$$

$$= \begin{bmatrix} B_{\ell_1}M_{i_1} \otimes B_{\ell_2}M_{i_2} \otimes \frac{1}{\sqrt{\ell_3}}\mathbf{j}'_{\ell_3}M_{i_3} \otimes \cdots \otimes \frac{1}{\sqrt{\ell_k}}\mathbf{j}'_{\ell_k}M_{i_k} \\ B_{\ell_1}M_{i_1} \otimes \frac{1}{\sqrt{\ell_2}}\mathbf{j}'_{\ell_2}M_{i_2} \otimes B_{\ell_3}M_{i_3} \otimes \cdots \otimes \frac{1}{\sqrt{\ell_k}}\mathbf{j}'_{\ell_k}M_{i_k} \\ \vdots \\ \frac{1}{\sqrt{\ell_1}}\mathbf{j}'_{\ell_1}M_{i_1} \otimes \cdots \otimes \frac{1}{\sqrt{\ell_{k-2}}}\mathbf{j}'_{\ell_{k-2}}M_{i_{k-2}} \otimes B_{\ell_{k-1}}M_{i_{k-1}} \otimes B_{\ell_k}M_{i_k} \end{bmatrix} B'_T$$

$$= \begin{bmatrix} B_{\ell_1}M_{i_1}B'_{\ell_1} \otimes B_{\ell_2}M_{i_2}B'_{\ell_2} \otimes \cdots \otimes \frac{1}{\sqrt{\ell_k}}\mathbf{j}'_{\ell_k}M_{i_k}\frac{1}{\sqrt{\ell_k}}\mathbf{j}_{\ell_k} \\ B_{\ell_1}M_{i_1}B'_{\ell_1} \otimes \frac{1}{\sqrt{\ell_2}}\mathbf{j}'_{\ell_2}M_{i_2}\frac{1}{\sqrt{\ell_2}}\mathbf{j}_{\ell_2} \otimes \cdots \otimes \frac{1}{\sqrt{\ell_k}}\mathbf{j}'_{\ell_k}M_{i_k}\frac{1}{\sqrt{\ell_k}}\mathbf{j}_{\ell_k} \\ \vdots \\ \frac{1}{\sqrt{\ell_1}}\mathbf{j}'_{\ell_1}M_{i_1}\frac{1}{\sqrt{\ell_1}}\mathbf{j}_{\ell_1} \otimes \cdots \otimes B_{\ell_{k-1}}M_{i_{k-1}}B'_{\ell_{k-1}} \otimes B_{\ell_k}M_{i_k}B'_{\ell_k} \end{bmatrix}.$$

Now the $(\ell_{q_1}-1)(\ell_{q_2}-1) \times (\ell_{q_1}-1)(\ell_{q_2}-1)$ block matrix of $B_T D_{\mathbf{d}} B'_T$ for attributes q_1 and q_2 contains $(k-2)$ terms of the form

$$\frac{1}{\sqrt{\ell_j}}\mathbf{j}'_{\ell_j}M_{i_j}\frac{1}{\sqrt{\ell_j}}\mathbf{j}_{\ell_j} = (\ell_j - 1)^{i_j} \text{ for } j = 1,\ldots,k \; j \neq q_1, q_2$$

and there are two terms

$$B_{\ell_{q1}}M_{i_{q1}}B'_{\ell_{q1}} = (-1)^{i_{q1}} I_{\ell_{q1}-1}$$

and

$$B_{\ell_{q2}}M_{i_{q2}}B'_{\ell_{q2}} = (-1)^{i_{q2}} I_{\ell_{q2}-1}.$$

Then the $(\ell_{q_1}-1)(\ell_{q_2}-1) \times (\ell_{q_1}-1)(\ell_{q_2}-1)$ block matrix of $B_T D_{\mathbf{d}} B'_T$ for attributes q_1 and q_2 is

$$\left[\prod_{j=1, j\neq q_1, q_2}^{k} (\ell_j - 1)^{i_j}\right](-1)^{i_{q1}}(-1)^{i_{q2}} I_{(\ell_{q1}-1)(\ell_{q2}-1)},$$

as required. □

In the following lemma we derive an expression for the determinant of $\det(C_{MT})$.

■ LEMMA 6.5.2.

Under the null hypothesis, the determinant of C_{MT} is given by

$$\det(C_{MT}) = \prod_{q=1}^{k}\left[\frac{1}{m^2}\sum_{\mathbf{d}}\left[y_{\mathbf{d}}\left(\prod_{j=1, j\neq q}^{k}(\ell_j-1)^{i_j}\right)\left[(\ell_q-1)^{i_q} - (-1)^{i_q}\right]\right]\right]^{\ell_q-1}$$

$$\times \prod_{q_1=1}^{k-1}\prod_{q_2=q_1+1}^{k}\left[\frac{1}{m^2}\sum_{\mathbf{d}}\left[y_{\mathbf{d}}L_{\mathbf{d}}\left(1-\frac{1}{(1-\ell_{q1})^{i_{q1}}(1-\ell_{q2})^{i_{q2}}}\right)\right]\right]^{(\ell_{q1}-1)(\ell_{q2}-1)},$$

where $L_{\mathbf{d}} = \prod_{j=1}^{k}(\ell_j - 1)^{i_j}$.

Proof. Recall that p is the total number of parameters to be estimated and is defined to be

$$p = \sum_{q=1}^{k}(\ell_q - 1) + \sum_{q_1=1}^{k-1} \sum_{q_2=q_1+1}^{k} (\ell_{q_1} - 1)(\ell_{q_2} - 1).$$

From Lemma 6.2.1, under the null hypothesis,

$$\begin{aligned}
C_{MT} &= B_{MT}\Lambda B'_{MT} \\
&= B_{MT}\left[\frac{m-1}{m^2} z I_L - \frac{1}{m^2}\sum_{\mathbf{d}} y_{\mathbf{d}} D_{\mathbf{d}}\right] B'_{MT} \\
&= \frac{1}{m^2}\sum_{\mathbf{d}}\left[y_{\mathbf{d}}\left(\prod_{j=1}^{k}(\ell_j - 1)^{i_j}\right)\right] B_{MT}B'_{MT} \\
&\quad - \frac{1}{m^2}\sum_{\mathbf{d}}\left[y_{\mathbf{d}} B_{MT} D_{\mathbf{d}} B'_{MT}\right] \\
&= \frac{1}{m^2}\sum_{\mathbf{d}}\left[y_{\mathbf{d}}\left(\prod_{j=1}^{k}(\ell_j - 1)^{i_j}\right)\right] I_p \\
&\quad - \frac{1}{m^2}\sum_{\mathbf{d}}\left\{y_{\mathbf{d}}\begin{bmatrix} B_M D_{\mathbf{d}} B'_M & \mathbf{0} \\ \mathbf{0} & B_T D_{\mathbf{d}} B'_T \end{bmatrix}\right\} \\
&= \frac{1}{m^2}\sum_{\mathbf{d}}\left\{y_{\mathbf{d}}\left[\left(\prod_{j=1}^{k}(\ell_j - 1)^{i_j}\right) I_p - \begin{bmatrix} B_M D_{\mathbf{d}} B'_M & \mathbf{0} \\ \mathbf{0} & B_T D_{\mathbf{d}} B'_T \end{bmatrix}\right]\right\}.
\end{aligned}$$

Thus C_{MT} is a block diagonal matrix with block matrices for the main effects of attributes $1, \ldots, k$ down the diagonal, then block matrices for the each of the interaction effects between all pairs of attributes down the rest of the diagonal.

Using Lemma 6.3.2, the block diagonal term of C_{MT} for main effects for factor q is given by

$$\frac{1}{m^2}\sum_{\mathbf{d}}\left\{y_{\mathbf{d}}\left(\prod_{j=1, j\neq q}^{k}(\ell_j - 1)^{i_j}\right)\left[(\ell_q - 1)^{i_q} - (-1)^{i_q}\right]\right\} I_{\ell_q - 1}$$

and the component of the determinant for the main effects for all attributes is

$$\prod_{q=1}^{k}\left[\frac{1}{m^2}\sum_{\mathbf{d}}\left\{y_{\mathbf{d}}\left(\prod_{j=1, j\neq q}^{k}(\ell_j - 1)^{i_j}\right)\left[(\ell_q - 1)^{i_q} - (-1)^{i_q}\right]\right\}\right]^{\ell_q - 1}. \tag{6.3}$$

Now consider the $(\ell_{q_1} - 1)(\ell_{q_2} - 1) \times (\ell_{q_1} - 1)(\ell_{q_2} - 1)$ block matrix of C_{MT} for the interaction effect of attributes q_1 and q_2. Using Lemma 6.5.1, we have

$$\frac{1}{m^2}\sum_{\mathbf{d}}\left\{y_{\mathbf{d}}\left[\left(\prod_{j=1}^{k}(\ell_j - 1)^{i_j}\right) I_{(\ell_{q_1}-1)(\ell_{q_2}-1)} - B_T D_{\mathbf{d}} B'_T\right]\right\}$$

$$= \frac{1}{m^2}\sum_{\mathbf{d}}\left\{y_{\mathbf{d}}\left[\left(\prod_{j=1}^{k}(\ell_j-1)^{i_j}\right)I_{(\ell_{q_1}-1)(\ell_{q_2}-1)} - \left(\prod_{j=1,j\neq q_1,q_2}^{k}(\ell_j-1)^{i_j}\right)(-1)^{i_{q_1}}(-1)^{i_{q_2}}I_{(\ell_{q_1}-1)(\ell_{q_2}-1)}\right]\right\}$$

$$= \frac{1}{m^2}\sum_{\mathbf{d}}\left\{y_{\mathbf{d}}\left[\left(\prod_{j=1}^{k}(\ell_j-1)^{i_j}\right) - \left(\prod_{j=1,j\neq q_1,q_2}^{k}(\ell_j-1)^{i_j}\right)(-1)^{i_{q_1}}(-1)^{i_{q_2}}\right]\right\}I_{(\ell_{q_1}-1)(\ell_{q_2}-1)}$$

$$= \frac{1}{m^2}\sum_{\mathbf{d}}\left\{y_{\mathbf{d}}\left(\prod_{j=1}^{k}(\ell_j-1)^{i_j}\right)\left[1-\frac{(-1)^{i_{q_1}}(-1)^{i_{q_2}}}{(\ell_{q_1}-1)^{i_{q_1}}(\ell_{q_2}-1)^{i_{q_2}}}\right]\right\}I_{(\ell_{q_1}-1)(\ell_{q_2}-1)}$$

$$= \frac{1}{m^2}\sum_{\mathbf{d}}\left\{y_{\mathbf{d}}L_{\mathbf{d}}\left[1-\frac{1}{(1-\ell_{q_1})^{i_{q_1}}(1-\ell_{q_2})^{i_{q_2}}}\right]\right\}I_{(\ell_{q_1}-1)(\ell_{q_2}-1)}.$$

Then the component of the determinant of C_{MT} for the two-factor interaction effects only is given by

$$\prod_{q_1=1}^{k-1}\prod_{q_2=q_1+1}^{k}\left[\frac{1}{m^2}\sum_{\mathbf{d}}\left\{y_{\mathbf{d}}L_{\mathbf{d}}\left(1-\frac{1}{(1-\ell_{q_1})^{i_{q_1}}(1-\ell_{q_2})^{i_{q_2}}}\right)\right\}\right]^{(\ell_{q_1}-1)(\ell_{q_2}-1)} \tag{6.4}$$

To obtain the determinant of C_{MT}, Equations (6.3) and (6.4) are multiplied together. □

■ EXAMPLE 6.5.2.

Let $m = 3$, $k = 2$, $\ell_1 = 2$, and $\ell_2 = 3$. Assume that the choice sets in the choice experiment are the ones with difference vectors $\mathbf{v}_1$ and $\mathbf{v}_3$ only. First, we calculate C_{MT} and hence $\det(C_{MT})$ by evaluating

$$C_{MT} = B_{MT}\Lambda B'_{MT}.$$

Then we also calculate $\det(C_{MT})$ by using Lemma 6.5.2.

The matrices Λ and B_{MT} are given in Examples 6.2.2 and 6.5.1, respectively. Thus

$$C_{MT} = B_{MT}\Lambda B'_{MT} = \begin{bmatrix} \frac{1}{9} & : & 0 & 0 & : & 0 & 0 \\ \cdots & & \cdots & \cdots & & \cdots & \cdots \\ 0 & : & \frac{1}{6} & 0 & : & 0 & 0 \\ 0 & : & 0 & \frac{1}{6} & : & 0 & 0 \\ \cdots & & \cdots & \cdots & & \cdots & \cdots \\ 0 & : & 0 & 0 & : & \frac{1}{9} & 0 \\ 0 & : & 0 & 0 & : & 0 & \frac{1}{9} \end{bmatrix}.$$

Hence $\det(C_{MT}) = \left(\frac{1}{9}\right)^3\left(\frac{1}{6}\right)^2$.

Alternatively, we can use the results from Lemma 6.5.2. The $y_{\mathbf{d}}$ given in Example 6.2.2 are

$$y_{01} = \frac{2}{8}, \quad y_{10} = 0, \quad \text{and} \quad y_{11} = \frac{2}{8}.$$

Then the main effects part of $\det(C_{MT})$ is

$$\prod_{q=1}^{2}\left[\frac{1}{3^2}\sum_{\mathbf{d}}\left\{y_{\mathbf{d}}\left(\prod_{j=1,j\neq q}^{2}(\ell_j-1)^{i_j}\right)\left[(\ell_q-1)^{i_q} - (-1)^{i_q}\right]\right\}\right]^{\ell_q-1}$$

$$\begin{aligned}
= & \left[\frac{1}{9}\left\{y_{01}(\ell_2-1)^1[(\ell_1-1)^0-(-1)^0]\right.\right.\\
& + y_{10}(\ell_2-1)^0[(\ell_1-1)^1-(-1)^1]\\
& \left.\left. + y_{11}(\ell_2-1)^1[(\ell_1-1)^1-(-1)^1]\right\}\right]^{\ell_1-1}\\
\times & \left[\frac{1}{9}\left\{y_{01}(\ell_1-1)^0[(\ell_2-1)^1-(-1)^1]\right.\right.\\
& + y_{10}(\ell_1-1)^1[(\ell_2-1)^0-(-1)^0]\\
& \left.\left. + y_{11}(\ell_1-1)^1[(\ell_2-1)^1-(-1)^1]\right\}\right]^{\ell_2-1}\\
= & \left[\frac{1}{9}\left\{\frac{2}{8}(3-1)^1[(2-1)^0-(-1)^0] + 0\right.\right.\\
& \left.\left. + \frac{2}{8}(3-1)^1[(2-1)^1-(-1)^1]\right\}\right]^{2-1}\\
\times & \left[\frac{1}{9}\left\{\frac{2}{8}(2-1)^0[(3-1)^1-(-1)^1] + 0\right.\right.\\
& \left.\left. + \frac{2}{8}(2-1)^1[(3-1)^1-(-1)^1]\right\}\right]^{3-1}\\
= & \left[\frac{1}{9}\left\{\frac{2}{8}\times 2\times 0+\frac{2}{8}\times 2\times 2\right\}\right] \times \left[\frac{1}{9}\left\{\frac{2}{8}\times 1\times 3+\frac{2}{8}\times 1\times 3\right\}\right]^2\\
= & \left(\frac{1}{9}\right)\left(\frac{1}{6}\right)^2.
\end{aligned}$$

The two-factor interaction part of $\det(C_{MT})$ is

$$\prod_{q_1=1}^{1}\prod_{q_2=2}^{2}\left[\frac{1}{3^2}\sum_{\mathbf{d}}\left\{y_{\mathbf{d}}L_{\mathbf{d}}\left(1-\frac{1}{(1-\ell_{q_1})^{i_{q_1}}(1-\ell_{q_2})^{i_{q_2}}}\right)\right\}\right]^{(\ell_{q_1}-1)(\ell_{q_2}-1)}$$

$$\begin{aligned}
= & \left[\frac{1}{9}\left\{y_{01}(\ell_1-1)^0(\ell_2-1)^1\left[1-\frac{1}{(1-\ell_1)^0(1-\ell_2)^1}\right]\right.\right.\\
& + y_{10}(\ell_1-1)^1(\ell_2-1)^0\left[1-\frac{1}{(1-\ell_1)^1(1-\ell_2)^0}\right]\\
& \left.\left. + y_{11}(\ell_1-1)^1(\ell_2-1)^1\left[1-\frac{1}{(1-\ell_1)^1(1-\ell_2)^1}\right]\right\}\right]^{(\ell_1-1)(\ell_2-1)}
\end{aligned}$$

$$\begin{aligned}
&= \left[\frac{1}{9}\left\{\frac{2}{8}(2-1)^0(3-1)^1\left[1-\frac{1}{(1-2)^0(1-3)^1}\right] + 0\right.\right.\\
&\quad \left.\left. + \frac{2}{8}(2-1)^1(3-1)^1\left[1-\frac{1}{(1-2)^1(1-3)^1}\right]\right\}\right]^{(2-1)(3-1)}\\
&= \left[\frac{1}{9}\left\{\frac{2}{8}\times 1\times 2\left[1+\frac{1}{2}\right] + \frac{2}{8}\times 1\times 2\left[1-\frac{1}{2}\right]\right\}\right]^2\\
&= \left(\frac{1}{9}\right)^2
\end{aligned}$$

Then

$$\det(C_{MT}) = \left(\frac{1}{9}\right)\left(\frac{1}{6}\right)^2 \times \left(\frac{1}{9}\right)^2 = \left(\frac{1}{9}\right)^3\left(\frac{1}{6}\right)^2.$$

All possible designs for $m = 3$, $k = 2$, $\ell_1 = 2$, and $\ell_2 = 3$ are shown in Table 6.8. In this example, the design consisting of choice sets from all three difference vectors has the maximum $\det(C_{MT})$ of

$$\det(C_{\text{opt},MT}) = \left(\frac{2}{15}\right)^5.$$

The efficiencies of the other designs are calculated relative to this design. Recall that the total number of parameters to be estimated is

$$\begin{aligned}
p &= \sum_{q=1}^{k}(\ell_q - 1) + \sum_{q_1=1}^{k-1}\sum_{q_2=q_1+1}^{k}(\ell_{q_1}-1)(\ell_{q_2}-1)\\
&= [(2-1)+(3-1)] + (2-1)(3-1)\\
&= 5.
\end{aligned}$$

Then, for example, the efficiency of the design consisting of choice sets with difference vector $\mathbf{v}_2$ only, is

$$\text{Eff}_D = 100\left[\frac{\det(C_{MT})}{\det(C_{\text{opt},MT})}\right]^{1/5} = 100\left[\frac{(\frac{4}{27})^3(\frac{1}{9})^2}{(\frac{2}{15})^5}\right]^{1/5} = 99.03\%.$$

□

We have been unable to get any general results, true for all m, for the form of the designs that are optimal for estimating main effects and two-factor interactions.

We can get specific results for fixed m and k. For instance, for particular m and k values, we can evaluate $\det(C_{MT})$ for all possible designs.

■ EXAMPLE 6.5.3.

Suppose $m = 2$ and $k = 2$. Then the main effects part of $\det(C_{MT})$ is

$$\prod_{q=1}^{2}\left[\frac{1}{2^2}\sum_{\mathbf{d}}\left\{y_{\mathbf{d}}\left(\prod_{j=1,j\neq q}^{2}(\ell_j-1)^{i_j}\right)\left[(\ell_q-1)^{i_q} - (-1)^{i_q}\right]\right\}\right]^{\ell_q-1}$$

Table 6.8 All Possible Choice Experiment Designs for $k = 2$, $\ell_1 = 2$, and $\ell_2 = 3$ when $m = 3$: Main Effects and Two-Factor Interactions

$\mathbf{v}_j$ in Design	N	$\det(C_{MT})$	Efficiency (%)
$\mathbf{v}_1$	2	$(0)(\frac{1}{6})^4$	0
$\mathbf{v}_2$	12	$(\frac{4}{27})^3(\frac{1}{9})^2$	99.03
$\mathbf{v}_3$	6	$(\frac{4}{27})(\frac{1}{6})^2(\frac{5}{54})^2$	96.51
$\mathbf{v}_1$ & $\mathbf{v}_2$	14	$(\frac{8}{63})(\frac{5}{42})^2(\frac{19}{126})^2$	99.42
$\mathbf{v}_1$ & $\mathbf{v}_3$	8	$(\frac{1}{9})^3(\frac{1}{6})^2$	98.01
$\mathbf{v}_2$ & $\mathbf{v}_3$	18	$(\frac{4}{27})(\frac{7}{54})^4$	99.85
$\mathbf{v}_1$, $\mathbf{v}_2$ & $\mathbf{v}_3$	20	$(\frac{2}{15})^5$	100

$$\begin{aligned}
= & \left[\frac{1}{4}\left\{y_{01}(\ell_2-1)^1[(\ell_1-1)^0-(-1)^0]\right.\right. \\
& + y_{10}(\ell_2-1)^0[(\ell_1-1)^1-(-1)^1] \\
& \left.\left. + y_{11}(\ell_2-1)^1[(\ell_1-1)^1-(-1)^1]\right\}\right]^{\ell_1-1} \\
\times & \left[\frac{1}{4}\left\{y_{01}(\ell_1-1)^0[(\ell_2-1)^1-(-1)^1]\right.\right. \\
& + y_{10}(\ell_1-1)^1[(\ell_2-1)^0-(-1)^0] \\
& \left.\left. + y_{11}(\ell_1-1)^1[(\ell_2-1)^1-(-1)^1]\right\}\right]^{\ell_2-1} \\
= & \left[\frac{1}{4}\left\{\ell_1 y_{10}+\ell_1(\ell_2-1)y_{11}\right\}\right]^{\ell_1-1} \times \left[\frac{1}{4}\left\{\ell_2 y_{01}+\ell_2(\ell_1-1)y_{11}\right\}\right]^{\ell_2-1} \\
= & \left[\frac{\ell_1}{4}\left\{y_{10}+(\ell_2-1)y_{11}\right\}\right]^{\ell_1-1} \times \left[\frac{\ell_2}{4}\left\{y_{01}+(\ell_1-1)y_{11}\right\}\right]^{\ell_2-1}.
\end{aligned}$$

The two-factor interaction part of $\det(C_{MT})$ is

$$\left[\frac{1}{2^2}\sum_{\mathbf{d}}\left\{y_{\mathbf{d}}(\ell_1-1)^{i_1}(\ell_2-1)^{i_2}\left(1-\frac{1}{(1-\ell_1)^{i_1}(1-\ell_2)^{i_2}}\right)\right\}\right]^{(\ell_1-1)(\ell_2-1)}$$

$$\begin{aligned}
= & \left[\frac{1}{4}\left\{y_{01}(\ell_1-1)^0(\ell_2-1)^1\left[1-\frac{1}{(1-\ell_1)^0(1-\ell_2)^1}\right]\right.\right. \\
& + y_{10}(\ell_1-1)^1(\ell_2-1)^0\left[1-\frac{1}{(1-\ell_1)^1(1-\ell_2)^0}\right] \\
& \left.\left. + y_{11}(\ell_1-1)^1(\ell_2-1)^1\left[1-\frac{1}{(1-\ell_1)^1(1-\ell_2)^1}\right]\right\}\right]^{(\ell_1-1)(\ell_2-1)}
\end{aligned}$$

$$= \left[\frac{1}{4}\left\{\frac{(\ell_2-1)\ell_2 y_{01}}{\ell_2-1} + \frac{(\ell_1-1)\ell_1 y_{10}}{\ell_1-1} + \frac{(\ell_1-1)(\ell_2-1)[(\ell_1-1)(\ell_2-1)-1]y_{11}}{(\ell_1-1)(\ell_2-1)}\right\}\right]^{(\ell_1-1)(\ell_2-1)}$$

$$= \left[\frac{1}{4}\left\{\ell_2 y_{01} + \ell_1 y_{10} + (\ell_1\ell_2-\ell_1-\ell_2)y_{11}\right\}\right]^{(\ell_1-1)(\ell_2-1)}.$$

Then

$$\det(C_{MT}) = \left[\frac{\ell_1}{4}\left\{y_{10}+(\ell_2-1)y_{11}\right\}\right]^{\ell_1-1} \times \left[\frac{\ell_2}{4}\left\{y_{01}+(\ell_1-1)y_{11}\right\}\right]^{\ell_2-1} \times \left[\frac{1}{4}\left\{\ell_2 y_{01} + \ell_1 y_{10} + (\ell_1\ell_2-\ell_1-\ell_2)y_{11}\right\}\right]^{(\ell_1-1)(\ell_2-1)}.$$

Now

$$\mathbf{v}_1 = 01, \quad c_{\mathbf{v}_1} = \ell_2 - 1,$$
$$\mathbf{v}_2 = 10, \quad c_{\mathbf{v}_2} = \ell_1 - 1,$$
$$\mathbf{v}_3 = 11, \quad c_{\mathbf{v}_3} = (\ell_1 - 1)(\ell_2 - 1).$$

So

$$y_{01} = \frac{2c_{\mathbf{v}_1}a_{\mathbf{v}_1}}{2(\ell_1-1)^0(\ell_2-1)^1} = a_{\mathbf{v}_1},$$

$$y_{10} = \frac{2c_{\mathbf{v}_2}a_{\mathbf{v}_2}}{2(\ell_1-1)^1(\ell_2-1)^0} = a_{\mathbf{v}_2},$$

$$y_{11} = \frac{2c_{\mathbf{v}_3}a_{\mathbf{v}_3}}{2(\ell_1-1)^1(\ell_2-1)^1} = a_{\mathbf{v}_3}.$$

Then

$$\det(C_{MT}) = \left[\frac{\ell_1}{4}\left\{a_{\mathbf{v}_2}+(\ell_2-1)a_{\mathbf{v}_3}\right\}\right]^{\ell_1-1} \times \left[\frac{\ell_2}{4}\left\{a_{\mathbf{v}_1}+(\ell_1-1)a_{\mathbf{v}_3}\right\}\right]^{\ell_2-1} \times \left[\frac{1}{4}\left\{\ell_2 a_{\mathbf{v}_1} + \ell_1 a_{\mathbf{v}_2} + (\ell_1\ell_2-\ell_1-\ell_2)a_{\mathbf{v}_3}\right\}\right]^{(\ell_1-1)(\ell_2-1)}.$$

Now we can evaluate explicitly the determinants for each of the possible designs for $m = 2$ and $k = 2$. Recall that

$$a_{\mathbf{v}_j} = \frac{i_{\mathbf{v}_j}}{N} = \frac{i_{\mathbf{v}_j} m}{L\sum_j c_{\mathbf{v}_j} i_{\mathbf{v}_j}}$$

since

$$N = \frac{L}{m} c_{\mathbf{v}_j} i_{\mathbf{v}_j}.$$

The values for $a_{\mathbf{v}_j}$ for all the possible designs are given in Table 6.9 and the determinants for each of these designs are given in Table 6.10. It is possible to investigate the relative magnitudes of these determinants. In all cases, the largest determinant is that obtained from the choice experiment containing all the pairs. But this design has a determinant that is only about 5% larger than that from the choice experiment with only the pairs with difference

Table 6.9 Values of $a_{\mathbf{v}_j}$ for $m = 2$ and $k = 2$

$\mathbf{v}_j$ in Design	$a_{\mathbf{v}_1}$	$a_{\mathbf{v}_2}$	$a_{\mathbf{v}_3}$
$\mathbf{v}_1 = (01)$	$\frac{2}{\ell_1\ell_2(\ell_2-1)}$	0	0
$\mathbf{v}_2 = (10)$	0	$\frac{2}{\ell_1\ell_2(\ell_1-1)}$	0
$\mathbf{v}_3 = (11)$	0	0	$\frac{2}{\ell_1\ell_2(\ell_1-1)(\ell_2-1)}$
$\mathbf{v}_1$ & $\mathbf{v}_2$	$\frac{2}{\ell_1\ell_2(\ell_1+\ell_2-2)}$	$\frac{2}{\ell_1\ell_2(\ell_1+\ell_2-2)}$	0
$\mathbf{v}_1$ & $\mathbf{v}_3$	$\frac{2}{\ell_1^2\ell_2(\ell_2-1)}$	0	$\frac{2}{\ell_1^2\ell_2(\ell_2-1)}$
$\mathbf{v}_2$ & $\mathbf{v}_3$	0	$\frac{2}{\ell_1\ell_2^2(\ell_1-1)}$	$\frac{2}{\ell_1\ell_2^2(\ell_1-1)}$
$\mathbf{v}_1$, $\mathbf{v}_2$ & $\mathbf{v}_3$	$\frac{2}{R}$	$\frac{2}{R}$	$\frac{2}{R}$

$R = \ell_1\ell_2\left[(\ell_1 - 1) + (\ell_2 - 1) + (\ell_1 - 1)(\ell_2 - 1)\right]$.

Table 6.10 Values of $\det(C_{MT})$ for $m = 2$ and $k = 2$

$\mathbf{v}_j$ in Design	$\det(C_{MT})$
$\mathbf{v}_1 = (01)$	0
$\mathbf{v}_2 = (10)$	0
$\mathbf{v}_3 = (11)$	$\left[\frac{1}{2\ell_2(\ell_1-1)}\right]^{\ell_1-1}\left[\frac{1}{2\ell_1(\ell_2-1)}\right]^{\ell_2-1}\left[\frac{(\ell_1-1)(\ell_2-1)-1}{2\ell_1\ell_2(\ell_1-1)(\ell_2-1)}\right]^{(\ell_1-1)(\ell_2-1)}$
$\mathbf{v}_1$ & $\mathbf{v}_2$	$\left[\frac{1}{2\ell_2(\ell_1+\ell_2-2)}\right]^{\ell_1-1}\left[\frac{1}{2\ell_1(\ell_1+\ell_2-2)}\right]^{\ell_2-1}\left[\frac{\ell_1+\ell_2}{2\ell_1\ell_2(\ell_1+\ell_2-2)}\right]^{(\ell_1-1)(\ell_2-1)}$
$\mathbf{v}_1$ & $\mathbf{v}_3$	$\left[\frac{1}{2\ell_1\ell_2}\right]^{\ell_1-1}\left[\frac{1}{2\ell_1(\ell_2-1)}\right]^{\ell_2-1}\left[\frac{1}{2\ell_1\ell_2}\right]^{(\ell_1-1)(\ell_2-1)}$
$\mathbf{v}_2$ & $\mathbf{v}_3$	$\left[\frac{1}{2\ell_2(\ell_1-1)}\right]^{\ell_1-1}\left[\frac{1}{2\ell_1\ell_2}\right]^{\ell_2-1}\left[\frac{1}{2\ell_1\ell_2}\right]^{(\ell_1-1)(\ell_2-1)}$
$\mathbf{v}_1$, $\mathbf{v}_2$ & $\mathbf{v}_3$	$\left[\frac{1}{2(\ell_1\ell_2-1)}\right]^{\ell_1-1}\left[\frac{1}{2(\ell_1\ell_2-1)}\right]^{\ell_2-1}\left[\frac{1}{2(\ell_1\ell_2-1)}\right]^{(\ell_1-1)(\ell_2-1)}$

vector (11) when $\ell_1 = \ell_2 = 20$, but it is 965 times larger than the determinant for the experiment with only the pairs with difference vectors (01) and (10) when $\ell_1 = \ell_2 = 20$. Indeed, as ℓ_1 and ℓ_2 tend to infinity, the limit of the ratio of the determinant of the design with all pairs to that of the design with pairs with difference (11) only is 1. □

Consider now the situation when $k = 3$ and $m = 2$. Here there are 7 difference vectors possible and so there are 127 possible choice experiments to consider. For $\ell_1 = 2, \ell_2 = 2, 3$ and $\ell_3 \leq 8$, the optimal design has all the pairs with difference vectors (011), (101), and (110). For $\ell_1 = 2$, $4 \leq \ell_2 \leq 8$, $\ell_2 \leq \ell_3 \leq 8$, the optimal design has all the pairs with difference vectors (101) and (110). When $\ell_1 = 3$, the optimal design has all the pairs with difference vectors (011), (101), and (110) when $\ell_2 = 3, 4, 5$ and $\ell_3 = 3, 4, 5, 6$; but it has all the pairs with difference vectors (101) and (110) when $5 \leq \ell_2 \leq 8$ and $\ell_2 \leq \ell_3 \leq 8$ (except $\ell_2 = 5, \ell_3 = 5, 6$). This situation continues for all the cases we have investigated. If we assume, without loss of generality, that $\ell_1 \leq \ell_2 \leq \ell_3$, then for fixed ℓ_1 all three difference vectors of weight 2 give the optimal design when ℓ_2 and ℓ_3 are "close enough" to ℓ_1. As ℓ_2 and ℓ_3 get larger, it is sufficient to have only those pairs with difference vectors (101) and (110). See Appendix 6. A.2 for details. When $\ell_1 = 2$ then the determinant of the design with pairs with difference vectors (110) and (101) is at most 11.3 times that of the determinant of the design with pairs with difference vectors (110), (101), and (011). When $\ell_1 = 5$ the same multiple is only 1.32.

When $m = 2$ and $k = 4$ we have considered all cases with $2 \leq \ell_1 \leq \ell_2 \leq \ell_3 \leq \ell_4 \leq 8$. There are 8 different sets of optimal difference vectors. The most common has three difference vectors of weight 3: (1011), (1101), and (1110). The second most common has two difference vectors of weight 3: (1101) and (1110). Details may be found in Appendix 6. A.3. These designs all have a large number of choice sets, and we can use the method in Section 6.4 to obtain near-optimal designs with a smaller number of choice sets.

■ EXAMPLE 6.5.4.

Suppose $m = 2$, $k = 4$ with $\ell_1 = 2$, $\ell_2 = 3$, $\ell_3 = 6$, and $\ell_4 = 6$. By investigating all possible designs, we found that the 2160 pairs with difference vectors (1101) and (1110) form the optimal design. By starting with the complete factorial and adding generators (1101), (1110), and (1011) we obtain a design in 648 pairs that is 94.4% efficient and the C_{MT} matrix is block diagonal. □

Results for $m = 2$ and $k = 3$ are given in Appendix 6. A.4. For $m = 3$ and $k = 2$ for $2 \leq \ell_1 \leq \ell_2 \leq 8$ see Appendix 6. A.5. For $m = 4$ and $k = 2$ for some values of ℓ_1 and ℓ_2 see Appendix 6. A.6.

6.5.1 Exercises

1. Let $k = 3$ and $m = 2$. Let $\ell_1 = \ell_2 = 2$ and let $\ell_3 = 3$. Give the pairs that are optimal for the estimation of main effects and two-factor interactions. Suppose that you only want to estimate the interaction between factors 1 and 2. What is the determinant of the appropriate information matrix? Can you find a set of pairs that do better than this?

2. Another type of construction that has been developed by Grasshoff et al. (2004) uses each row of an OA to construct a choice set of size 2 by letting the symbols of the OA refer to ordered pairs of attribute levels (for the two options in each choice set). Consider an OA[18; 3^6, 6; 2] and let $\ell_1 = \ldots = \ell_6 = 3$ and $\ell_7 = 4$. Then equate the

levels of the factors in the OA with pairs of levels for each attribute in some way; perhaps $0 \mapsto 0, 1$; $1 \mapsto 0, 2$ and $2 \mapsto 1, 2$ for the 3-level attributes and $0 \mapsto 0, 1$; $1 \mapsto 0, 2$, $2 \mapsto 0, 3$, $3 \mapsto 1, 2$, $4 \mapsto 1, 3$ and $5 \mapsto 2, 3$ for the 4-level attribute. So if one row in the OA was 0221014, then the corresponding choice set would be (0110001, 1222123).

(a) Construct all 18 pairs associated with an OA[18; 3^6, 6; 2].

(b) What is the efficiency of these pairs for the estimation of main effects? main effects plus two-factor interactions?

(c) What pairs would you get for these situations if you used the approach outlined in this chapter and how efficient would they be?

3. Hadamard matrices are also used in Grasshoff et al. (2004) to construct paired comparison designs. Let A_ℓ have as its rows the pairs in an optimal design with $k = 1$ and $m = 2$. Let $H_{n,k}$ be k columns of a Hadamard matrix of order n, where $k \leq n$. Then the pairs are obtained from the matrix $H_{n,k} \otimes A_\ell$. For instance, if $\ell = 2 = k$ then $A_2 = [(0, 1)]$ and $H_{2,2} = \begin{bmatrix} 1 & 1 \\ 1 & -1 \end{bmatrix}$. Thus we get $H_{2,2} \otimes A_2 = \begin{bmatrix} (0,1) & (0,1) \\ (0,1) & -(0,1) \end{bmatrix}$. We get the choice set (00, 11) from the first row and the choice set (01, 10) from the second row. Give the 12 choice sets of size 2 that you get from this approach when $\ell = 3$ and $k = 4$. How efficient is the design for the estimation of main effects? main effects plus two-factor interactions?

6.6 REFERENCES AND COMMENTS

Most of the results in this chapter originally appeared in Burgess and Street (2005). Readers can find software to construct choice sets from an initial factorial design and sets of generators, as well as calculate the corresponding information matrix and variance-covariance matrix, at the following website: http://maths.science.uts.edu.au/maths/wiki/SPExpts.

The choice of which of the different possible B matrices to use depends very much on the effects that are of interest. If you are interested in estimating the linear, quadratic, and so on, effects for an attribute with discrete quantitative levels, then the B matrix with contrasts that arise from the appropriate orthogonal polynomial contrasts are natural. If, on the other hand, you have chosen to represent a pair of binary attributes by one 4-level attribute because you want to be able to estimate the main effects and the two-factor interaction effect between these two attributes, then it makes sense to use the contrast matrix for the 2^2 factorial design. The choice of B will not affect the structure of the C^{-1} matrix if the choice experiment is optimal: C^{-1} will be a diagonal matrix. On the other hand, if the choice experiment is not optimal, then it is sometimes possible for one B matrix to result in a diagonal C^{-1} matrix and for another choice of B to give a C^{-1} matrix which is block diagonal (so the effects of interest are estimated independently of each other but the particular components within an effect that are determined by the choice of B matrix are not estimated independently).

Bunch et al. (1996) give a construction in which an OMEP is used for the treatment combinations in the first option, and subsequent options are constructed by shifting the levels of each attribute by adding a constant, using modulo arithmetic. This method is really a subset of the methods described in this chapter.

Appendix

6. A.1 Optimal Designs for $m = 2$ and $k = 2$

6. A.2 Optimal Designs for $m = 2$ and $k = 3$

6. A.3 Optimal Designs for $m = 2$ and $k = 4$

6. A.4 Optimal Designs for $m = 2$ and $k = 5$

6. A.5 Optimal Designs for $m = 3$ and $k = 2$

6. A.6 Optimal Designs for $m = 4$ and $k = 2$

6. A.7 Optimal Designs for Symmetric Attributes for $m = 2$

6. A.1 OPTIMAL DESIGNS FOR $m = 2$ AND $k = 2$

ℓ_1	ℓ_2	Difference Vectors	$\det(C_{opt})$
2	2	(01), (10) & (11)	$(\frac{1}{6})^3$
2	3	(01), (10) & (11)	$(\frac{1}{10})^5$
2	4	(01), (10) & (11)	$(\frac{1}{14})^7$
2	5	(01), (10) & (11)	$(\frac{1}{18})^9$
2	6	(01), (10) & (11)	$(\frac{1}{22})^{11}$
2	7	(01), (10) & (11)	$(\frac{1}{26})^{13}$
2	8	(01), (10) & (11)	$(\frac{1}{30})^{15}$
3	3	(01), (10) & (11)	$(\frac{1}{16})^8$
3	4	(01), (10) & (11)	$(\frac{1}{22})^{11}$
3	5	(01), (10) & (11)	$(\frac{1}{28})^{14}$
3	6	(01), (10) & (11)	$(\frac{1}{34})^{17}$
3	7	(01), (10) & (11)	$(\frac{1}{40})^{20}$
3	8	(01), (10) & (11)	$(\frac{1}{46})^{23}$
4	4	(01), (10) & (11)	$(\frac{1}{30})^{15}$
4	5	(01), (10) & (11)	$(\frac{1}{38})^{19}$
4	6	(01), (10) & (11)	$(\frac{1}{46})^{23}$
4	7	(01), (10) & (11)	$(\frac{1}{54})^{27}$
4	8	(01), (10) & (11)	$(\frac{1}{62})^{31}$
5	5	(01), (10) & (11)	$(\frac{1}{48})^{24}$
5	6	(01), (10) & (11)	$(\frac{1}{58})^{29}$
5	7	(01), (10) & (11)	$(\frac{1}{68})^{34}$
5	8	(01), (10) & (11)	$(\frac{1}{78})^{39}$
6	6	(01), (10) & (11)	$(\frac{1}{70})^{35}$
6	7	(01), (10) & (11)	$(\frac{1}{82})^{41}$
6	8	(01), (10) & (11)	$(\frac{1}{94})^{47}$
7	7	(01), (10) & (11)	$(\frac{1}{96})^{48}$
7	8	(01), (10) & (11)	$(\frac{1}{110})^{55}$
8	8	(01), (10) & (11)	$(\frac{1}{126})^{63}$

6. A.2 OPTIMAL DESIGNS FOR $m = 2$ AND $k = 3$

ℓ_1	ℓ_2	ℓ_3	Difference Vectors	$\det(C_{\text{opt}})$
2	2	2	(011),(101) & (110)	$(\frac{1}{12})^6$
2	2	3	(011),(101) & (110)	$(\frac{1}{15})^1(\frac{1}{20})^8$
2	2	4	(011),(101) & (110)	$(\frac{3}{56})^1(\frac{1}{28})^{11}$
2	2	5	(011),(101) & (110)	$(\frac{1}{36})^1(\frac{1}{36})^1(\frac{1}{36})^4(\frac{2}{45})^1(\frac{1}{36})^4(\frac{1}{36})^4$
2	2	6	(011),(101) & (110)	$(\frac{1}{44})^1(\frac{1}{44})^1(\frac{1}{44})^5(\frac{5}{132})^1(\frac{1}{44})^5(\frac{1}{44})^5$
2	2	7	(011),(101) & (110)	$(\frac{1}{52})^1(\frac{1}{52})^1(\frac{1}{52})^6(\frac{3}{91})^1(\frac{1}{52})^6(\frac{1}{52})^6$
2	2	8	(011),(101) & (110)	$(\frac{1}{60})^1(\frac{1}{60})^1(\frac{1}{60})^7(\frac{7}{240})^1(\frac{1}{60})^7(\frac{1}{60})^7$
2	3	3	(011), (101) & (110)	$(\frac{1}{36})^1(\frac{1}{32})^2(\frac{1}{32})^2(\frac{11}{288})^2(\frac{11}{288})^2(\frac{1}{32})^4$
2	3	4	(011), (101) & (110)	$(\frac{5}{264})^1(\frac{1}{44})^2(\frac{1}{44})^3(\frac{1}{33})^2(\frac{7}{264})^3(\frac{1}{44})^6$
2	3	5	(011), (101) & (110)	$(\frac{1}{70})^1(\frac{1}{56})^2(\frac{1}{56})^4(\frac{1}{40})^2(\frac{17}{840})^{11}(\frac{1}{56})^8$
2	3	6	(011), (101) & (110)	$(\frac{7}{612})^1(\frac{1}{68})^2(\frac{1}{68})^5(\frac{13}{612})^2(\frac{5}{306})^5(\frac{1}{68})^{10}$
2	3	7	(011), (101) & (110)	$(\frac{1}{105})^1(\frac{1}{80})^2(\frac{1}{80})^6(\frac{31}{1680})^2(\frac{23}{1680})^6(\frac{1}{80})^{12}$
2	3	8	(011), (101) & (110)	$(\frac{3}{368})^1(\frac{1}{92})^2(\frac{1}{92})^7(\frac{3}{184})^2(\frac{13}{1104})^7(\frac{1}{92})^{14}$
2	4	4	(101) & (110)	$(\frac{1}{32})^1(\frac{1}{96})^3(\frac{1}{96})^3(\frac{}{48})^3(\frac{1}{48})^3(\frac{1}{48})^9$
2	4	5	(101) & (110)	$(\frac{1}{40})^1(\frac{1}{140})^3(\frac{1}{112})^4(\frac{1}{56})^3(\frac{9}{560})^4(\frac{9}{560})^{12}$
2	4	6	(101) & (110)	$(\frac{1}{48})^1(\frac{1}{192})^3(\frac{1}{128})^5(\frac{1}{64})^3(\frac{5}{384})^5(\frac{5}{384})^{15}$
2	4	7	(101) & (110)	$(\frac{1}{56})^1(\frac{1}{252})^3(\frac{1}{144})^6(\frac{1}{72})^3(\frac{11}{1008})^6(\frac{11}{1008})^{18}$
2	4	8	(101) & (110)	$(\frac{1}{64})^1(\frac{1}{320})^3(\frac{1}{160})^7(\frac{1}{80})^3(\frac{3}{320})^7(\frac{3}{320})^{21}$
2	5	5	(101) & (110)	$(\frac{1}{50})^1(\frac{1}{160})^4(\frac{1}{160})^4(\frac{11}{800})^4(\frac{11}{800})^4(\frac{1}{80})^{16}$
2	5	6	(101) & (110)	$(\frac{1}{60})^1(\frac{1}{216})^4(\frac{1}{180})^5(\frac{13}{1080})^4(\frac{1}{90})^5(\frac{11}{1080})^{20}$
2	5	7	(101) & (110)	$(\frac{1}{70})^1(\frac{1}{280})^4(\frac{1}{200})^6(\frac{3}{280})^4(\frac{13}{1400})^6(\frac{3}{350})^{24}$

ℓ_1	ℓ_2	ℓ_3	**Difference Vectors**	$\det(C_{\text{opt}})$
2	5	8	(101) & (110)	$(\frac{1}{80})^1(\frac{1}{352})^4(\frac{1}{220})^7(\frac{17}{1760})^4(\frac{7}{880})^7(\frac{13}{1760})^{28}$
2	6	6	(101) & (110)	$(\frac{1}{72})^1(\frac{1}{240})^5(\frac{1}{240})^5(\frac{7}{720})^5(\frac{7}{720})^5(\frac{1}{120})^{25}$
2	6	7	(101) & (110)	$(\frac{1}{84})^1(\frac{1}{308})^5(\frac{1}{264})^6(\frac{2}{231})^5(\frac{5}{616})^6(\frac{13}{1848})^{30}$
2	6	8	(101) & (110)	$(\frac{1}{96})^1(\frac{1}{384})^5(\frac{1}{288})^7(\frac{1}{128})^5(\frac{1}{144})^7(\frac{7}{1152})^{35}$
2	7	8	(101) & (110)	$(\frac{1}{112})^1(\frac{1}{416})^6(\frac{1}{364})^7(\frac{19}{2912})^6(\frac{9}{1456})^7(\frac{15}{2912})^{42}$
2	8	8	(101) & (110)	$(\frac{1}{128})^1(\frac{1}{448})^7(\frac{1}{448})^7(\frac{5}{896})^7(\frac{5}{896})^7(\frac{1}{224})^{49}$
3	3	3	(011), (101) & (110)	$(\frac{1}{54})^2(\frac{1}{54})^2(\frac{1}{54})^2(\frac{5}{216})^4(\frac{5}{216})^4(\frac{5}{216})^4$
3	3	4	(011), (101) & (110)	$(\frac{5}{384})^2(\frac{5}{384})^2(\frac{1}{72})^3(\frac{7}{384})^4(\frac{19}{1152})^6(\frac{19}{1152})^6$
3	3	5	(011), (101) & (110)	$(\frac{1}{100})^2(\frac{1}{100})^2(\frac{1}{90})^4(\frac{3}{200})^4(\frac{23}{1800})^8(\frac{23}{1800})^8$
3	3	6	(011), (101) & (110)	$(\frac{7}{864})^2(\frac{7}{864})^2(\frac{1}{108})^5(\frac{11}{864})^4(\frac{1}{96})^{10}(\frac{1}{96})^{10}$
3	3	7	(011), (101) & (110)	$(\frac{1}{147})^2(\frac{1}{147})^2(\frac{1}{126})^6(\frac{13}{1176})^4(\frac{31}{3528})^{12}(\frac{31}{3528})^{12}$
3	3	8	(011), (101) & (110)	$(\frac{3}{512})^2(\frac{3}{512})^2(\frac{1}{144})^7(\frac{5}{512})^4(\frac{35}{4608})^{14}(\frac{35}{4608})^{14}$
3	4	4	(011), (101) & (110)	$(\frac{1}{112})^2(\frac{5}{504})^3(\frac{5}{504})^3(\frac{13}{1008})^6(\frac{13}{1008})^6(\frac{1}{84})^9$
3	4	5	(011), (101) & (110)	$(\frac{7}{1040})^2(\frac{1}{130})^3(\frac{5}{624})^4(\frac{11}{1040})^6(\frac{31}{3120})^8(\frac{29}{3120})^{12}$
3	4	6	(011), (101) & (110)	$(\frac{1}{186})^2(\frac{7}{1116})^3(\frac{5}{744})^5(\frac{5}{558})^6(\frac{1}{124})^{10}(\frac{17}{2232})^{15}$
3	4	7	(011), (101) & (110)	$(\frac{1}{224})^2(\frac{1}{189})^3(\frac{5}{864})^6(\frac{47}{6048})^6(\frac{41}{6048})^{12}(\frac{13}{2016})^{18}$
3	4	8	(011), (101) & (110)	$(\frac{5}{1312})^2(\frac{3}{656})^3(\frac{5}{984})^7(\frac{9}{1312})^6(\frac{23}{3936})^{14}(\frac{11}{1968})^{21}$
3	5	5	(011), (101) & (110)	$(\frac{1}{200})^2(\frac{1}{160})^4(\frac{1}{160})^4(\frac{13}{1600})^8(\frac{13}{1600})^8(\frac{7}{960})^{16}$
3	5	6	(011), (101) & (110)	$(\frac{3}{760})^2(\frac{7}{1368})^4(\frac{1}{190})^5(\frac{47}{6840})^8(\frac{1}{152})^{10}(\frac{41}{6840})^{20}$
3	5	7	(101) & (110)	$(\frac{1}{140})^2(\frac{1}{420})^4(\frac{1}{300})^6(\frac{1}{168})^8(\frac{23}{4200})^{12}(\frac{1}{175})^{24}$
3	5	8	(101) & (110)	$(\frac{1}{160})^2(\frac{1}{528})^4(\frac{1}{330})^7(\frac{7}{1320})^8(\frac{5}{1056})^{14}(\frac{13}{2640})^{28}$
3	6	6	(101) & (110)	$(\frac{1}{144})^2(\frac{1}{360})^5(\frac{1}{360})^5(\frac{1}{180})^{10}(\frac{1}{180})^{10}(\frac{1}{180})^{25}$

ℓ_1	ℓ_2	ℓ_3	**Difference Vectors**	$\det(C_{opt})$
3	6	7	(101) & (110)	$(\frac{1}{168})^2(\frac{1}{462})^5(\frac{1}{396})^6(\frac{3}{616})^{10}(\frac{13}{2772})^{12}(\frac{13}{2772})^{30}$
3	6	8	(101) & (110)	$(\frac{1}{192})^2(\frac{1}{576})^5(\frac{1}{432})^7(\frac{5}{1152})^{10}(\frac{7}{1728})^{14}(\frac{7}{1728})^{35}$
3	7	7	(101) & (110)	$(\frac{1}{196})^2(\frac{1}{504})^6(\frac{1}{504})^6(\frac{29}{7056})^{12}(\frac{29}{7056})^{12}(\frac{1}{252})^{36}$
3	7	8	(101) & (110)	$(\frac{1}{224})^2(\frac{1}{624})^6(\frac{1}{546})^7(\frac{1}{273})^{12}(\frac{31}{8736})^{14}(\frac{5}{1456})^{42}$
3	8	8	(101) & (110)	$(\frac{1}{256})^2(\frac{1}{672})^7(\frac{1}{672})^7(\frac{17}{5376})^{14}(\frac{17}{5376})^{14}(\frac{1}{336})^{49}$
4	4	4	(011), (101) & (110)	$(\frac{1}{144})^3(\frac{1}{144})^3(\frac{1}{144})^3(\frac{1}{108})^9(\frac{1}{108})^9(\frac{1}{108})^9$
4	4	5	(011), (101) & (110)	$(\frac{7}{1320})^3(\frac{7}{1320})^3(\frac{1}{176})^4(\frac{1}{132})^9(\frac{19}{2640})^{12}(\frac{19}{2640})^{12}$
4	4	6	(011), (101) & (110)	$(\frac{1}{234})^3(\frac{1}{234})^3(\frac{1}{208})^5(\frac{1}{156})^9(\frac{11}{1872})^{15}(\frac{11}{1872})^{15}$
4	4	7	(011), (101) & (110)	$(\frac{1}{280})^3(\frac{1}{280})^3(\frac{1}{240})^6(\frac{1}{180})^9(\frac{5}{1008})^{18}(\frac{5}{1008})^{18}$
4	4	8	(011), (101) & (110)	$(\frac{5}{1632})^3(\frac{5}{1632})^3(\frac{1}{272})^7(\frac{1}{204})^9(\frac{7}{1632})^{21}(\frac{7}{1632})^{21}$
4	5	5	(011), (101) & (110)	$(\frac{1}{250})^3(\frac{7}{1600})^4(\frac{7}{1600})^4(\frac{47}{8000})^{12}(\frac{47}{8000})^{12}(\frac{9}{1600})^{16}$
4	5	6	(011), (101) & (110)	$(\frac{3}{940})^3(\frac{1}{282})^4(\frac{7}{1880})^5(\frac{7}{1410})^{12}(\frac{9}{1880})^{15}(\frac{13}{2820})^{20}$
4	5	7	(011), (101) & (110)	$(\frac{1}{378})^3(\frac{1}{336})^4(\frac{7}{2160})^6(\frac{13}{3024})^{12}(\frac{61}{15120})^{18}(\frac{59}{15120})^{24}$
4	5	8	(011), (101) & (110)	$(\frac{11}{4880})^3(\frac{5}{1952})^4(\frac{7}{2440})^7(\frac{37}{9760})^{12}(\frac{17}{4880})^{21}(\frac{33}{9760})^{28}$
4	6	6	(011), (101) & (110)	$(\frac{1}{396})^3(\frac{1}{330})^5(\frac{3}{330})^5(\frac{2}{495})^{15}(\frac{2}{495})^{15}(\frac{1}{264})^{25}$
4	6	7	(011), (101) & (110)	$(\frac{11}{5292})^3(\frac{1}{392})^5(\frac{1}{378})^6(\frac{37}{10584})^{15}(\frac{1}{294})^{18}(\frac{17}{5292})^{30}$
4	6	8	(011), (101) & (110)	$(\frac{1}{568})^3(\frac{5}{2272})^5(\frac{1}{426})^7(\frac{7}{2272})^{15}(\frac{5}{1704})^{21}(\frac{19}{6816})^{35}$
4	7	8	(101) & (110)	$(\frac{1}{336})^3(\frac{1}{832})^6(\frac{1}{728})^7(\frac{15}{5824})^{18}(\frac{11}{4368})^{21}(\frac{15}{5824})^{42}$
4	8	8	(101) & (110)	$(\frac{1}{384})^3(\frac{1}{896})^7(\frac{1}{896})^7(\frac{1}{448})^{21}(\frac{1}{448})^{21}(\frac{1}{448})^{49}$
5	5	5	(011), (101) & (110)	$(\frac{1}{300})^4(\frac{1}{300})^4(\frac{1}{300})^4(\frac{11}{2400})^{16}(\frac{11}{2400})^{16}(\frac{11}{2400})^{16}$
5	5	6	(011), (101) & (110)	$(\frac{3}{1120})^4(\frac{3}{1120})^4(\frac{1}{350})^5(\frac{13}{3360})^{16}(\frac{3}{800})^{20}(\frac{3}{800})^{20}$
5	5	7	(011), (101) & (110)	$(\frac{1}{448})^4(\frac{1}{448})^4(\frac{1}{400})^6(\frac{3}{896})^{16}(\frac{71}{22400})^{24}(\frac{71}{22400})^{24}$

ℓ_1	ℓ_2	ℓ_3	**Difference Vectors**	$\det(C_{\mathbf{opt}})$
5	5	8	(011), (101) & (110)	$(\frac{11}{5760})^4(\frac{11}{5760})^4(\frac{1}{450})^7(\frac{17}{5760})^{16}(\frac{79}{28800})^{28}(\frac{79}{28800})^{28}$
5	6	6	(011), (101) & (110)	$(\frac{1}{468})^4(\frac{3}{1300})^5(\frac{3}{1300})^5(\frac{37}{11700})^{20}(\frac{37}{11700})^{20}(\frac{1}{325})^{25}$
5	6	7	(011), (101) & (110)	$(\frac{11}{6216})^4(\frac{1}{518})^5(\frac{3}{1480})^6(\frac{17}{6216})^{20}(\frac{83}{31080})^{24}(\frac{27}{10360})^{30}$
5	6	8	(011), (101) & (110)	$(\frac{1}{664})^4(\frac{11}{6640})^5(\frac{3}{1660})^7(\frac{1}{415})^{20}(\frac{23}{9960})^{28}(\frac{3}{1328})^{35}$
5	7	8	(011), (101) & (110)	$(\frac{13}{10528})^4(\frac{11}{7520})^6(\frac{1}{658})^7(\frac{107}{52640})^{24}(\frac{3}{1504})^{28}(\frac{101}{52640})^{42}$
5	8	8	(011), (101) & (110)	$(\frac{1}{960})^4(\frac{11}{8400})^7(\frac{11}{8400})^7(\frac{59}{33600})^{28}(\frac{59}{33600})^{28}(\frac{1}{600})^{49}$
6	6	6	(011), (101) & (110)	$(\frac{1}{540})^5(\frac{1}{540})^5(\frac{1}{540})^5(\frac{7}{2700})^{25}(\frac{7}{2700})^{25}(\frac{7}{2700})^{25}$
6	6	7	(011), (101) & (110)	$(\frac{11}{7140})^5(\frac{11}{7140})^5(\frac{1}{612})^6(\frac{4}{1785})^{25}(\frac{47}{21420})^{30}(\frac{47}{21420})^{30}$
6	6	8	(011), (101) & (110)	$(\frac{1}{760})^5(\frac{1}{760})^5(\frac{1}{684})^7(\frac{3}{1520})^{25}(\frac{13}{6840})^{35}(\frac{13}{6840})^{35}$
6	7	7	(011), (101) & (110)	$(\frac{1}{784})^5(\frac{11}{8064})^6(\frac{11}{8064})^6(\frac{107}{56448})^{30}(\frac{107}{56448})^{30}(\frac{5}{2688})^{36}$
6	7	8	(011), (101) & (110)	$(\frac{13}{11984})^5(\frac{1}{856})^6(\frac{11}{8988})^7(\frac{5}{2996})^{30}(\frac{59}{35952})^{35}(\frac{29}{17976})^{42}$
6	8	8	(011), (101) & (110)	$(\frac{1}{1088})^5(\frac{1}{952})^7(\frac{1}{952})^7(\frac{11}{7616})^{35}(\frac{11}{7616})^{35}(\frac{1}{714})^{49}$
7	7	7	(011), (101) & (110)	$(\frac{1}{882})^6(\frac{1}{882})^6(\frac{1}{882})^6(\frac{17}{10584})^{36}(\frac{17}{10584})^{36}(\frac{17}{10584})^{36}$
7	7	8	(011), (101) & (110)	$(\frac{13}{13440})^6(\frac{13}{13440})^6(\frac{1}{980})^7(\frac{19}{13440})^{36}(\frac{131}{94080})^{42}(\frac{131}{94080})^{42}$
7	8	8	(011), (101) & (110)	$(\frac{1}{1216})^6(\frac{13}{14896})^7(\frac{13}{14896})^7(\frac{73}{59584})^{42}(\frac{73}{59584})^{42}(\frac{9}{7448})^{49}$
8	8	8	(011), (101) & (110)	$(\frac{1}{1344})^7(\frac{1}{1344})^7(\frac{1}{1344})^7(\frac{5}{4704})^{49}(\frac{5}{4704})^{49}(\frac{5}{4704})^{49}$

6. A.3 OPTIMAL DESIGNS FOR $m = 2$ AND $k = 4$

ℓ_1	ℓ_2	ℓ_3	ℓ_4	Difference Vectors
2	2	2	2	(0011), (0101), (0110), (0111), (1001), (1010), (1011), (1100), (1101) & (1110)
2	2	2	3–8	(0111), (1011), (1101) & (1110)
2	2	3	3	(0111), (1011), (1101) & (1110)
2	2	3	4–8	(0111), (1101) & (1110) or (1011), (1101) & (1110)
2	2	4	4–5	(0111), (1101) & (1110) or (1011), (1101) & (1110)
2	2	4	6–8	(0110), (1101) & (1110) or (1010), (1101) & (1110)
2	2	5	5	(0111), (1101) & (1110) or (1011), (1101) & (1110)
2	2	5	6–8	(0110), (1101) & (1110) or (1010), (1101) & (1110)
2	2	6	6	(0101), (1101) & (1110) or (0110), (1101) & (1110) or (1001), (1101) & (1110) or (1010), (1101) & (1110)
2	2	6	7–8	(0110), (1101) & (1110) or (1010), (1101) & (1110)
2	2	7	7	(0101), (1101) & (1110) or (0110), (1101) & (1110) or (1001), (1101) & (1110) or (1010), (1101) & (1110)
2	2	7	8	(0110), (1101) & (1110) or (1010), (1101) & (1110)
2	2	8	8	(0101), (1101) & (1110) or (0110), (1101) & (1110) or (1001), (1101) & (1110) or (1010), (1101) & (1110)
2	3	3	3–8	(1011), (1101) & (1110)
2	3	4	4–8	(1011), (1101) & (1110)
2	3	5	5–8	(1101) & (1110)
2	3	6	6–8	(1101) & (1110)
2	3	7	7–8	(1101) & (1110)
2	3	8	8	(1101) & (1110)
2	4	4	4–8	(1011), (1101) & (1110)
2	4	5	5–8	(1011), (1101) & (1110)
2	4	6	6	(1011), (1101) & (1110)

ℓ_1	ℓ_2	ℓ_3	ℓ_4	**Difference Vectors**
2	4	6	7–8	(1101) & (1110)
2	4	7	7–8	(1101) & (1110)
2	4	8	8	(1101) & (1110)
2	5	5	5–8	(1011), (1101) & (1110)
2	5	6	6–8	(1011), (1101) & (1110)
2	5	7	7–8	(1011), (1101) & (1110)
2	5	8	8	(1101) & (1110)
2	6	6	6–8	(1011), (1101) & (1110)
2	6	7	7–8	(1011), (1101) & (1110)
2	6	8	8	(1011), (1101) & (1110)
2	7	7	7–8	(1011), (1101) & (1110)
2	7	8	8	(1011), (1101) & (1110)
2	8	8	8	(1011), (1101) & (1110)
3	3	3	3–7	(0111), (1011), (1101) & (1110)
3	3	3	8	(0111), (1011) & (1110) or (0111), (1101) & (1110) or (1011), (1101) & (1110)
3	3	4	4	(0111), (1011), (1101) & (1110)
3	3	4	5–8	(1101) & (1110)
3	3	5	5–8	(1101) & (1110)
3	3	6	6–8	(1101) & (1110)
3	3	7	7–8	(1101) & (1110)
3	3	8	8	(1101) & (1110)
3	4	4	4–8	(1011), (1101) & (1110)
3	4	5	5–6	(1011), (1101) & (1110)
3	4	5	7–8	(1101) & (1110)
3	4	6	6–8	(1101) & (1110)
3	4	7	7–8	(1101) & (1110)
3	4	8	8	(1101) & (1110)
3	5	5	5–8	(1011), (1101) & (1110)
3	5	6	6–8	(1011), (1101) & (1110)
3	5	7	7	(1011), (1101) & (1110)
3	5	7	8	(1101) & (1110)
3	5	8	8	(1101) & (1110)
3	6	6	6–8	(1011), (1101) & (1110)
3	6	7	7–8	(1011), (1101) & (1110)
3	6	8	8	(1011), (1101) & (1110)
3	7	7	7–8	(1011), (1101) & (1110)
3	7	8	8	(1011), (1101) & (1110)
3	8	8	8	(1011), (1101) & (1110)

ℓ_1	ℓ_2	ℓ_3	ℓ_4	**Difference Vectors**
4	4	4	4–8	(0111), (1011), (1101) & (1110)
4	4	5	5	(0111), (1011), (1101) & (1110)
4	4	5	6	(0111), (1101) & (1110) or (1011), (1101) & (1110)
4	4	5	7–8	(1101) & (1110)
4	4	6	6–8	(1101) & (1110)
4	4	7	7–8	(1101) & (1110)
4	4	8	8	(1101) & (1110)
4	5	5	5–8	(1011), (1101) & (1110)
4	5	6	6–8	(1011), (1101) & (1110)
4	5	7	7–8	(1101) & (1110)
4	5	8	8	(1101) & (1110)
4	6	6	6–8	(1011), (1101) & (1110)
4	6	7	7–8	(1011), (1101) & (1110)
4	6	8	8	(1011), (1101) & (1110)
4	7	7	7–8	(1011), (1101) & (1110)
4	7	8	8	(1011), (1101) & (1110)
4	8	8	8	(1011), (1101) & (1110)
5	5	6	7–8	(0111), (1101) & (1110) or (1011), (1101) & (1110)
5	5	5	5–8	(0111), (1011), (1101) & (1110)
5	5	6	6	(0111), (1011), (1101) & (1110)
5	5	7	7–8	(1101) & (1110)
5	5	8	8	(1101) & (1110)
5	6	6	6–8	(1011), (1101) & (1110)
5	6	7	7–8	(1011), (1101) & (1110)
5	6	8	8	(1011), (1101) & (1110)
5	7	7	7–8	(1011), (1101) & (1110)
5	7	8	8	(1011), (1101) & (1110)
5	8	8	8	(1011), (1101) & (1110)
6	6	6	6–8	(0111), (1011), (1101) & (1110)
6	6	7	7–8	(0111), (1011), (1101) & (1110)
6	6	8	8	(0111), (1101) & (1110) or (1011), (1101) & (1110)
6	7	7	7–8	(1011), (1101) & (1110)
6	7	8	8	(1011), (1101) & (1110)
6	8	8	8	(1011), (1101) & (1110)
7	7	7	7–8	(0111), (1011), (1101) & (1110)
7	7	8	8	(0111), (1011), (1101) & (1110)
7	8	8	8	(0111), (1011), (1101) & (1110)
8	8	8	8	(0111), (1011), (1101) & (1110)

6. A.4 OPTIMAL DESIGNS FOR $m = 2$ AND $k = 5$

ℓ_1 ℓ_2 ℓ_3 ℓ_4 ℓ_5	Difference Vectors
2 2 2 2 2	(00111), (01011), (01101), (01110), (10011), (10101), (10110), (11001), (11010) & (11100)
2 2 2 3 3	(00111), (01011), (01101), (01110), (10011), (10101), (10110), (11001), (11010), (11101) & (11110)
2 2 4 4 4	(10011), (10101), (10110), (11011), (11101) & (11110)
2 2 4 4 8	(01110), (10110), (11011), (11101) & (11110)
2 2 4 8 8	(11101), (11110) & one of (01101), (01110), (10101) or (10110)
2 3 3 4 5	(10011), (11101) & (11110)
2 4 4 4 4	(10111), (11011), (11101) & (11110)
2 4 4 4 8	(01111), (10111), (11011), (11101) & (11110)
2 4 4 8 8	(11101) & (11110)
2 4 8 8 8	(11011), (11101) & (11110)
2 8 8 8 8	(10111), (11011), (11101) & (11110)
3 3 3 3 3	(00111), (01011), (01101), (01110), (01111), (10011), (10101), (10110), (10111), (11001), (11010), (11011), (11100), (11101) & (11110)
4 4 4 4 4	(01111), (10111), (11011), (11101) & (11110)
4 4 4 4 8	(11110) and any two of (01111), (10111), (11011) & (11101)
4 4 4 8 8	(11101) & (11110)
4 4 8 8 8	(11011), (11101) & (11110)
4 8 8 8 8	(10111), (11011), (11101) & (11110)
5 5 5 5 5	(01111), (10111), (11011), (11101) & (11110)
6 6 6 6 6	(01111), (10111), (11011), (11101) & (11110)
7 7 7 7 7	(01111), (10111), (11011), (11101) & (11110)
8 8 8 8 8	(01111), (10111), (11011), (11101) & (11110)

6. A.5 OPTIMAL DESIGNS FOR $m = 3$ AND $k = 2$

ℓ_1	ℓ_2	Difference Vectors	$\det(C_{\text{opt}})$
2	2	(01,10,11)	$(\frac{2}{9})^3$
2	3	(01,01,01), (01,10,11) & (01,11,11)	$(\frac{2}{15})^5$
2	4	(01,01,01), (01,10,11) & (01,11,11)	$(\frac{2}{21})^7$
2	5	(01,01,01), (01,10,11) & (01,11,11)	$(\frac{2}{27})^9$
2	6	(01,01,01), (01,10,11) & (01,11,11)	$(\frac{2}{33})^{11}$
2	7	(01,01,01), (01,10,11) & (01,11,11)	$(\frac{2}{39})^{13}$
2	8	(01,01,01), (01,10,11) & (01,11,11)	$(\frac{2}{45})^{15}$
3	3	(01,01,01), (10,10,10) & (11,11,11) or (01,10,11), (01,11,11) & (10,11,11) or (01,01,01), (01,10,11), (01,11,11), (10,10,10), (10,11,11) & (11,11,11)	$(\frac{1}{12})^8$
3	4	(01,01,01), (01,10,11), (01,11,11), (10,10,10), (10,11,11) & (11,11,11)	$(\frac{2}{33})^{11}$
3	5	(01,01,01), (01,10,11), (01,11,11), (10,10,10), (10,11,11) & (11,11,11)	$(\frac{1}{21})^{14}$
3	6	(01,01,01), (01,10,11), (01,11,11), (10,10,10), (10,11,11) & (11,11,11)	$(\frac{2}{51})^{17}$
3	7	(01,01,01), (01,10,11), (01,11,11), (10,10,10), (10,11,11) & (11,11,11)	$(\frac{1}{30})^{20}$
3	8	(01,01,01), (01,10,11), (01,11,11), (10,10,10), (10,11,11) & (11,11,11)	$(\frac{2}{69})^{23}$
4	4	(01,10,11) & (11,11,11) or (01,01,01), (01,11,11) & (10,10,10) or (01,01,01), (01,10,11), (01,11,11), (10,10,10), (10,11,11) & (11,11,11)	$(\frac{2}{45})^{15}$
4	5	(01,01,01), (01,10,11), (01,11,11), (10,10,10), (10,11,11) & (11,11,11)	$(\frac{2}{57})^{19}$

ℓ_1	ℓ_2	Difference Vectors	$\det(C_{\text{opt}})$
4	6	(01,01,01), (01,10,11) & (11,11,11) or (01,11,11), (10,10,10) & (10,11,11) or (01,01,01), (01,10,11), (01,11,11), (10,10,10), (10,11,11) & (11,11,11)	$(\frac{2}{69})^{23}$
4	7	(01,01,01), (01,10,11), (01,11,11), (10,10,10), (10,11,11) & (11,11,11)	$(\frac{2}{81})^{27}$
4	8	(01,01,01), (01,10,11), (01,11,11), (10,10,10), (10,11,11) & (11,11,11)	$(\frac{2}{93})^{31}$
5	5	(01,11,11) & (10,11,11) or (01,01,01), (01,10,11), (10,10,10) & (11,11,11) or (01,01,01), (01,10,11), (01,11,11), (10,10,10), (10,11,11) & (11,11,11)	$(\frac{1}{36})^{24}$
5	6	(01,01,01), (01,10,11), (01,11,11), (10,10,10), (10,11,11) & (11,11,11)	$(\frac{2}{87})^{29}$
5	7	(01,01,01), (01,10,11), (01,11,11), (10,10,10), (10,11,11) & (11,11,11)	$(\frac{1}{51})^{34}$
5	8	(01,01,01), (01,10,11), (01,11,11), (10,10,10), (10,11,11) & (11,11,11)	$(\frac{2}{117})^{39}$
6	6	(01,01,01), (01,10,11), (01,11,11), (10,10,10), (10,11,11) & (11,11,11)	$(\frac{2}{105})^{35}$
6	7	(01,01,01), (01,10,11), (01,11,11), (10,10,10), (10,11,11) & (11,11,11)	$(\frac{2}{123})^{41}$
6	8	(01,01,01), (01,10,11), (01,11,11), (10,10,10), (10,11,11) & (11,11,11)	$(\frac{2}{141})^{47}$
7	7	(01,01,01), (01,10,11), (01,11,11), (10,10,10), (10,11,11) & (11,11,11)	$(\frac{1}{72})^{48}$
7	8	(01,01,01), (01,10,11), (01,11,11), (10,10,10), (10,11,11) & (11,11,11)	$(\frac{2}{165})^{55}$
8	8	(01,01,01), (01,10,11), (01,11,11), (10,10,10), (10,11,11) & (11,11,11)	$(\frac{2}{189})^{63}$

6. A.6 OPTIMAL DESIGNS FOR $m = 4$ AND $k = 2$

ℓ_1	ℓ_2	Difference Vectors	$\det(C_{\text{opt}})$
2	3	(01,01,01,10,11,11), (01,01,10,10,11,11) & (01,01,10,11,11,11)	$(\frac{3}{50})^5$
2	4	(01,01,01,01,01,01), (01,01,10,10,11,11) & (01,01,11,11,11,11) or (01,01,01,01,01,01), (01,01,01,10,11,11), (01,01,10,11,11,11) & (01,01,11,11,11,11)	$(\frac{3}{28})^7$
2	5	(01,01,01,10,11,11) & (01,01,11,11,11,11)	$(\frac{3}{36})^9$
2	6	(01,01,01,10,11,11) & (01,01,11,11,11,11)	$(\frac{3}{44})^{11}$
2	7	(01,01,01,01,01,01), (01,01,10,10,11,11), (01,01,10,11,11,11) & (01,01,11,11,11,11)	$(\frac{3}{52})^{13}$
2	8	(01,01,01,01,01,01), (01,01,10,10,11,11), (01,01,10,11,11,11) & (01,01,11,11,11,11)	$(\frac{3}{60})^{15}$
3	3	(01,01,10,11,11,11) & (01,10,10,11,11,11) or (01,01,01,10,11,11), (01,01,10,10,11,11), (01,10,10,10,11,11) & (01,10,11,11,11,11) or (01,01,01,01,01,01), (01,01,01,10,11,11), (01,01,01,11,11,11), (01,01,10,10,11,11), (01,01,10,11,11,11), (01,10,10,10,11,11), (01,10,10,11,11,11), (01,01,11,11,11,11), (01,10,11,11,11,11) & (01,11,11,11,11,11)	$(\frac{3}{32})^8$

ℓ_1	ℓ_2	Difference Vectors	det(C_{opt})
3	4	(01,01,01,01,01,01), (01,01,01,11,11,11), (01,10,10,10,11,11) & (01,11,11,11,11,11) or (01,01,01,01,01,01), (01,01,01,10,11,11), (01,01,10,11,11,11), (01,10,10,11,11,11) & (01,10,11,11,11,11) or (01,01,01,01,01,01), (01,01,10,10,11,11), (01,01,10,11,11,11), (01,10,10,11,11,11) & (01,11,11,11,11,11) or (01,01,01,01,01,01), (01,01,10,11,11,11), (01,10,10,10,11,11), (01,01,11,11,11,11) & (01,11,11,11,11,11) or (01,01,01,10,11,11), (01,01,01,11,11,11), (01,01,10,10,11,11), (01,10,10,11,11,11) & (01,10,11,11,11,11) or (01,01,01,10,11,11), (01,01,01,11,11,11), (01,10,10,10,11,11),(01,01,11,11,11,11) & (01,10,11,11,11,11) or (01,01,01,11,11,11), (01,01,10,10,11,11), (01,10,10,10,11,11), (01,01,11,11,11,11) & (01,11,11,11,11,11) or (01,01,01,10,11,11), (01,01,10,10,11,11), (01,01,10,11,11,11), (01,10,10,11,11,11), (01,01,11,11,11,11) & (01,10,11,11,11,11) or (01,01,01,01,01,01), (01,01,01,10,11,11), (01,01,01,11,11,11),(01,01,10,10,11,11), (01,01,10,11,11,11), (01,10,10,10,11,11), (01,10,10,11,11,11), (01,01,11,11,11,11), (01,10,11,11,11,11)& (01,11,11,11,11,11)	$(\frac{3}{44})^{11}$

6. A.7 OPTIMAL DESIGNS FOR SYMMETRIC ATTRIBUTES FOR $m = 2$

Attributes	Levels	Difference Vectors
2	2–12	All difference vectors, i.e., (01), (10) & (11)
3	2–12	All difference vectors of weight 2 i.e. (011), (101) & (110)
4	2	All difference vectors of weight 2 & 3
	3–12	All difference vectors of weight 3
5	2	All difference vectors of weight 3
	3	All difference vectors of weight 3 & 4
	4–12	All difference vectors of weight 4
6	2	All difference vectors of weight 3 & 4
	3	All difference vectors of weight 4
	4	All difference vectors of weight 4 & 5
	5–12	All difference vectors of weight 5
7	2	All difference vectors of weight 4
	3–4	All difference vectors of weight 5
	5	All difference vectors of weight 5 & 6
	6–12	All difference vectors of weight 6
8	2	All difference vectors of weight 4 & 5
	3	All difference vectors of weight 5 & 6
	4–5	All difference vectors of weight 6
	6–12	All difference vectors of weight 7
9	2	All difference vectors of weight 5
	3	All difference vectors of weight 6
	4–6	All difference vectors of weight 7
	8–12	All difference vectors of weight 8
10	2	All difference vectors of weight 5 & 6
	3	All difference vectors of weight 7
	4	All difference vectors of weight 7 & 8
	5–7	All difference vectors of weight 8
	8–12	All difference vectors of weight 9

CHAPTER 7

VARIOUS TOPICS

We have now seen how to construct optimal and near-optimal designs for estimating main effects and main effects plus two-factor interactions for any number of attributes each with any number of levels and with choice sets of any (constant) size using the MNL model. In this chapter we touch on a number of other important topics in the design of choice experiments.

The first topic we consider is the design of optimal choice experiments when all choice sets contain a none option or an option which is common to all choice sets, a base alternative, or both a none option and a base alternative.

Next we discuss how to determine the best choice set size, in terms of the number of levels of each of the attributes, as well as how to compare choice sets of different sizes.

So far in this book we have not placed any restrictions on the number of attributes that can differ between the options in a choice set. Yet there is some evidence that respondents do not perform as consistently when there are many features to trade-off. We show how to construct optimal choice experiments when we limit the number of attributes which can be different between the options in each choice set.

All of the designs that have been developed in the earlier chapters have been optimized when we have no prior information about the values of the π_i. If we do have such prior information, then it can be used to calculate a different Λ matrix, as indicated in Chapter 3, and this matrix can be used to determine a modified C matrix (the matrix of contrasts is, of course, unaltered by prior information). We consider the use of prior information for two specific examples.

7.1 OPTIMAL STATED CHOICE EXPERIMENTS WHEN ALL CHOICE SETS CONTAIN A SPECIFIC OPTION

In this section we consider the construction of choice experiments in which all choice sets contain either a none option, a common base option, or both a none option and a common base option.

The none option is sometimes viewed as an option to defer choice and is sometimes included so that participants do not have to choose between two or more options that they find equally attractive (although Davidson (1970) has described an MNL model in which ties are allowed). A common base option, described by some combination of the attribute levels, is often included in choice experiments to investigate treatment options, where you want to compare the standard treatment option with a proposed new treatment option in each choice set. Finally we consider the design of choice experiments in which both a common base and a none option is included in every choice set.

7.1.1 Choice Experiments with a None Option

In this section we consider what happens when we adjoin a none option to each choice set in a stated preference choice experiment. It turns out that there is a simple relationship between the matrices for a forced choice stated preference experiment and those from the same choice sets with a none option adjoined to each choice set.

For a discussion of how to design experiments to avoid consumers choosing to defer choice (another way of viewing the none option) see Haaijer et al. (2001) and references cited therein. There are references in Chapter 1 to other non-mathematical issues.

We let B_f be the contrast matrix for the forced choice experiment (so $B_f = B_h$ and may contain contrasts for main effects and perhaps interaction effects), let Λ_f be the Λ matrix for the forced choice experiment, and let $C_f = B_f \Lambda_f B_f'$ be the information matrix for the forced choice experiment. We assume that B_f is $p \times L$ where there are p contrasts of interest, and note that Λ_f will contain rows and columns of 0s if not all treatment combinations appear in the choice experiment. We will use B_n, Λ_n, and $C_n = B_n \Lambda_n B_n'$ for the corresponding matrices when a none option has been included in each choice set. We assume that each choice set in the forced choice experiment has m options in it and that there are N such choice sets.

The next result establishes the relationship between B_f and B_n and between Λ_f and Λ_n. As usual we let $L = \prod_{q=1}^{k} \ell_q$.

■ LEMMA 7.1.1.

1. *Let* $d^2 = L(L+1)$. *Then*

$$B_n = \begin{bmatrix} B_f & \mathbf{0}_p \\ \frac{1}{d}\mathbf{j}_L' & \frac{-L}{d} \end{bmatrix}$$

where $\mathbf{0}_p$ *is a* $p \times 1$ *vector of zeroes and* none *is the final treatment combination. Thus* $B_n B_n' = I_{p+1}$.

2. *Let* r_i *be the number of times the ith treatment combination appears in the stated preference experiment and let* $\mathbf{r} = (r_1, \ldots, r_L)'$. *Let* D *be a matrix with these*

replication numbers on the diagonal. Then

$$\Lambda_n = \frac{1}{(m+1)^2 N} \begin{bmatrix} m^2 N \Lambda_f + D & -\mathbf{r} \\ -\mathbf{r}' & mN \end{bmatrix}.$$

Proof.

1. The only additional contrast is between none and some treatment combination. The divisor d ensures that the contrast is of unit length.

2. The none option will appear with each treatment combination as many times as that treatment combination appears in the design so

$$r_{\text{none}} = mN = \sum_i r_i.$$

Again only one additional row and column need to be adjoined to Λ. □

Now we are in a position to evaluate C_n.

■ THEOREM 7.1.1.

Consider a forced choice stated preference experiment in which a none option has been adjoined to each choice set. Then

$$C_n = \frac{1}{(m+1)^2 N} \begin{bmatrix} m^2 N C_f + B_f D B_f' & \frac{L+1}{d} B_f \mathbf{r} \\ \frac{L+1}{d} \mathbf{r}' B_f' & \frac{mN(L+1)}{L} \end{bmatrix}.$$

Proof. This result follows directly from the definition, noting that $\mathbf{j}'_L \mathbf{r} = mN$, $\Lambda_f \mathbf{j}_L = \mathbf{0}_L$ and $\mathbf{j}'_L D = \mathbf{r}$. □

If all the treatment combinations in the complete factorial appear r times in the choice experiment, then we can establish the following result.

■ COROLLARY 7.1.1.

Suppose that all the treatment combinations in the complete factorial appear r times in the choice experiment. Then

$$C_n = \frac{1}{(m+1)^2 N} \begin{bmatrix} m^2 C_f + \frac{m}{L} I_p & \mathbf{0}_p \\ \mathbf{0}_p' & \frac{mN(L+1)}{L} \end{bmatrix}.$$

Proof. Since all the treatment combinations appear r times we have that

$$\mathbf{r} = r\mathbf{j}_L = \frac{mN}{L}\mathbf{j}_L \text{ and } D = rI_L = \frac{mN}{L} I_L.$$

Thus we know that

$$B_f \mathbf{r} = \mathbf{0}_p \text{ and that } B_f D B_f' = \frac{mN}{L} I_p.$$

The result follows. □

■ COROLLARY 7.1.2.

The optimal designs for estimating main effects when a none option is included in each choice set have

$$\det(C_{\text{opt},M,n}) = \frac{m(L+1)}{L(m+1)^2} \times \prod_{q=1}^{k} \left[\frac{2\ell_q S_q + m(\ell_q - 1)}{(m+1)^2 (\ell_q - 1) L} \right]^{\ell_q - 1},$$

where S_q is the least upper bound for the sum of the differences for a particular attribute q.

Proof. The optimal designs for estimating main effects given in Chapters 4, 5, and 6 satisfy the conditions of Corollary 7.1.1. So if we adjoin a none option to each choice set in one of these optimal designs, then

$$C_{M,n} = \frac{1}{L(m+1)^2}\begin{bmatrix} \frac{2S_1\ell_1+m(\ell_1-1)}{\ell_1-1}I_{\ell_1-1} & \cdots & 0 & 0 \\ \vdots & \vdots & \vdots & \vdots \\ 0 & \cdots & \frac{2S_k\ell_k+m(\ell_k-1)}{\ell_k-1}I_{\ell_k-1} & 0 \\ 0 & \cdots & 0 & m(L+1) \end{bmatrix}.$$

The result follows. □

Hence the same designs that are optimal for the estimation of main effects in the forced choice setting are optimal for the estimation of main effects when a none option has been adjoined to each choice set. We can also calculate the effect of estimating just the main effects (i.e., C_n without the row and column for the none option) relative to the maximum determinant of C_M for the forced choice setting. We find that the efficiency of the design for the estimation of main effects only is reduced from 100% to

$$\frac{m^2}{(m+1)^2}\left[\prod_{q=1}^{k}\left(\frac{2\ell_q S_q + m(\ell_q-1)}{2\ell_q S_q}\right)^{(\ell_q-1)}\right]^{1/p}.$$

But adjoining a none option does change the properties of the design. If there is no none option then the optimal design for main effects cannot be used to give any information about two-factor interactions. The inclusion of a none option may make it possible to estimate two-factor interactions as well, although the efficiency may not be very high. The component of the information matrix corresponding to main effects is given by

$$\frac{1}{(m+1)^2N}(m^2NC_f + \frac{m}{L}I_p),$$

and so correlated effects will remain correlated after the introduction of a none option.

■ **EXAMPLE 7.1.1.**

Suppose that we have $k = 3$ attributes with $\ell_1 = \ell_2 = 2$ and $\ell_3 = 3$. If $m = 2$, the optimal design for estimating main effects in a forced choice setting is obtained by taking the complete factorial and adding the generator (1,1,1) to get the following 12 choice sets:

(000, 111), (001, 112), (002, 110), (010, 101), (011, 102), (012, 100),
(100, 011), (101, 012), (102, 010), (110, 001), (111, 002), (112, 000).

If we adjoin a none option to each of these choice sets then these 12 choice sets are still optimal for the estimation of main effects but the component of the information matrix corresponding to main effects has changed.

Let B_M be the normalized main effects contrast matrix. Then the normalized contrast matrix for the stated preference experiment when the none option is included is given by

$$B_n = \frac{1}{\sqrt{156}}\begin{bmatrix} & & & & & \sqrt{156}B_M & & & & & & & \mathbf{0}'_4 \\ 1 & 1 & 1 & 1 & 1 & 1 & 1 & 1 & 1 & 1 & 1 & 1 & -12 \end{bmatrix}.$$

For main effects only for the forced choice design we have

$$C_{M,f} = \frac{1}{48}\begin{bmatrix} 4 & 0 & 0 & 0 \\ 0 & 4 & 0 & 0 \\ 0 & 0 & 3 & 0 \\ 0 & 0 & 0 & 3 \end{bmatrix}.$$

For main effects only for the "12 plus none" design we have

$$C_{M,n} = \frac{1}{108}\begin{bmatrix} 6 & 0 & 0 & 0 & 0 \\ 0 & 6 & 0 & 0 & 0 \\ 0 & 0 & 5 & 0 & 0 \\ 0 & 0 & 0 & 5 & 0 \\ 0 & 0 & 0 & 0 & 26 \end{bmatrix}.$$

Thus we see that

$$\det(C_{\text{opt},M}) = \left(\frac{1}{12}\right)^2\left(\frac{1}{16}\right)^2 \text{ and that } \det(C_{\text{opt},M,n}) = \left(\frac{1}{18}\right)^2\left(\frac{5}{108}\right)^2\frac{13}{54}.$$

The determinant for the main effects only if the none option is included is $\left(\frac{1}{18}\right)^2\left(\frac{5}{108}\right)^2$. When we compare the efficiencies of the designs with none adjoined to forced choice designs, we use the determinant for the main effects only (or for main effects plus two-factor interactions only) so that we can see what we gain (or lose) by adjoining the none. In this case we see that if we adjoin a none option to the 12 choice sets which are optimal for estimating main effects, then the design is 67.2% efficient for estimating main effects plus two-factor interactions. The same set of 12 choice sets is now 70.3% efficient for estimating main effects only, however. This is calculated by

$$\left(\frac{\left(\frac{1}{18}\right)^2\left(\frac{5}{108}\right)^2}{\left(\frac{1}{12}\right)^2\left(\frac{1}{16}\right)^2}\right)^{1/4}.$$

If $m = 2$, the optimal design for estimating main effects plus two-factor interactions is obtained by taking the complete factorial and adding the generators (0,1,1), (1,0,1) and (1,1,0) and has the following 30 choice sets:

(000, 011), (001, 012), (002, 010), (010, 001), (011, 002), (012, 000),
(100, 111), (101, 112), (102, 110), (110, 101), (111, 102), (112, 100),
(000, 101), (001, 102), (002, 100), (010, 111), (011, 112), (012, 110),
(100, 001), (101, 002), (102, 000), (110, 011), (111, 012), (112, 010),
(000, 110), (001, 111), (002, 112), (010, 100), (011, 101), (012, 102).

In this example all effects are independently estimated in all of the designs discussed.

Table 7.1 gives the efficiencies of all the designs for estimating both main effects only and main effects plus two-factor interactions relative to the best forced choice experiment and considering only the effects of interest. □

In the next example adjoining a none option does not make the two-factor interaction effects estimable.

■ **EXAMPLE 7.1.2.**

Suppose that we have $k = 3$ attributes with $\ell_1 = \ell_2 = 2$ and $\ell_3 = 4$ levels. Suppose that we

Table 7.1 Efficiencies of the Four Designs Discussed in Example 7.1.1

Design	Estimating Main Effects	Estimating ME plus 2fi
12 forced choice	100%	0%
12 plus none	70.3%	67.2%
30 forced choice	69.3%	100%
30 plus none	56.5%	80.4%

use the choice sets given in Table 7.2 to estimate main effects plus two-factor interactions. Then we get that

$$C_{MT} = \begin{bmatrix} \frac{1}{16} & 0 & 0 & 0 & 0 & 0 & 0 & 0 & 0 & \frac{-1}{16\sqrt{5}} & \frac{-1}{32} & \frac{1}{32\sqrt{5}} \\ 0 & \frac{1}{16} & 0 & 0 & 0 & 0 & \frac{-1}{16\sqrt{5}} & \frac{-1}{32} & \frac{1}{32\sqrt{5}} & 0 & 0 & 0 \\ 0 & 0 & \frac{3}{80} & 0 & \frac{1}{80} & 0 & 0 & 0 & 0 & 0 & 0 & 0 \\ 0 & 0 & 0 & \frac{1}{32} & 0 & 0 & 0 & 0 & 0 & 0 & 0 & 0 \\ 0 & 0 & \frac{1}{80} & 0 & \frac{9}{160} & 0 & 0 & 0 & 0 & 0 & 0 & 0 \\ 0 & 0 & 0 & 0 & 0 & 0 & 0 & 0 & 0 & 0 & 0 & 0 \\ 0 & \frac{-1}{16\sqrt{5}} & 0 & 0 & 0 & 0 & \frac{1}{40} & 0 & \frac{-1}{80} & 0 & 0 & 0 \\ 0 & \frac{-1}{32} & 0 & 0 & 0 & 0 & 0 & \frac{1}{32} & 0 & 0 & 0 & 0 \\ 0 & \frac{1}{32\sqrt{5}} & 0 & 0 & 0 & 0 & \frac{-1}{80} & 0 & \frac{1}{160} & 0 & 0 & 0 \\ \frac{-1}{16\sqrt{5}} & 0 & 0 & 0 & 0 & 0 & 0 & 0 & 0 & \frac{1}{40} & 0 & \frac{-1}{80} \\ \frac{-1}{32} & 0 & 0 & 0 & 0 & 0 & 0 & 0 & 0 & 0 & \frac{1}{32} & 0 \\ \frac{1}{32\sqrt{5}} & 0 & 0 & 0 & 0 & 0 & 0 & 0 & 0 & \frac{-1}{80} & 0 & \frac{1}{160} \end{bmatrix}.$$

We see that $\det(C_{MT}) = 0$ and so no effects can be estimated. If we now assume that each choice set has a none option adjoined, then a row and column is adjoined to C_{MT} for the none option and we have

$$C_{MT,n} = \begin{bmatrix} \frac{1}{24} & 0 & 0 & 0 & 0 & 0 & 0 & 0 & 0 & \frac{-1}{24\sqrt{5}} & \frac{-1}{48} & \frac{1}{48\sqrt{5}} & 0 \\ 0 & \frac{1}{24} & 0 & 0 & 0 & 0 & \frac{-1}{24\sqrt{5}} & \frac{-1}{48} & \frac{1}{48\sqrt{5}} & 0 & 0 & 0 & 0 \\ 0 & 0 & \frac{11}{360} & 0 & \frac{1}{180} & \frac{-1}{72\sqrt{5}} & 0 & 0 & 0 & 0 & 0 & 0 & 0 \\ 0 & 0 & 0 & \frac{1}{36} & 0 & \frac{-1}{144} & 0 & 0 & 0 & 0 & 0 & 0 & 0 \\ 0 & 0 & \frac{1}{180} & 0 & \frac{7}{180} & \frac{1}{144\sqrt{5}} & 0 & 0 & 0 & 0 & 0 & 0 & 0 \\ 0 & 0 & \frac{-1}{72\sqrt{5}} & \frac{-1}{144} & \frac{1}{144\sqrt{5}} & \frac{1}{72} & 0 & 0 & 0 & 0 & 0 & 0 & 0 \\ 0 & \frac{-1}{24\sqrt{5}} & 0 & 0 & 0 & 0 & \frac{1}{40} & 0 & \frac{-1}{180} & \frac{-1}{180} & \frac{-1}{144\sqrt{5}} & \frac{-1}{240} & 0 \\ 0 & \frac{-1}{48} & 0 & 0 & 0 & 0 & 0 & \frac{1}{36} & 0 & \frac{-1}{144\sqrt{5}} & 0 & \frac{-1}{72\sqrt{5}} & 0 \\ 0 & \frac{1}{48\sqrt{5}} & 0 & 0 & 0 & 0 & \frac{-1}{180} & 0 & \frac{1}{60} & \frac{-1}{240} & \frac{-1}{72\sqrt{5}} & \frac{1}{180} & 0 \\ \frac{-1}{24\sqrt{5}} & 0 & 0 & 0 & 0 & 0 & \frac{-1}{180} & \frac{-1}{144\sqrt{5}} & \frac{-1}{240} & \frac{1}{40} & 0 & \frac{-1}{180} & 0 \\ \frac{-1}{48} & 0 & 0 & 0 & 0 & 0 & \frac{-1}{144\sqrt{5}} & 0 & \frac{-1}{72\sqrt{5}} & 0 & \frac{1}{36} & 0 & 0 \\ \frac{1}{48\sqrt{5}} & 0 & 0 & 0 & 0 & 0 & \frac{-1}{240} & \frac{-1}{72\sqrt{5}} & \frac{1}{180} & \frac{-1}{180} & 0 & \frac{1}{60} & 0 \\ 0 & 0 & 0 & 0 & 0 & 0 & 0 & 0 & 0 & 0 & 0 & 0 & \frac{17}{72} \end{bmatrix}$$

as we would expect from Theorem 7.1.1. Thus we see that the matrix $B_f D B_f'$ is not a diagonal matrix and in this case adjoining the none option has not improved the properties of the design since we have $\det(C_{MT,n}) = 0$. □

The result in Theorem 7.1.1 holds for any stated preference choice experiment in which a none option has been adjoined to each choice set. Obviously the expression for C_n is easier to work with if $B_f \mathbf{r} = \mathbf{0}_p$. This is true if the treatment combinations that appear

Table 7.2 Choice Sets with $k = 3$ Attributes, $\ell_1 = \ell_2 = 2$ and $\ell_3 = 4$

Option 1	Option 2
0 0 0	1 1 1
1 1 0	0 0 1
1 1 1	0 0 2
0 0 1	1 1 2
0 1 2	1 0 3
1 0 2	0 1 3
1 0 3	0 1 0
0 1 3	1 0 0

in the choice experiment form a fractional factorial design in which each level of each attribute is equally represented. $B_f DB'_f$ will then be a multiple of the identity matrix if the treatment combinations form a regular symmetric fractional factorial design. In this case effects which are not independently estimated in the original forced choice experiment will not be independently estimated when the none option is adjoined but sometimes effects which can not be estimated in the original forced choice can be estimated in the extended design as we have seen for the complete factorial in Example 7.1.1.

7.1.2 Optimal Binary Response Experiments

In a binary response design respondents are shown a set of options, in turn, and asked for each option whether or not they would choose it. Thus this is really the simplest example of having a none option in each choice set. The options themselves might be actual products for sale or they might be treatments in a medical setting, for instance. Hall et al. (2002) used a binary response experiments to investigate chickenpox vaccination.

In the next result we prove that showing the products determined by a resolution 3 fraction results in a diagonal C matrix with the largest possible determinant amongst designs with a diagonal C matrix. Thus each choice set is of size 2: the product under investigation and the "no" option.

■ THEOREM 7.1.2.
Suppose that the main effects of the attributes for options described by k attributes, where the qth attribute has ℓ_q levels, are to be estimated using a binary response design. Then a resolution 3, equi-replicate fraction gives rise to a diagonal information matrix for the estimation of main effects and it has the largest determinant amongst designs with diagonal information matrices. Thus

$$\det(C_{\text{opt}}) = \left(\frac{1}{4N}\right)^p \frac{(L+1)N}{L}.$$

Proof. Let us assume that N of the L products are to be presented to respondents. Then we can order the products so that these N products are the first N products in some fixed, but arbitrary, ordering of the L products. We use this ordering to label the first L rows and columns of Λ and the first L columns of B. The final row and column of Λ and the final column of B correspond to the "no" option. Thus we obtain the Λ_n and B_n matrices as we did in the previous section. Indeed B_n is exactly the same as it was in Theorem 7.1.1.

Since each product is shown individually to each respondent the number of choice sets is N and we can write

$$\Lambda_n = \frac{1}{4N}\begin{bmatrix} I_N & \mathbf{0}_{N,L-N} & -\mathbf{j}_N \\ \mathbf{0}_{L-N,N} & \mathbf{0}_{L-N,L-N} & \mathbf{0}_{L-N} \\ -\mathbf{j}'_N & \mathbf{0}'_{L-N} & N \end{bmatrix}.$$

We then get that

$$C_n = \frac{1}{4N}\begin{bmatrix} I_p & \mathbf{0}_p \\ \mathbf{0}'_p & \frac{(L+1)N}{L} \end{bmatrix},$$

which is a special case of the result in Theorem 7.1.1. □

7.1.3 Common Base Option

If a particular combination of attribute levels appears in all choice sets then this combination is called the *common base option*. It may represent the current situation or the current treatment for a particular health condition (examples in Ryan and Hughes (1997), Ryan and Farrar (2000), and Longworth et al. (2001)) or the common base may be randomly chosen from the main effects plan and have all the other scenarios from the plan compared to it pairwise (examples in Ryan (1999), Ryan et al. (2000), and Scott (2002)).

Our first optimality result gives the determinant of the information matrix for any resolution 3 fractional factorial design in which any one of the treatment combinations may be used as the common base and only one other option appears in each choice set. To ensure that the matrix of contrasts for main effects for the treatment combinations that appear in the choice experiment is unambiguously defined, we will insist that the treatment combinations that appear in the choice experiment form a fractional factorial design of resolution 3.

From Theorem 6.3.1, we know that the maximum possible determinant for the information matrix for the estimation of main effects only, when choice sets are of size 2, is given by

$$C_{\text{opt},M,m=2} = \prod_{q=1}^{k}\left[\frac{\ell_q}{2(\ell_q-1)L}\right]^{\ell_q-1}.$$

We will use C_c for the information matrix when there is a common base.

■ **THEOREM 7.1.3.**

Let F be a resolution 3 fractional factorial design with equal replication of levels within each attribute and with N treatment combinations in total. Choose any treatment combination in F to be the common base. Then

$$\det(C_c) = \frac{1+p}{(4(N-1))^p}\left(\frac{N}{L}\right)^p.$$

The optimal design arises from the smallest resolution 3 design and so

$$\det(C_{\text{opt},c}) = \frac{1+p}{(4(N_{\min}-1))^p}\left(\frac{N_{\min}}{L}\right)^p.$$

Proof. Without loss of generality, order the treatment combinations so that the first treatment combination is the common base, the next $N-1$ treatment combinations are the other treatment combinations in F and as usual let B_M be the normalized main effects contrast matrix. Then

$$4(N-1)\Lambda_c = \begin{bmatrix} \begin{bmatrix} (N-1) & -1 & -1 & \dots & -1 \\ -1 & 1 & 0 & \dots & 0 \\ -1 & 0 & 1 & \dots & 0 \\ \vdots & \vdots & \vdots & \dots & \vdots \\ -1 & 0 & 0 & \dots & 1 \end{bmatrix} & \mathbf{0}_{N,L-N} \\ \mathbf{0}_{L-N,N} & \mathbf{0}_{L-N,L-N} \end{bmatrix}.$$

To get a neat expression for the information matrix, it is helpful to partition B_M in the same way that Λ has been partitioned. We let $B_M = [B_1\ B_2]$. Then

$$\begin{aligned} C_c &= B_M \Lambda B_M' \\ &= \frac{1}{4(N-1)}[B_1\ B_2] \begin{bmatrix} \begin{bmatrix} (N-1) & -1 & -1 & \dots & -1 \\ -1 & 1 & 0 & \dots & 0 \\ -1 & 0 & 1 & \dots & 0 \\ \vdots & \vdots & \vdots & \dots & \vdots \\ -1 & 0 & 0 & \dots & 1 \end{bmatrix} & \mathbf{0}_{N,L-N} \\ \mathbf{0}_{L-N,N} & \mathbf{0}_{L-N,L-N} \end{bmatrix} \begin{bmatrix} B_1' \\ B_2' \end{bmatrix} \\ &= \frac{1}{4(N-1)} B_1 \begin{bmatrix} (N-1) & -1 & -1 & \dots & -1 \\ -1 & 1 & 0 & \dots & 0 \\ -1 & 0 & 1 & \dots & 0 \\ \vdots & \vdots & \vdots & \dots & \vdots \\ -1 & 0 & 0 & \dots & 1 \end{bmatrix} B_1' \\ &= \frac{1}{4(N-1)} B_1 \begin{bmatrix} (N-2) & -1 & -1 & \dots & -1 \\ -1 & 0 & 0 & \dots & 0 \\ -1 & 0 & 0 & \dots & 0 \\ \vdots & \vdots & \vdots & \dots & \vdots \\ -1 & 0 & 0 & \dots & 0 \end{bmatrix} B_1' + \frac{1}{4(N-1)} B_1 B_1' \\ &= \frac{1}{4(N-1)} \left(\frac{N}{L} J_p + \frac{N}{L} I_p \right). \end{aligned}$$

Thus we have that

$$\det(C_c) = \left(\frac{N}{4(N-1)L}\right)^p \det(I_p + J_p) = (1+p)\left(\frac{N}{4(N-1)L}\right)^p.$$

(Recall that $\det(I_p + J_p) = (p+1)$.) □

The efficiency of the design, relative to the optimal forced choice stated preference experiment with choice sets of size 2, is given by

$$\left(\frac{\det(C_c)}{\det(C_{\text{opt},M,m=2})}\right)^{\frac{1}{p}}.$$

Using Theorem 7.1.3, we can see that, from the point of view of statistical efficiency, all resolution 3 designs with the same number of treatment combinations are equally good for a particular number of attributes with given number of levels. It is also immaterial which of the treatment combinations is used as the common base.

■ EXAMPLE 7.1.3.

Consider an experiment where there are four attributes describing each option. Three of the attributes have 2 levels and one has 4 levels. The smallest fractional factorial design of resolution 3 has 8 treatment combinations. Then using any $2 \times 2 \times 2 \times 4//16$ and using any treatment combination as a common base gives a design which is 93.3% efficient relative to a design with 8 treatment combinations and 45.17% efficient relative to a forced choice stated preference design. Using the complete factorial and using any treatment combination as a common base gives a design which is 90.3% efficient relative to a design with 8 treatment combinations and 43.71% efficient relative to a forced choice stated preference design. Using a $2 \times 2 \times 2 \times 4//8$ gives an efficiency of 48.40% efficient relative to a forced choice stated preference design. □

In Scott (2002) a similar experiment was carried out with four attributes (two with 2 levels, one with 3 levels, and one with 4 levels) and using 16 choice sets. In this case the efficiency depends on the treatment combination chosen to be the common base. If the common base has the level of the 3-level attribute which appears 8 times, then the efficiency is 44.49%; and if the common base has either of the other levels of the 3-level attribute, then the efficiency is 46.12%.

If we want to use this approach to estimate main effects plus two-factor interactions, then we need to start with a resolution 5 fractional factorial design. Unfortunately, in this setting we can only know that all effects are estimable and compare particular designs. We cannot calculate the efficiency relative to the optimal design since only a general expression for $\det(C)$ is available; see Chapter 6.

7.1.4 Common Base and None Option

The earlier results can be combined to give an expression for the information matrix for choice sets when there is both a common base and a none option in each choice set. Although in theory the choice sets could be of any size, we will only consider the situation when the choice sets are of size $m = 3$ and contain the common base, the neither of these option and one other option which will be different for each choice set.

In this case the correct contrast matrix is B_n, defined in Theorem 7.1.1. By re-ordering the treatment combinations if necessary, and assuming there are N treatment combinations in the fraction, we can write the Λ matrix, Λ_{cn}, as

$$\Lambda_{cn} = \frac{1}{9(N-1)}\begin{bmatrix} 2(N-1) & -1 & -1 & \dots & -1 & 0 & \dots & 0 & -(N-1) \\ -1 & 2 & 0 & \dots & 0 & 0 & \dots & 0 & -1 \\ -1 & 0 & 2 & \dots & 0 & 0 & \dots & 0 & -1 \\ \vdots & \vdots & \vdots & \ddots & \vdots & \vdots & \ddots & \vdots & \vdots \\ -1 & 0 & 0 & \dots & 2 & 0 & \dots & 0 & -1 \\ 0 & 0 & 0 & \dots & 0 & 0 & \dots & 0 & 0 \\ \vdots & \vdots & \vdots & \ddots & \vdots & \vdots & \ddots & \vdots & \vdots \\ 0 & 0 & 0 & \dots & 0 & 0 & \dots & 0 & 0 \\ -(N-1) & -1 & -1 & \dots & -1 & 0 & \dots & 0 & 2(N-1) \end{bmatrix}.$$

If we assume that the treatment combinations involved in the experiment form a fractional factorial design, it is straightforward to show that the information matrix when there is a common base and a none option is C_{cn}, given by

$$C_{cn} = \frac{1}{9(N-1)} \begin{bmatrix} \frac{2N}{L} I_p + 2(N-1)\mathbf{b}_1\mathbf{b}_1' & \frac{(L+1)(N-2)}{d}\mathbf{b}_1 \\ \frac{(L+1)(N-2)}{d}\mathbf{b}_1' & \frac{2(L+1)(N-1)}{L} \end{bmatrix},$$

where $\mathbf{b}_1$ is the first column of B_M.

7.2 OPTIMAL CHOICE SET SIZE

In the previous chapters in the book we have assumed that we want optimal designs for a particular choice set size m. In this section we show that the statistically optimal choice set size is a function of the number of levels of each of the attributes. We find that although pairs are the most common choice set size used they are rarely the most efficient size to use. Whether the choice sets in these designs may prove to be too large for respondents to manage comfortably is something that potential users need to consider. The results that we give can be used to determine the relative efficiency of practical values of m.

7.2.1 Main Effects Only for Asymmetric Attributes

In this section we consider choice experiments in which each of the attributes can have any number of levels and in which main effects only are to be estimated. From the results in Chapter 6 we know the maximum value of the determinant of C_M for a particular value of m. This allows the efficiency of any design to be calculated for particular values of m, k and ℓ_q, $q = 1, \ldots, k$.

The following theorem establishes an upper bound for the maximum value of $\det(C_M)$ when m is not fixed and gives the values of m for which this bound is attained.

■ THEOREM 7.2.1.
For fixed values of k and ℓ_q, $q = 1, \ldots, k$, when estimating main effects only, the upper bound for the maximum value of $\det(C_M)$ over all values of m is

$$\det(C_{\text{opt},M,\text{any }m}) = \left[\frac{1}{L}\right]^p,$$

where $p = \sum_{q=1}^{k}(\ell_q - 1)$. The upper bound is attained when m is a multiple of the least common multiple of $\ell_1, \ell_2, \ldots, \ell_q$.

Proof. Let $\det(C_{\text{opt},M})_q$ be the maximum possible contribution of attribute q towards $\det(C_{\text{opt},M,\text{any }m})$. Clearly

$$\det(C_{\text{opt},M,\text{any }m}) = \prod_{q=1}^{k} \det(C_{\text{opt},M})_q,$$

so we can examine $\det(C_{\text{opt},M})_q$ for each attribute separately. For each attribute q there are four possible cases to consider.

1. Suppose $m = c\ell_q$, where c is a positive integer and $\ell_q \leq m$. If $\ell_q = 2$, then m is even and $S_q = m^2/4$. Then for attribute q

$$\det(C_{\text{opt},M,\text{case1}})_q = \left[\frac{m^2\ell_q}{2m^2L}\right]^{\ell_q-1} = \frac{1}{L}.$$

If $\ell_q \neq 2$, then $m = \ell_q x + y$, where $x = c$ and $y = 0$. Thus

$$S_q = (m^2 - (\ell_q x^2 + 2xy + y))/2 = c^2\ell_q(\ell_q - 1)/2$$

and for attribute q

$$\det(C_{\text{opt},M,\text{case1}})_q = \left[\frac{c^2\ell_q^2(\ell_q - 1)}{c^2\ell_q^2(\ell_q - 1)L}\right]^{\ell_q - 1} = \left[\frac{1}{L}\right]^{\ell_q - 1}.$$

2. Suppose $m \neq c\ell_q$, where c is a positive integer and $2 < \ell_q < m$. Then we have $S_q = (m^2 - (\ell_q x^2 + 2xy + y))/2$ for $0 < y < \ell_q$ and the determinant for that attribute is given by

$$\det(C_{\text{opt},M,\text{case2}})_q = \left[\frac{\ell_q(m^2 - (\ell_q x^2 + 2xy + y))}{m^2(\ell_q - 1)L}\right]^{\ell_q - 1}.$$

We need to show that this determinant is less than the upper bound. That is, we need to show that

$$\left[\frac{\ell_q(m^2 - (\ell_q x^2 + 2xy + y))}{m^2(\ell_q - 1)L}\right]^{\ell_q - 1} < \left[\frac{1}{L}\right]^{\ell_q - 1}$$

or

$$[\ell_q m^2 - \ell_q(\ell_q x^2 + 2xy + y)] - m^2(\ell_q - 1) < 0.$$

This is true because $\ell_q(\ell_q x^2 + 2xy + y) > m^2$ for all possible values of ℓ_q, x, and y. Thus the determinant for attribute q in this case will be less than the upper bound.

3. Suppose that $2 \leq m < \ell_q$. In this case $S_q = m(m-1)/2$ and therefore

$$\det(C_{\text{opt},M,\text{case3}})_q = \left[\frac{\ell_q(m-1)}{m(\ell_q - 1)L}\right]^{\ell_q - 1} < \left[\frac{1}{L}\right]^{\ell_q - 1}$$

since $(m-1)/m < (\ell_q - 1)/\ell_q$.

4. Suppose $\ell_q = 2$ and m is odd. Now $S_q = (m^2 - 1)/4$ and for attribute q

$$\det(C_{\text{opt},M,\text{case4}})_q = \left[\frac{\ell_q(m^2 - 1)}{2m^2(\ell_q - 1)L}\right]^{\ell_q - 1} = \frac{m^2 - 1}{m^2 L}.$$

This determinant is less than the upper bound since $(m^2 - 1)/m^2 < 1$.

In cases 2, 3, and 4 the maximum value of $\det(C_M)$ is always less than the upper bound, which is only ever attained for case 1 attributes. Thus the upper bound can only be realized if all attributes belong to case 1. □

In the following examples we determine $\det(C_{\text{opt},M,m})$ for values of m from 2 to 16. The D-efficiency for each of these values of m is calculated relative to the maximum determinant over all values of m, $\det(C_{\text{opt},M})$; that is,

$$\text{Eff}_D = \left(\frac{\det(C_{\text{opt},M,m})}{\det(C_{\text{opt},M,\text{any } m})}\right)^{1/p}.$$

■ EXAMPLE 7.2.1.

Suppose there are five attributes ($k = 5$), each with four levels ($\ell_q = 4$ for $q = 1, \ldots, 5$) and that the choice sets each have four options ($m = 4$). From Theorem 7.2.1 $\det(C_{\text{opt},M,\text{any } m})$ is $(1/4^5)^{15}$ which is attained when m is a multiple of 4. In Table 7.3 the determinants and efficiencies are given for different values of m. Note that the choice set size of $m = 4$ is optimal, as is the choice set size of $m = 8$ and other multiples of 4, and that pairs ($m = 2$) are not nearly as efficient as other values of m. □

Table 7.3 Efficiency of Different Values of m for Five Attributes, Each with Four Levels for Main Effects Only.

m	$\det(C_{\text{opt}.M.m})$	Efficiency (%)
2	1.60004×10^{-48}	66.67
3	1.19733×10^{-46}	88.89
4	7.00649×10^{-46}	100
5	3.79812×10^{-46}	96.00
6	3.97781×10^{-46}	96.30
7	5.14254×10^{-46}	97.96
8	7.00649×10^{-46}	100
9	5.81534×10^{-46}	98.77
10	5.72872×10^{-46}	98.67
11	6.18641×10^{-46}	99.17
12	7.00649×10^{-46}	100
13	6.40972×10^{-46}	99.41
14	6.32461×10^{-46}	99.32
15	6.55365×10^{-46}	99.56
16	7.00649×10^{-46}	100

The following example illustrates using the theorem to choose values of m.

■ EXAMPLE 7.2.2.

Consider four attributes where $\ell_1 = 2$, $\ell_2 = 3$, and $\ell_3 = \ell_4 = 4$. The least common multiple $\text{lcm}(2, 3, 4)$ is 12 and this is the smallest value of m for which $\det(C_{\text{opt},M,\text{any } m})$ is attained. When $m = 12$,

$$\det(C_{\text{opt},M,\text{any } m}) = \left[\frac{1}{L}\right]^{\sum_{q=1}^{k}(\ell_q - 1)} = \left[\frac{1}{96}\right]^9 = 1.44397 \times 10^{-18}$$

since all four attributes belong to case 1. Now choice sets of size 12 are too large for respondents in practice but knowing the efficiency of other values of m relative to this upper bound is useful when deciding what size the choice sets should be.

Suppose $m = 3$. For $\ell_1 = 2$ and m odd (case 4),

$$\det(C_{\text{opt},M,3,\text{case4}})_1 = \left[\frac{m^2 - 1}{m^2 L}\right]^{\ell_q - 1} = \frac{1}{108}.$$

For $\ell_2 = 3$ (case 1),

$$\det(C_{\text{opt},M,3,\text{case1}})_2 = \left[\frac{1}{L}\right]^{\ell_q - 1} = \left[\frac{1}{96}\right]^2$$

and for $\ell_3 = \ell_4 = 4$ (case 2),

$$\det(C_{\text{opt},M,3,\text{case2}})_3 = \det(C_{\text{opt},M,3,\text{case2}})_4 = \left[\frac{\ell_q(m-1)}{m(\ell_q-1)L}\right]^{\ell_q-1} = \left[\frac{1}{108}\right]^3.$$

Hence

$$\det(C_M) = \left(\frac{1}{108}\right)\left(\frac{1}{96}\right)^2\left(\frac{1}{108}\right)^3\left(\frac{1}{108}\right)^3 = 6.33128 \times 10^{-19}.$$

The efficiency compared with using $m = 12$ is

$$\left(\frac{6.33128 \times 10^{-19}}{1.44397 \times 10^{-18}}\right)^{1/9} \times 100 = 91.25\%.$$

Now consider $m = 4$. Attributes 1, 3, and 4 all belong to case 1 and will have determinants equal to the upper bound, while attribute 2 belongs to case 3 since $2 < \ell_2 < m$. Now $m = \ell_2 x + y$ so $x = y = 1$ and for attribute 2

$$\det(C_{\text{opt},M,4,\text{case3}})_2 = \left[\frac{\ell_q(m^2 - (\ell_q x^2 + 2xy + y))}{m^2(\ell_q - 1)L}\right]^{\ell_q-1} = \left[\frac{5}{512}\right]^2.$$

Thus

$$\det(C_{\text{opt},M,4}) = \left(\frac{1}{96}\right)\left(\frac{5}{512}\right)^2\left(\frac{1}{96}\right)^3\left(\frac{1}{96}\right)^3 = 1.26912 \times 10^{-18}$$

with an efficiency of 98.58%.

Table 7.4 gives the maximum determinant and the efficiency for values of m from 2 to 16. Observe that as long as $m > 2$ the efficiencies are greater than 90%. □

7.2.2 Main Effects and Two-Factor Interactions for Binary Attributes

In Chapter 5 the upper bound for the determinant of the information matrix, C_{MT} is established, when estimating main effects and all two-factor interactions, when all k attributes are binary. The maximum possible value of $\det(C_{MT})$ for fixed values of m and k is given by

$$\det(C_{\text{opt},MT,m}) = \begin{cases} \left(\frac{(m-1)(k+2)}{m(k+1)2^k}\right)^{k+k(k-1)/2}, & k \text{ even}, \\ \left(\frac{(m-1)(k+1)}{mk2^k}\right)^{k+k(k-1)/2}, & k \text{ odd}. \end{cases}$$

Clearly, as $m \to \infty$, $(m-1)/m \to 1$ giving an upper bound for the maximum value of $\det(C_{MT})$, which is

$$\det(C_{\text{opt},MT,\text{any } m}) = \begin{cases} \left(\frac{(k+2)}{(k+1)2^k}\right)^{k+k(k-1)/2}, & k \text{ even}, \\ \left(\frac{(k+1)}{k2^k}\right)^{k+k(k-1)/2}, & k \text{ odd}. \end{cases}$$

Obviously the larger the choice set, the larger the value of $\det(C_{\text{opt},MT})$ and hence the more efficient the design. For example, suppose there are 6 binary attributes. In Table 7.5 the values of $\det(C_{\text{opt},MT,m})$ and efficiencies are given for choice sets of size 2 to 10. The efficiency is calculated with respect to $\det(C_{\text{opt},MT})$ with $p = k + k(k-1)/2$. These results show that in this setting the size of the choice sets should be as large as the respondents can cope with.

Table 7.4 Efficiency of Different Values of m for Four Asymmetric Attributes for Main Effects Only.

m	$\det(C_{\text{opt},M,m})$	Efficiency (%)
2	7.13073×10^{-20}	71.59
3	6.33128×10^{-19}	91.25
4	1.26912×10^{-18}	98.58
5	1.00000×10^{-18}	96.00
6	1.15137×10^{-18}	97.52
7	1.19940×10^{-18}	97.96
8	1.39920×10^{-18}	99.65
9	1.32371×10^{-18}	99.04
10	1.30573×10^{-18}	98.89
11	1.34005×10^{-18}	99.17
12	1.44397×10^{-18}	100
13	1.36887×10^{-18}	99.41
14	1.37192×10^{-18}	99.43
15	1.39964×10^{-18}	99.65
16	1.43271×10^{-18}	99.91

Table 7.5 Efficiency of Different Values of m for 6 Binary Attributes when Estimating Main Effects and Two-factor Interactions.

m	Max $\det(C_{MT})$	Efficiency (%)
2	9.25596×10^{-44}	50.00
3	3.89166×10^{-41}	66.67
4	4.61677×10^{-40}	75.00
5	1.79036×10^{-39}	80.00
6	4.21935×10^{-39}	83.33
7	7.62375×10^{-39}	85.71
8	1.17549×10^{-38}	87.50
9	1.63625×10^{-38}	88.89
10	2.12395×10^{-38}	90.00

7.2.3 Choice Experiments with Choice Sets of Various Sizes

Although all the work on optimal choice experiments that has been done to date has assumed that all choice sets in a particular experiment are of the same size, there is in fact no mathematical requirement for that to be true. In this section we derive the information matrix for a forced choice experiment in which there are N_s choice sets of size m_s, $1 \leq s \leq r$. We then show that the optimal choice set sizes are the ones that we have found in the previous sections.

Extending the notation in the earlier chapters, we let $n^{(s)}_{i_1,i_2}$ be the number of times that options T_{i_1} and T_{i_2} appear together in the N_s choice sets of size m_s. Let $N = \sum_s N_s$ be the total number of choice sets. Under the null hypothesis that all items are equally attractive, Λ contains the proportion of choice sets in which pairs of profiles appear together and hence

$$\Lambda = \frac{1}{N}\sum_{s=1}^{r} N_s\Lambda_s.$$

Substituting, we get

$$C = B\Lambda B' = \sum_s \frac{N_s}{N} B\Lambda_s B' = \sum_s \frac{N_s}{N} C_s.$$

Since the entries in the matrix for one choice set size have no effect on the entries in the matrix for another choice set size, we need to maximize the coefficients for each choice set size independently.

If we are only estimating main effects, then from Theorem 7.2.1 we know that each m_s needs to be a multiple of the least common multiple of $\ell_1, \ell_2, \ldots, \ell_k$ to get the largest determinant.

■ **EXAMPLE 7.2.3.**

Suppose that there are $k = 2$ binary attributes and that we have choice sets of three different sizes: $N_1 = 2$ choice sets, (00, 11) and (01, 10), each of size $m_1 = 2$ (design 1); $N_2 = 4$ choice sets, (00, 01, 10), (00, 01, 11), (00, 10, 11) and (01, 10, 11), each of size $m_2 = 3$ (design 2); and $N_3 = 1$ choice set, (00, 01, 10, 11) of size $m_3 = 4$ (design 3). If we use just the choice sets of size 2 and 3 in the choice experiment (designs 1 & 2), then we have that $n^{(1)}_{00,01} = n^{(1)}_{00,10} = n^{(1)}_{01,11} = n^{(1)}_{10,11} = 0$ and $n^{(1)}_{00,11} = n^{(1)}_{01,10} = 1$. (Recall that $n^{(s)}_{i_1,i_2} = n^{(s)}_{i_2,i_1}$.) For the sets of size 3, $n^{(2)}_{i_1,i_2} = 2 \;\forall\; i_1 \neq i_2$. The Λ matrix for these 6 choice sets is

$$\Lambda = \begin{bmatrix} \frac{1}{24}+\frac{1}{9} & -\frac{1}{27} & -\frac{1}{27} & -\frac{1}{24}-\frac{1}{27} \\ -\frac{1}{27} & \frac{1}{24}+\frac{1}{9} & -\frac{1}{24}-\frac{1}{27} & -\frac{1}{27} \\ -\frac{1}{27} & -\frac{1}{24}-\frac{1}{27} & \frac{1}{24}+\frac{1}{9} & -\frac{1}{27} \\ -\frac{1}{24}-\frac{1}{27} & -\frac{1}{27} & -\frac{1}{27} & \frac{1}{24}+\frac{1}{9} \end{bmatrix}.$$

If we wish to estimate main effects only, these 6 choice sets have an efficiency of 92.52% compared to the optimal experiments which are those that contain choice sets of size 2, or 4, or 2 and 4 together (see Table 7.6). Only using the 4 choice sets of size 3 (design 2) gives a design with an efficiency of 88.89% however; so including some choice sets of an optimal size improves the efficiency of the triples. □

Table 7.6 Choice Sets of Up to Three Different Sizes for 6 Binary Attributes for Main Effects Only.

Design(s)	m_s Values	N	Max det(C_M)	Efficiency (%)
1	2	2	0.0625	100
2	3	4	0.0494	88.89
3	4	1	0.0625	100
1 & 2	2 & 3	6	0.0536	92.52
1 & 3	2 & 4	3	0.0625	100
2 & 3	3 & 4	5	0.0519	91.04
1, 2 & 3	2, 3 & 4	7	0.0548	93.58

7.2.4 Concluding Comments on Choice Set Size

We have given results on the best choice set sizes to use to maximize the statistical efficiency of the choice experiment. Some authors, like DeShazo and Fermo (2002), have used larger choice set size as an indicator of a more complex task and so might argue that choice experiments in which several different choice set sizes are used have different values for respondent efficiency. Thus practitioners will need to decide how to trade-off gains in statistical efficiency with potential losses in respondent efficiency.

7.3 PARTIAL PROFILES

If the number of attributes used to describe each option is of the order of 16 and if all the attributes are allowed to differ between the options in each choice set then there is evidence that respondents will not perform as consistently as they might; see, for instance, Swait and Adamowicz (1996), Severin (2000), and Maddala et al. (2002). Holling et al. (1998) provide a summary of a number of studies in which both the number of attributes to be used and the the number of choice sets to be answered was compared. More choice sets with more options slowed down respondents and decreased the predictive ability of the model. In Maddala et al. (2002), for example, the authors compared a choice experiment in which 6 attributes were allowed to differ between options in a choice set with a choice experiment in which only 4 of the 6 attributes were allowed to differ between the options in a choice set. They found no significant differences in consistency, perceived difficulty or fatigue but they did observe differences in the stated preferences which these authors think may link with the importance of context to a stated preference experiment.

Some authors have focused on finding optimal designs when there is a limit on the number of attributes that are different between options within a choice set. If only some of the attributes may differ between the options within a choice set, then the options are said to be described by a *partial profile*.

One way to do this is to show all the attributes but have several attributes with the same level for all the options in a choice set. This was the approach taken by Severin (2000) using pairs. Another way to limit the number of attributes different is to choose subsets of the k attributes and find a choice experiment on the subset of attributes only; the remaining attributes are not shown. Various such subsets are used in turn to get the complete choice experiment. This approach was introduced by Green (1974).

If $m = 2$, then allowing only f of the k attributes to be different, but showing all k attributes, means that the optimal design for estimating main effects is the design in which all choice sets with f attributes different are included. The corresponding $\det(C_{\text{opt},M,pp})$ is given by putting $a_f = 1/(2^{k-1}\binom{k}{f})$ in the expression for $\det(C_M)$ in Theorem 4.1.1. This gives an optimal determinant of

$$\det(C_{\text{opt},M,pp}) = \left[\frac{f}{k} \times \frac{1}{2^k}\right]^k.$$

Green (1974) used balanced incomplete block designs (BIBDs) (defined in Section 2.4) to determine which attributes would be allowed to vary in a set of choice sets. He used all the blocks of the BIBD in turn to get all of the choice sets in an experiment.

■ **EXAMPLE 7.3.1.**

Suppose that pizzas are described by the 16 attributes, each with two levels, as given in Table 7.7. Then a (16,20,5,4,1) BIBD (Table 7.8, from Mathon and Rosa (2006)) could be used to determine which four attributes were varied in the choice sets associated with each block. Each block of the BIBD would give rise to four choice sets, as illustrated in Table 7.9. Using all $20 \times 4 = 80$ choice sets gives a design which is 100% efficient if only four attributes are allowed to have different levels between the two options in a choice set and is 25% efficient if there are no restrictions on the number of attributes which may have different levels in the two options in a choice set. □

Table 7.7 Sixteen Attributes Used to Describe Pizzas

Attribute	Level 0	Level 1
Quality of ingredients	All fresh	Some tinned
Price	\$13	\$17
Pizza temperature	Steaming hot	warm
Manners of operator	Friendly	Unfriendly
Delivery charge	Free	\$2
Delivery time	30 minutes	45 minutes
Vegetarian available	Yes	No
Pizza type	Traditional	Gourmet
Type of outlet	Chain	Local
Baking method	Woodfire oven	Traditional oven
Range	Large menu	Restricted menu
Distance to outlet	Same suburb	Next suburb
Type of crust	Thin	Thick
Available sizes	Single size	Three sizes
Opening hours	Till 10 pm	Till 1 am

Green's idea has been extended by using designs in which each pair of treatment combinations appears in at least one block by Grasshoff et al. (2004) rather than insisting on equality of pair replication as in the incomplete block design. Usually this relaxation means that the choice experiment is smaller but the relative efficiency of these two approaches has yet to be determined.

Table 7.8 (16,20,5,4,1) BIBD

0 1 2 3	0 4 5 6	0 7 8 9	0 a b c
0 d e f	1 4 7 a	1 5 b d	1 6 8 e
1 9 c f	2 4 c e	2 5 7 f	2 6 9 b
2 8 a d	3 4 9 d	3 5 8 c	3 6 a f
3 7 b e	4 8 b f	5 9 a e	6 7 c d

Table 7.9 The Four Choice Sets from the First Block of the (16, 20, 5, 4, 1)BIBD

Option 1				Option 2			
Fresh	\$13	Hot	Friendly	Tinned	\$17	Warm	Unfriendly
Fresh	\$13	Warm	Unfriendly	Tinned	\$17	Hot	Friendly
Fresh	\$17	Hot	Unfriendly	Tinned	\$13	Warm	Friendly
Fresh	\$17	Warm	Friendly	Tinned	\$13	Hot	Unfriendly

7.4 CHOICE EXPERIMENTS USING PRIOR POINT ESTIMATES

So far in this book we have constructed optimal designs under the null hypothesis of no difference between the options. However the expression for the Λ matrix given in Equations (3.2) and (3.3) includes the π_i values and so in fact we could find the optimal designs for any prior values of the π_i that we thought were appropriate. We discuss this idea in this section.

Additional information about the relative magnitudes of the π_i values may be available from earlier studies or from taking a sequential approach to experimentation. We give two examples below. In the first there is a wide range of values of π where the optimal designs for the null hypothesis perform well. In the second case the optimal designs for the null hypothesis are at best about 63% efficient.

In the first example we consider two binary attributes and let the class of competing designs be any set of pairs. This means that we have extended the class of competing designs from what we have considered in the remainder of the book. This is possible because the computations are not overwhelming in this small case. We have assumed that the two-factor interaction effect is 0 so that we can work with the simplest form of the C matrix.

■ EXAMPLE 7.4.1.

Suppose that there are $k = 2$ binary attributes and that the choice sets are of size $m = 2$. There are 4 treatment combinations and 6 pairs of distinct treatment combinations. Each such pair could be a choice set in the experiment. Suppose that all $2^6 - 1 = 63$ non-empty sets of distinct pairs form the class of competing designs. Suppose also that $\pi_1 = 1/h$, $\pi_2 = \pi_3 = 1$ and $\pi_4 = h$. When $h = 1$ the optimal design consists of the two foldover pairs (Design 1). This is still the optimal design for $0.25 < h < 4.7$ while outside that range the optimal design contains the three choice sets (00, 01), (00, 10), and (01, 10) (Design 2). Intuitively this makes sense: For extreme values of h, there is no information to be gained from comparing 00 and 11 since one or the other is certain to be chosen. The values of $\det(C_M)$ for these two designs for various values of h between 0 and 10 are shown in Figure 7.1. □

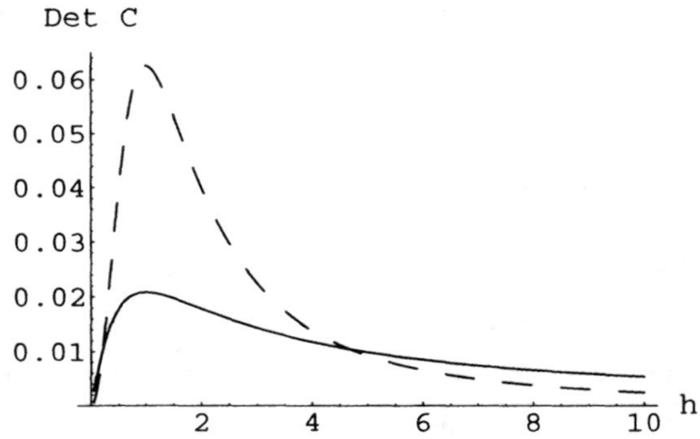

Figure 7.1 $\det(C_M)$ for Design 1 (dashed) and Design 2 (solid)

In the second example we consider the performance of the designs we have found to be optimal under the null hypothesis for the situation described in Carlsson and Martinsson (2003) for 4 attributes when prior estimates of π_i are assumed.

■ **EXAMPLE 7.4.2.**

Carlsson and Martinsson (2003) look at designs for $k = 4$ attributes where $\ell_1 = \ell_2 = \ell_4 = 3$ and $\ell_3 = 2$. They specify a linear utility which we have translated into π_i values, and these are used to give values consistent with the prior point estimates in the Λ matrix. The 54 treatment combinations and the corresponding π_i are given in Table 7.10.

It is their third case that corresponds to the situation that we have been using with generic options and an MNL model. The design that they call "cyclical" is an example of the design that is optimal under the null hypothesis. The particular design that they have given has an efficiency of about 54% relative to the efficiency of the optimal design that they give using the prior π_i values for the estimation of the linear components of the main effects. A different choice of generator could increase the efficiency of the cyclical design to about 63% in 18 choice sets but clearly in this case about 60% more respondents will be required to get the same accuracy from a design optimal under the null hypothesis relative to one of the designs advocated in Carlsson and Martinsson (2003) for these prior estimates, assuming that this prior information is correct. □

7.5 REFERENCES AND COMMENTS

Most of the material in Section 7.1 first appeared in Street and Burgess (2004b). The results in Section 7.2 first appeared in Burgess and Street (2006).

There are a number of authors who have chosen to find optimal designs using a Bayesian approach, where a prior distribution for the π_i, rather than point values, are assumed. Sandor and Wedel (2001) and Kessels et al. (2006) provide a good entry to this literature.

Table 7.10 The Treatment Combinations and the Corresponding Assumed π_i

Treatment Combination	π_i	Treatment Combination	π_i
0 0 0 0	0.045	0 0 0 1	0.025
0 0 0 2	0.014	0 0 1 0	0.067
0 0 1 1	0.037	0 0 1 2	0.020
0 1 0 0	0.549	0 1 0 1	0.301
0 1 0 2	0.166	0 1 1 0	0.819
0 1 1 1	0.450	0 1 1 2	0.247
0 2 0 0	6.686	0 2 0 1	3.669
0 2 0 2	2.014	0 2 1 0	9.974
0 2 1 1	5.474	0 2 1 2	3.004
1 0 0 0	0.122	1 0 0 1	0.067
1 0 0 2	0.037	1 0 1 0	0.183
1 0 1 1	0.100	1 0 1 2	0.055
1 1 0 0	1.491	1 1 0 1	0.819
1 1 0 2	0.449	1 1 1 0	2.226
1 1 1 1	1.221	1 1 1 2	0.670
1 2 0 0	18.174	1 2 0 1	9.974
1 2 0 2	5.474	1 2 1 0	27.113
1 2 1 1	14.880	1 2 1 2	8.167
2 0 0 0	0.333	2 0 0 1	0.183
2 0 0 2	0.100	2 0 1 0	0.497
2 0 1 1	0.273	2 0 1 2	0.150
2 1 0 0	4.055	2 1 0 1	2.226
2 1 0 2	1.221	2 1 1 0	6.050
2 1 1 1	3.320	2 1 1 2	1.822
2 2 0 0	49.402	2 2 0 1	27.113
2 2 0 2	14.880	2 2 1 0	73.700
2 2 1 1	40.447	2 2 1 2	22.198

To date, we have assumed that each attribute has some fixed number of levels, even if the attribute describes a continuous measure. Kanninen (2002) has discussed how the levels of one continuous attribute should be chosen to get better designs than are available using pre-specified levels. Kanninen (2002) makes the assumption that the utility for each option will be a linear function of the attributes describing that option. Thus, in keeping with the results from simple linear regression, she finds that it is best to have the attributes set at the upper and lower endpoints. But if this assumption is wrong, then there is no protection hence most people take some measurements in the interval.

One extension to stated preference choice experiments that has been considered by various people is the extension of the choice task to indicate not only the "best" option in each choice set but also the "worst" option in each choice set; see Flynn et al. (2007) for an example. The task has on occasion been extended to ask respondents to indicate the best and worst of the remaining options and so on until at most one option remains. Optimal designs for this situation are yet to be developed.

CHAPTER 8

PRACTICAL TECHNIQUES FOR CONSTRUCTING CHOICE EXPERIMENTS

In this chapter we discuss some techniques that are commonly used to construct optimal or near-optimal choice experiments. In Chapters 4, 5, and 6 we have given constructions for optimal or near-optimal choice experiments, but these constructions are not necessarily the easiest way to construct choice experiments in practice. Before thinking about the design of the choice experiment, the experimenter will have already decided on the number of attributes and the number of levels for each attribute, the number of options in each choice set (see Section 7.2 for results about the optimal size of the choice sets), the effects that are to be estimated, and the number of choice sets that each respondent can complete.

The basic idea is to find a suitable starting design, and then add sets of generators to obtain the choice sets. We discuss various methods to obtain a fractional factorial design to use as a starting design if an appropriate one is not readily available. These methods, which are defined in Chapter 2, include collapsing levels of attributes, expansive replacement, contractive replacement, and adding one more attribute. There are various ways to get a starting design: Find one in Sloane (2006b) or Kuhfeld (2006); use the techniques in Chapter 2; or use the tables and constructions in Dey (1985) and Hedayat et al. (1999). We also give some techniques for choosing the sets of generators to be used to construct the choice sets.

The ideal choice experiment would have the following desirable properties:

1. A manageable number of choice sets for the respondents;

2. Choice sets which are optimal or near-optimal;

3. The variance-covariance matrix, C^{-1}, should be diagonal or at least block diagonal;

4. Equal replication of levels of each attribute across all the choice sets;
5. Equal replication of levels of each attribute within each option;
6. All combinations of levels of an attribute should appear equally often over all the choice sets;
7. Avoidance of any predictable pattern in the choice sets.

Unfortunately, it is usually impossible to satisfy all of these properties simultaneously, and trade-offs must be made. The properties that are the most important may vary from one choice experiment to another. For example, properties 2 to 7 can easily be achieved by creating additional choice sets. Then property 1 can only be achieved if the choice sets can be split into blocks which consist of a manageable number of choice sets.

The number of choice sets that each respondent can complete will depend on the complexity of the choice sets, including the number of attributes and the number of options in the choice set (see Holling et al. (1998) and Brazell and Louviere (1995)). Usually we have each of the respondents completing all of the choice sets but if the design has more choice sets than a respondent can complete then the choice sets can be split into blocks either randomly or using a spare attribute, if there is one available.

It is desirable for the C^{-1} matrix to be diagonal or block diagonal because that means that all of the effects of interest are orthogonal or uncorrelated and can therefore be estimated independently of each other. This is discussed in Section 3.4.1. Thus the structure of the C^{-1} matrix should be checked for each design constructed. Note that if the C matrix is diagonal (or block diagonal) for the effects of interest, then C^{-1} is also diagonal (or block diagonal).

For the analysis it is desirable that, for each attribute, each level appears equally often over all of the choice sets. For example, if there are 4 binary attributes and 8 choice sets of size 2 in the experiment, then for the first attribute we would expect 0 to appear 8 times over all the choice sets. However, sometimes having each level of an attribute appearing equally often within each option is preferable, so that 0 appears 4 times in the first option and 4 times in the second option. It is also desirable to have as many combinations of levels of an attribute as possible appearing in the choice experiment. For an attribute with 4 levels, we would ideally want levels 0 and 1, 0 and 2, 0 and 3, 1 and 2, 1 and 3, and 2 and 3, appearing the same number of times in the choice experiment.

From a psychological point of view, it is often preferable to avoid any predictable pattern in the order of presentation of the choice sets, as well as in the options within the choice sets themselves. The choice sets should be placed in a random order before being presented to the respondents. In addition, if there is a discernable pattern for an attribute in the options within the choice sets, then this can be overcome by either reordering the options in some of the choice sets or, if that is not possible, by creating additional choice sets with a different pattern.

In this chapter we discuss various techniques that we use to construct designs which satisfy as many of these desirable properties as possible. We discuss various ways to obtain a starting design from readily available designs and how to generate the choice sets in a fairly simple way. We also compare our construction with other commonly used strategies for constructing choice experiments. Software that allows readers to construct choice sets from a starting design by adding sets of generators is available at the following website: http://maths.science.uts.edu.au/maths/wiki/SPExpts.

Except in one example (Example 8.1.7), in this chapter we have assumed that the levels of the attributes are not ordered in any way that would result in the possibility of dominated alternatives. We also assume that there are no unrealistic treatment combinations to be avoided, again except in Example 8.1.7.

8.1 SMALL NEAR-OPTIMAL DESIGNS FOR MAIN EFFECTS ONLY

Designs which estimate main effects only are appropriate when all interactions between the attributes are believed to be negligible. The formal construction of designs for estimating main effects only in Theorem 6.4.1 is complicated to use and in practice it is often easier to follow the approach used in Theorem 5.2.1. This approach requires F, a starting design of resolution at least 3, and a set of generators G, which have a difference vector that satisfies the upper bound for the sum of the differences (S_q), given in Theorem 6.1.1. It should be noted that just choosing sets of generators that have difference vectors satisfying the upper bound for S_q does not guarantee an optimal design. Trial and error may be needed in the selection of F and G to achieve optimality, or to get choice sets which satisfy as many as possible of the desirable properties above.

8.1.1 Smaller Designs for Examples in Section 6.4

We now look at each of the examples in Section 6.4 and use various methods to construct choice sets with as many of the desirable properties as possible.

■ EXAMPLE 8.1.1.

Let $m = 3$, $k = 2$, $\ell_1 = 2$, and $\ell_2 = 3$. In Example 6.3.5 the 6 choice sets with difference vector $\mathbf{v}_3 = (01, 11, 11)$ are given. These choice sets are optimal. In this example we are not going to be able to construct a smaller design, but we will still use it to illustrate the steps involved and to construct a design that is equally good. In this case F is the complete 2×3 factorial design, and so has treatment combinations 00, 01, 02, 10, 11 and 12, since no fraction has resolution 3 as $k \leq 2$. We now need $G = (\mathbf{g}_1 = \mathbf{0}, \mathbf{g}_2, \mathbf{g}_3)$ which has a difference vector equal to $\mathbf{v}_3$. One set of generators which has such a difference vector is $G = (00, 11, 12)$, but $G = (00, 01, 12)$ or $G = (00, 02, 11)$ also satisfy the condition. Then the choice sets are given by

$$(F, F + \mathbf{g}_2, F + \mathbf{g}_3) = (F, F + 11, F + 12),$$

where the addition is done modulo 2 for the first attribute and modulo 3 for the second attribute. These 6 choice sets, shown in Table 8.1, are 100% efficient. If we use the set of generators $G = (00, 01, 02)$ with difference vector $\mathbf{v}_1 = (01, 01, 01)$, then the efficiency is zero. This is because the level of the first attribute does not change across the options of the choice sets and therefore the main effect of that attribute cannot be estimated. Sets of generators such as $G = (00, 01, 10)$ with difference vector $\mathbf{v}_2 = (01, 10, 11)$ will result in choice sets which have an efficiency of 76.31%.

The design in Table 8.1 satisfies properties 1 to 6. We can satisfy property 7 by swapping the treatment combinations in options 2 and 3 for the last three choice sets. Alternatively, we can create an additional 6 choice sets by adding another set of generators, such as $G = (00, 02, 11)$. □

■ EXAMPLE 8.1.2.

Let $m = 6$, $k = 2$, and $\ell_1 = \ell_2 = 4$. In Example 6.4.3 we found that an optimal design could be constructed in 24 choice sets. We now look at constructing a (near-)optimal design

Table 8.1 Optimal Choice Sets for $k = 2$, $\ell_1 = 2$, $\ell_2 = 3$ when $m = 3$ for Main Effects Only

F	$F + 11$	$F + 12$
00	11	12
01	12	10
02	10	11
10	01	02
11	02	00
12	00	01

in fewer choice sets. Since $k = 2$, the starting design F must be the complete 4×4 factorial with the 16 treatment combinations 00, 01, 02, 03, 10, 11, 12, 13, 20, 21, 22, 23, 30, 31, 32, and 33. We need to choose a set of generators of the form

$$G = (\mathbf{g}_1 = \mathbf{0}, \mathbf{g}_2, \ldots, \mathbf{g}_m) = (00, \mathbf{g}_2, \mathbf{g}_3, \mathbf{g}_4, \mathbf{g}_5, \mathbf{g}_6).$$

First we choose $\mathbf{g}_2$ to be any generator with all attribute levels different from those in $\mathbf{g}_1 = \mathbf{0}$. Every subsequent generator has as many attribute levels as different as possible from all preceding generators. In this case we choose

$$G = (00, 11, 22, 33, 01, 10),$$

which has a difference vector

$$\mathbf{v} = (01, 01, 10, 10, 11, 11, 11, 11, 11, 11, 11, 11, 11, 11, 11),$$

but many other sets of generators would also be suitable.

We can check if the difference vector satisfies the upper bound for the sum of the differences given in Theorem 6.1.1. First we calculate S_1 and S_2 using Theorem 6.1.1. Now $m = 6 = \ell_q x + y = 4 \times 1 + 2$ so $x = 1$ and $y = 2$. Then

$$S_1 = S_2 = \frac{m^2 - (\ell_q x^2 + 2xy + y)}{2} = \frac{6^2 - (4 \times 1^2 + 2 \times 1 \times 2 + 2)}{2} = 13.$$

Clearly the difference vector above satisfies this upper bound.

The 16 choice sets given in Table 8.2 were constructed by adding the set of generators G to F. The addition is done modulo 4 for both attributes.

These choice sets are 99.96% efficient and have the following block diagonal information matrix:

$$C_M = \left[\begin{array}{ccc:ccc} \frac{43}{720} & 0 & \frac{1}{720} & 0 & 0 & 0 \\ 0 & \frac{17}{288} & 0 & 0 & 0 & 0 \\ \frac{1}{720} & 0 & \frac{89}{1440} & 0 & 0 & 0 \\ \hdashline 0 & 0 & 0 & \frac{43}{720} & 0 & \frac{1}{720} \\ 0 & 0 & 0 & 0 & \frac{17}{288} & 0 \\ 0 & 0 & 0 & \frac{1}{720} & 0 & \frac{89}{1440} \end{array}\right].$$

The main effects of both attributes can be estimated independently of each other, but the components of the main effects are not independent within an attribute. There are many sets of generators G that will result in choice experiments with 16 choice sets with similar properties. In fact there are 459 different sets of generators G that have a block diagonal C_M matrix and are at least 98% efficient.

For this example there are three ways in which we could get a diagonal C_M matrix. We could use the 24 or 48 choice sets that are optimal from Example 6.4.3, we could use different contrasts, or we could use a different set of generators that gives a diagonal matrix, but with some reduction in the efficiency. If we use different contrasts, such as those from a 2^2 design,

0	1	2	3
$-\frac{1}{2}$	$-\frac{1}{2}$	$\frac{1}{2}$	$\frac{1}{2}$
$-\frac{1}{2}$	$\frac{1}{2}$	$-\frac{1}{2}$	$\frac{1}{2}$
$\frac{1}{2}$	$-\frac{1}{2}$	$-\frac{1}{2}$	$\frac{1}{2}$

then we get the same efficiency but the C matrix is now diagonal:

$$C_M = \left[\begin{array}{ccc:ccc} \frac{17}{288} & 0 & 0 & 0 & 0 & 0 \\ 0 & \frac{1}{16} & 0 & 0 & 0 & 0 \\ 0 & 0 & \frac{17}{288} & 0 & 0 & 0 \\ \hdashline 0 & 0 & 0 & \frac{17}{288} & 0 & 0 \\ 0 & 0 & 0 & 0 & \frac{1}{16} & 0 \\ 0 & 0 & 0 & 0 & 0 & \frac{89}{1440} \end{array}\right].$$

If we want a diagonal C_M matrix with the orthogonal polynomial contrasts and are willing to have a less efficient choice experiment then we can use the set of generators

$$G = (00, 01, 02, 10, 20, 33)$$

with difference vector

$$(01, 01, 01, 10, 10, 10, 11, 11, 11, 11, 11, 11, 11, 11, 11),$$

where $S_1 = S_2 = 12$ to construct 16 choice sets that are 92.30% efficient with

$$C_M = \tfrac{1}{18} I_6.$$

There are 168 different sets of generators G that will result in the same C_M matrix. □

In the following example we show that the use of different fractional factorial designs F may result in choice experiments with different properties.

■ EXAMPLE 8.1.3.
Let $k = 3$, $\ell_1 = 2$, $\ell_2 = 3$, and $\ell_3 = 6$.

It is often the case that all seven of the desirable properties cannot be achieved and trade-offs must be made depending on the nature of the choice experiment. For example,

Table 8.2 Near-Optimal Choice Sets for $k = 2$, $\ell_1 = \ell_2 = 4$ when $m = 6$ for Main Effects Only

F	$F + 11$	$F + 22$	$F + 33$	$F + 01$	$F + 10$
0 0	1 1	2 2	3 3	0 1	1 0
0 1	1 2	2 3	3 0	0 2	1 1
0 2	1 3	2 0	3 1	0 3	1 2
0 3	1 0	2 1	3 2	0 0	1 3
1 0	2 1	3 2	0 3	1 1	2 0
1 1	2 2	3 3	0 0	1 2	2 1
1 2	2 3	3 0	0 1	1 3	2 2
1 3	2 0	3 1	0 2	1 0	2 3
2 0	3 1	0 2	1 3	2 1	3 0
2 1	3 2	0 3	1 0	2 2	3 1
2 2	3 3	0 0	1 1	2 3	3 2
2 3	3 0	0 1	1 2	2 0	3 3
3 0	0 1	1 2	2 3	3 1	0 0
3 1	0 2	1 3	2 0	3 2	0 1
3 2	0 3	1 0	2 1	3 3	0 2
3 3	0 0	1 1	2 2	3 0	0 3

Table 8.3 Difference Vectors and Sets of Generators for $k = 3$, $\ell_1 = 2$, $\ell_2 = 3$, and $\ell_3 = 6$

m	Difference Vector	G	S_1	S_2	S_3
2	(111)	(000,111)	1	1	1
3	(011,111,111)	(000,111,123)	2	3	3
4	(011,011,101,111,111,111)	(000,111,123,012)	4	5	6
5	(011,011,011,011,101,101,111,111,111,111)	(000,111,123,012,024)	6	8	10

minimizing the number of choice sets might be the first priority and a design with somewhat lower efficiency is perfectly acceptable. In this example, for each of three starting designs we will use the same set of generators G for the particular value of m and these, with the corresponding difference vectors, are given in Table 8.3. Note that there are many sets of generators G that have the difference vector given in Table 8.3 for the particular choice set size. For instance, $G = (000, 111, 022)$ will be just as efficient for $m = 3$ as $G = (000, 111, 123)$. We now discuss three different ways to obtain a starting design.

1. There is no $2 \times 3 \times 6$ OMEP readily available, so the complete factorial in 36 treatment combinations is used as the starting design F. In Example 6.4.4 we saw that adding one set of generators G to the complete factorial resulted in near-optimal or optimal designs. In Table 8.3 the best designs in 36 choice sets are shown for $m = 2, 3, 4, 5$ with efficiencies 90.56%, 99.39%, 99.78%, and 100% respectively. In all of these designs there is equal replication of levels in the choice experiment and, for the orthogonal polynomial contrasts, for $m = 2, 3, 4$, the C_M matrix is block diagonal for the attribute with 6 levels and diagonal for the other two attributes. When $m = 5$ the C_M matrix is diagonal for all three attributes.

2. Even though there is no $2 \times 3 \times 6$ OMEP readily available, we can obtain one by starting with a $3 \times 3 \times 6//18$ design and transforming a 3-level attribute into one with 2 levels by collapsing the levels. In Sloane (2006b) there is a $3^6 \times 6//18$ design. We delete the first four columns and then we collapse the three levels of the first of the remaining 3-level column to 2 levels by replacing all 2s by 1s and leaving the 1s and 0s unchanged. The original $3 \times 3 \times 6//18$ design, along with the $2 \times 3 \times 6$ design obtained by collapsing one column, is shown in Table 8.4.

Table 8.4 $2 \times 3 \times 6//18$ Fractional Factorial Design Obtained from the $3 \times 3 \times 6//18$ by Collapsing One Attribute

$3 \times 3 \times 6//18$			$2 \times 3 \times 6//18$		
0	0	0	0	0	0
0	1	1	0	1	1
1	0	2	1	0	2
2	2	3	1	2	3
2	1	4	1	1	4
1	2	5	1	2	5
1	1	0	1	1	0
1	2	1	1	2	1
2	1	2	1	1	2
0	0	3	0	0	3
0	2	4	0	2	4
2	0	5	1	0	5
2	2	0	1	2	0
2	0	1	1	0	1
0	2	2	0	2	2
1	1	3	1	1	3
1	0	4	1	0	4
0	1	5	0	1	5

We can now add the sets of generators for the different choice set sizes given in Table 8.3 to get 18 choice sets and we end up with the same efficiencies and structure of the

C_M matrix as we had in part 1. However, we do not always have equal replication of levels of the collapsed attribute over the choice experiment. In particular when $m = 3$, level 0 of attribute 1 appears 30 times overall, while level 1 appears only 24 times. Similarly, when $m = 5$, level 0 of attribute 1 appears 42 times overall, while level 1 appears 48 times. For $m = 2$ and $m = 4$ we have equal replication of the levels of all attributes.

3. We can also start with a fractional factorial $2 \times 2 \times 6//12$ design which has only two levels for the second attribute; this design is the F of Table 8.5. When we add the set of generators the addition, for this attribute, will be done modulo 3 so there will be three levels for this attribute in the choice experiment. However, in this case, the challenge will be in achieving equal replication of levels for this attribute. For $m = 2, 4, 5$ equal replication is not possible in 12 choice sets, but for $m = 3$ we can construct a near-optimal design (99.39%) which is given in Table 8.5. This design has equal replication of levels in the choice experiment and the C_M matrix has the same structure as before. It should be noted that not all levels of the second attribute appear in each option. □

Table 8.5 Near-Optimal Choice Sets for $k = 3$, $\ell_1 = 2$, $\ell_2 = 3$, and $\ell_3 = 6$ when $m = 3$ for Main Effects Only

F	$F + 111$	$F + 123$
0 0 0	1 1 1	1 2 3
0 0 1	1 1 2	1 2 4
0 0 2	1 1 3	1 2 5
0 1 3	1 2 4	1 0 0
0 1 4	1 2 5	1 0 1
0 1 5	1 2 0	1 0 2
1 1 0	0 2 1	0 0 3
1 1 1	0 2 2	0 0 4
1 1 2	0 2 3	0 0 5
1 0 3	0 1 4	0 2 0
1 0 4	0 1 5	0 2 1
1 0 5	0 1 0	0 2 2

8.1.2 Getting a Starting Design

As previously noted, many fractional factorial designs are available in Sloane (2006b) and Kuhfeld (2006). From these "parent" designs it is possible to obtain many other designs by discarding one or more attributes. Several constructions are given in Chapter 2, as well as some techniques for obtaining new fractional factorial designs from existing ones. In this section we illustrate various ways to get the fractional factorial designs, and then construct the choice sets. In the final example, we show how we can avoid unrealistic treatment combinations, and choice sets in which one treatment combination is dominated by another.

When the levels of an attribute need to be collapsed, the collapsing must be done in the starting design before the choice sets are constructed. In addition, collapsing levels of an attribute in the starting design, from ℓ_q levels to ℓ_s levels is straightforward as long as $\ell_s < \ell_q$ and $\ell_s \mid \ell_q$. This will ensure that the levels of the collapsed attribute are equally replicated. The most common collapsing of levels of this nature are given in the first part of Table 8.9.

When $\ell_s \nmid \ell_q$ there will be unequal replication of levels in the fractional factorial design which may or may not be able to be corrected when constructing the choice sets. Moreover, collapsing the levels of more than one attribute, where $\ell_s \nmid \ell_q$, can mean that the main effects of the collapsed attributes are correlated. In Example 6.4.4 collapsing levels of one attribute worked well in terms of obtaining choice sets which have equal replication and a C_M^{-1} matrix with an orthogonal structure for the main effects. However, collapsing more than one attribute usually results in a C_M^{-1} matrix in which the main effects of the collapsed attributes are not independent of each other. We illustrate this in the following example.

■ EXAMPLE 8.1.4.

Let $k = 4$, $\ell_1 = \ell_2 = 2$, $\ell_3 = 3$, and $\ell_4 = 6$. Using the $3^6 \times 6//18$ OMEP from Sloane (2006b), we delete the first three columns and then we collapse the three levels of the first two of the remaining 3-level columns to 2 levels by replacing all 2s by 1s for the first attribute and replacing all 2s by 0s for the second attribute. There is now unequal replication of the levels of the 2-level attributes in the OMEP and we wish to construct choice sets so that there is equal replication of levels of the attributes overall. In this example we use the B matrix from Example 6.3.3. Consider $m = 3$. The choice sets in Table 8.6 were constructed with the set of generators $G = (0000, 1111, 0123)$. These choice sets are 99.43% efficient and the variance-covariance matrix is given by

$$C_M^{-1} = \left[\begin{array}{c:c:cc:ccccc} \frac{26244}{323} & \frac{1458}{323} & 0 & 0 & 0 & 0 & 0 & 0 & 0 \\ \hdashline \frac{1458}{323} & \frac{26244}{323} & 0 & 0 & 0 & 0 & 0 & 0 & 0 \\ \hdashline 0 & 0 & 72 & 0 & 0 & 0 & 0 & 0 & 0 \\ 0 & 0 & 0 & 72 & 0 & 0 & 0 & 0 & 0 \\ \hdashline 0 & 0 & 0 & 0 & 108 & 0 & 0 & 0 & 0 \\ 0 & 0 & 0 & 0 & 0 & 108 & 0 & 0 & 0 \\ 0 & 0 & 0 & 0 & 0 & 0 & 81 & 0 & 0 \\ 0 & 0 & 0 & 0 & 0 & 0 & 0 & 81 & 0 \\ 0 & 0 & 0 & 0 & 0 & 0 & 0 & 0 & 81 \end{array}\right].$$

Note that the main effects of the two attributes with collapsed levels are correlated. We get similar results for $m = 2$ and $m = 5$ with the set of generators $G = (0000, 1111)$ with efficiency of 91.43% and $G = (0000, 1111, 0123, 1012, 0114)$ with efficiency of 97.07%, respectively. In both cases the main effects of the two collapsed attributes are correlated and there is unequal replication of the levels of those attributes in the choice sets. It is also worth noting that using a different collapsing scheme does not avoid this problem.

However, when $m = 4$, equal replication of levels of the attributes can be attained, with an efficiency of 99.81%. Furthermore, C_M, and therefore C_M^{-1}, are diagonal, and consequently all pairs of main effects are uncorrelated. The choice sets, constructed with the set of generators $G = (0000, 1111, 0123, 1014)$, are shown in Table 8.7 and the C_M matrix is given by

Table 8.6 Near-Optimal Choice Sets for $k = 4$, $\ell_1 = \ell_2 = 2$, $\ell_3 = 3$, and $\ell_4 = 6$ when $m = 3$ for Main Effects Only

$3^3 \times 6//18$	F	$F + 1111$	$F + 0123$
0 0 0 0	0 0 0 0	1 1 1 1	0 1 2 3
2 0 1 1	1 0 1 1	0 1 2 2	1 1 0 4
2 1 0 2	1 1 0 2	0 0 1 3	1 0 2 5
0 2 2 3	0 0 2 3	1 1 0 4	0 1 1 0
1 2 1 4	1 0 1 4	0 1 2 5	1 1 0 1
1 1 2 5	1 1 2 5	0 0 0 0	1 0 1 2
1 1 1 0	1 1 1 0	0 0 2 1	1 0 0 3
0 1 2 1	0 1 2 1	1 0 0 2	0 0 1 4
0 2 1 2	0 0 1 2	1 1 2 3	0 1 0 5
1 0 0 3	1 0 0 3	0 1 1 4	1 1 2 0
2 0 2 4	1 0 2 4	0 1 0 5	1 1 1 1
2 2 0 5	1 0 0 5	0 1 1 0	1 1 2 2
2 2 2 0	1 0 2 0	0 1 0 1	1 1 1 3
1 2 0 1	1 0 0 1	0 1 1 2	1 1 2 4
1 0 2 2	1 0 2 2	0 1 0 3	1 1 1 5
2 1 1 3	1 1 1 3	0 0 2 4	1 0 0 0
0 1 0 4	0 1 0 4	1 0 1 5	0 0 2 1
0 0 1 5	0 0 1 5	1 1 2 0	0 1 0 2

$$\left[\begin{array}{c|c|cc|ccccc}
\frac{1}{72} & 0 & 0 & 0 & 0 & 0 & 0 & 0 & 0 \\
\hline
0 & \frac{1}{72} & 0 & 0 & 0 & 0 & 0 & 0 & 0 \\
\hline
0 & 0 & \frac{5}{384} & 0 & 0 & 0 & 0 & 0 & 0 \\
0 & 0 & 0 & \frac{5}{384} & 0 & 0 & 0 & 0 & 0 \\
\hline
0 & 0 & 0 & 0 & \frac{1}{96} & 0 & 0 & 0 & 0 \\
0 & 0 & 0 & 0 & 0 & \frac{1}{96} & 0 & 0 & 0 \\
0 & 0 & 0 & 0 & 0 & 0 & \frac{1}{72} & 0 & 0 \\
0 & 0 & 0 & 0 & 0 & 0 & 0 & \frac{1}{72} & 0 \\
0 & 0 & 0 & 0 & 0 & 0 & 0 & 0 & \frac{1}{72}
\end{array}\right].$$

□

The following example looks at collapsing levels of attributes that results in equal replication of levels of the attributes. This example also illustrates that caution should be exercised when constructing pairs when the number of levels of a particular attribute is not a prime number. We also show that it may be possible to choose sets of generators so that the number of level changes is equal to S_q, but that this does not guarantee that the main effect of attribute q can be estimated.

■ **EXAMPLE 8.1.5.**

Let $k = 5$, $\ell_1 = \ell_2 = \ell_3 = 2$, and $\ell_4 = \ell_5 = 4$. Using the $4^5//16$ OMEP from Sloane (2006b), we collapse the levels of the first three attributes from 4 levels to 2 levels by changing all 2s to 1s and all 3s to 0s. These OMEPs are given in the first two columns of Table 8.8; note that all the levels of each of the 2-level attributes appear equally often.

Table 8.7 Near-Optimal Choice Sets for $k = 4$, $\ell_1 = \ell_2 = 2$, $\ell_3 = 3$, and $\ell_4 = 6$ when $m = 4$ for Main Effects Only

F	$F + 1111$	$F + 0123$	$F + 1014$
0 0 0 0	1 1 1 1	0 1 2 3	1 0 1 4
1 0 1 1	0 1 2 2	1 1 0 4	0 0 2 5
1 1 0 2	0 0 1 3	1 0 2 5	0 1 1 0
0 0 2 3	1 1 0 4	0 1 1 0	1 0 0 1
1 0 1 4	0 1 2 5	1 1 0 1	0 0 2 2
1 1 2 5	0 0 0 0	1 0 1 2	0 1 0 3
1 1 1 0	0 0 2 1	1 0 0 3	0 1 2 4
0 1 2 1	1 0 0 2	0 0 1 4	1 1 0 5
0 0 1 2	1 1 2 3	0 1 0 5	1 0 2 0
1 0 0 3	0 1 1 4	1 1 2 0	0 0 1 1
1 0 2 4	0 1 0 5	1 1 1 1	0 0 0 2
1 0 0 5	0 1 1 0	1 1 2 2	0 0 1 3
1 0 2 0	0 1 0 1	1 1 1 3	0 0 0 4
1 0 0 1	0 1 1 2	1 1 2 4	0 0 1 5
1 0 2 2	0 1 0 3	1 1 1 5	0 0 0 0
1 1 1 3	0 0 2 4	1 0 0 0	0 1 2 1
0 1 0 4	1 0 1 5	0 0 2 1	1 1 1 2
0 0 1 5	1 1 2 0	0 1 0 2	1 0 2 3

Suppose that $m = 4$. The choice sets in Table 8.8 were constructed with the set of generators $G = (00000, 11111, 10122, 01033)$ and are 100% efficient. This set of generators was chosen so that $S_1 = S_2 = S_3 = 4$ and $S_4 = S_5 = 6$. The C_M matrix is given by

$$C_M = \frac{1}{128} I_9.$$

Suppose that $m = 2$. A design which is 96.29% efficient can be constructed by adding 1 (mod 2) to the levels of the 2-level attributes and adding 1 or 3 (mod 4) to the levels of the 4-level attributes to obtain the treatment combinations for the second option (see Design 1 in Table 8.10), where $G = (00000, 11111)$. However, if we add 1 (mod 2) to the levels of the 2-level attributes and add 2 (mod 4) to the levels of the 4-level attributes to obtain the treatment combinations for the second option (see Design 2 in Table 8.10), where $G = (00000, 11122)$, we get $\det(C_M) = 0$ since the main effects of the 4-level attributes cannot be estimated. This is because, for each 4-level attribute, we have 0 paired with 2, 1 with 3, 2 with 0, and 3 with 1. Thus, only two (02, 13) of the six possible ordered pairs (01, 02, 03, 12, 13, 23) result, compared to four of the six (01, 12, 23, 30) if we add either 1 or 3 (mod 4). This situation arises because 2 + 2 = 0 (mod 4) and it is always an issue when constructing pairs when the number of levels of an attribute is not prime.

It is also worth noting that Design 1 in Table 8.10 does not satisfy the desirable properties 6 and 7 given at the beginning of this chapter. For the 4-level attributes 0 in option 1 is always paired with 1 in option 2, 1 in option 1 is always paired with 2 in option 2, 2 in option 1 is always paired with 3 in option 2, and 3 in option 1 is always paired with 0 in option 2. Furthermore, the only combinations of levels of each of the 4-level attributes that appear in the choice sets are 0 and 1, 1 and 2, 2 and 3, 3 and 0. By adding the sets of generators $G_1 = (00000, 11111)$, $G_2 = (00000, 11122)$, and $G_3 = (00000, 11133)$, properties 6 and 7 can be satisfied, but at a cost of an increase in the number of choice sets from 16 to 48. This larger design is 100% efficient. □

Table 8.8 Starting Design and Near-Optimal Choice Sets for $k = 5$, $\ell_1 = \ell_2 = \ell_3 = 2$, and $\ell_4 = \ell_5 = 4$ when $m = 4$ for Main Effects Only

$4^5//16$	$F = 2^3 \times 4^2//16$	$F + 11111$	$F + 10122$	$F + 01033$
00000	00000	11111	10122	01033
01111	01111	10022	11033	00100
02222	01122	10033	11000	00111
03333	00033	11100	10111	01022
10123	10123	01030	00001	11112
11032	11032	00103	01110	10021
12301	11001	00112	01123	10030
13210	10110	01021	00032	11103
00231	00131	11002	10013	01120
01320	01020	10131	11102	00013
02013	01013	10120	11131	00002
03102	00102	11013	10020	01131
10312	10012	01123	00130	11001
11203	11103	00010	01021	10132
12130	11130	00001	01012	10123
13021	10021	01132	00103	11010

In the following example we illustrate the use of both expansive replacement of one attribute by more than one attribute (see Construction 2.3.4) and contractive replacement of several attributes by one attribute (see Construction 2.3.5). The most commonly used expansive and contractive replacements of attributes are given in the middle and last sections, respectively, of Table 8.9. Contractive replacement is not always possible due to the conditions on the structure of the original array.

■ EXAMPLE 8.1.6.

Suppose $k = 13$, $\ell_q = 2$, for $q = 1, \ldots, 12$, $\ell_{13} = 4$, and $m = 3$. Now there are two ways in which we can obtain the starting design: a $2^{12} \times 4//16$ OMEP. The first way is to begin with the $4^5//16$ and use expansive replacement to get the starting design that we need. Alternatively, we can begin with the $2^{15}//16$ and use contractive replacement to get the desired starting design. In this example we will illustrate both methods.

Starting with the $4^5//16$, which is given in the first column in Table 8.11, we replace the 4 levels of the first attribute by the 4 treatment combinations of the $2^3//4$, which is shown in Table 2.25. That is, we replace the first 4-level attribute by three 2-level attributes. We replace 0 in the first attribute with 000, 1 with 011, 2 with 101, and 3 with 110. We repeat this procedure for the second, third and fourth 4-level attributes to get the $2^{12} \times 4//16$ design shown in the second column in Table 8.11. If we add the set of generators

$$G = (0000000000000, 1111111111111, 1010101010102)$$

to this starting design then we get the choice sets which are also given in Table 8.11. This design is 100% efficient and the information matrix is

$$C_M = \frac{1}{18432} I_{15}.$$

Table 8.9 Common Collapsing/Replacement of Attribute Levels

Starting Column(s)		Replace with Column(s)		Use
# Columns	# Levels	# Columns	# Levels	OMEP
1	16	1	8, 4 or 2	
1	8	1	4 or 2	
1	4	1	2	
1	9	1	3	
1	6	1	3 or 2	
1	4	3	2	$2^3//4$
1	8	7	2	$2^7//8$
1	8	{ 1 4	4 2	$2^4 \times 4//8$
1	16	5	4	$4^5//16$
1	16	15	2	$2^{15}//16$
1	16	{ 6 3	2 4	$2^6 \times 4^3//16$
1	9	4	3	$3^4//9$
3	2	1	8	$2^3//8$
3	2	1	4	$2^3//4$
2	4	1	16	$4^2//16$
2	3	1	9	$3^2//9$

In the second method, we start with the fourth $2^{15}//16$ OMEP, which is called the oa.16.15.2.2.3, from Sloane (2006b). We denote this OMEP by A and it is shown in the first column in Table 8.12. Let B be the $2^3//4$ which is displayed in Table 2.25. Then B is tight since it satisfies the condition given in Construction 2.3.5 for a tight orthogonal array: $N_1 = 4 = 1 + 3 \times (2 - 1)$. We need three columns of A that contain only the entries in the $2^3//4$, that is, treatment combinations 000, 011, 101 and 110. There are seven possibilities from which we can choose: columns 1, 8, and 9; columns 2, 8, and 10; columns 3, 8, and 11; columns 4, 8, and 12; columns 5, 8, and 13; columns 6, 8, and 14; or columns 7, 8, and 15. Suppose we choose columns 1, 8, and 9 and replace them with one 4-level column. We can do this since those $c = 3$ columns of A form $N/N_1 = 16/4 = 4$ copies of B. To get the 4-level column in the new design F, we replace 000 in the columns 1, 8, and 9 of A with 0, 011 with 1, 101 with 2, and 110 with 3. This new 4-level column is then combined with columns 2 to 7 and 10 to 15 of A to get the $2^{12} \times 4//16$ design. This design is given in the second column in Table 8.12.

The choice sets are then constructed with the same set of generators as the first method and the resulting design is 100% efficient with the same C_M matrix as before. The choice sets are also given in Table 8.12. □

As we discussed in Sections 1.6.2, 4.2.4, and 5.2.3, it is possible to have choice sets in which one treatment combination may dominate the other treatment combinations in the same choice set. This can happen when the levels of all of the attributes are ordered in some way, from best to worst or vice versa. In this situation we need to avoid starting designs which contain the treatment combination 000...0 or the treatment combination with all attributes at the high level. The best way of choosing sets of generators is to have

Table 8.10 Choice Sets for $k = 5$, $\ell_1 = \ell_2 = \ell_3 = 2$, and $\ell_4 = \ell_5 = 4$ when $m = 2$ for Main Effects Only

Design 1		Design 2	
F	$F + 11111$	F	$F + 11122$
0 0 0 0 0	1 1 1 1 1	0 0 0 0 0	1 1 1 2 2
0 1 1 1 1	1 0 0 2 2	0 1 1 1 1	1 0 0 3 3
0 1 1 2 2	1 0 0 3 3	0 1 1 2 2	1 0 0 0 0
0 0 0 3 3	1 1 1 0 0	0 0 0 3 3	1 1 1 1 1
1 0 1 2 3	0 1 0 3 0	1 0 1 2 3	0 1 0 0 1
1 1 0 3 2	0 0 1 0 3	1 1 0 3 2	0 0 1 1 0
1 1 0 0 1	0 0 1 1 2	1 1 0 0 1	0 0 1 2 3
1 0 1 1 0	0 1 0 2 1	1 0 1 1 0	0 1 0 3 2
1 0 1 3 1	0 1 0 0 2	1 0 1 3 1	0 1 0 1 3
1 1 0 2 0	0 0 1 3 1	1 1 0 2 0	0 0 1 0 2
1 1 0 1 3	0 0 1 2 0	1 1 0 1 3	0 0 1 3 1
1 0 1 0 2	0 1 0 1 3	1 0 1 0 2	0 1 0 2 0
0 0 0 1 2	1 1 1 2 3	0 0 0 1 2	1 1 1 3 0
0 1 1 0 3	1 0 0 1 0	0 1 1 0 3	1 0 0 2 1
0 1 1 3 0	1 0 0 0 1	0 1 1 3 0	1 0 0 1 2
0 0 0 2 1	1 1 1 3 2	0 0 0 2 1	1 1 1 0 3

Table 8.11 $2^{12} \times 4//16$ Obtained from $4^5//16$ by Expansive Replacement, and Choice Sets for $m = 3$

$4^5//16$	$F = 2^{12} \times 4//16$	$F + 1111111111111$	$F + 1010101010102$
00000	0000000000000	1111111111111	1010101010102
01111	0000110110111	1111001001002	1010011100013
02222	0001011011012	1110100100103	1011110001110
03333	0001101101103	1110010010010	1011000111001
10123	0110000111013	1001111000100	1100101101111
11032	0110110001102	1001001110013	1100011011000
12301	0111011100001	1000100011112	1101110110103
13210	0111101010110	1000010101001	1101000000012
20231	1010001011101	0101110100012	0000100001003
21320	1010111101010	0101000010101	0000010111112
22013	1011010000113	0100101111000	0001111010011
23102	1011100110002	0100011001113	0001001100100
30312	1100001100112	0011110011003	0110100110010
31203	1100111010003	0011000101110	0110010000101
32130	1101010111100	0010101000011	0111111101002
33021	1101100001011	0010011110102	0111001011113

a mix of levels so that in some attributes a low value is added, in some attributes middle values are added, and in the remaining attributes high values are added. Thus no choice set will contain treatment combinations that dominate other treatment combinations. In the following example we show how to obtain a starting design for the previous example. We also discuss avoiding unrealistic treatment combinations. In both cases we are using the ideas in Construction 2.3.9.

Table 8.12 $2^{12} \times 4//16$ Obtained from $2^{15}//16$ by Contractive Replacement, and Choice Sets for $m = 3$

$A = 2^{15}//16$	$F = 2^{12} \times 4//16$	$F + 1111111111111$	$F + 1010101010102$
000000000000000	0000000000000	1111111111111	1010101010102
101010101010101	0101010101012	1010101010103	1111111111110
011001100110011	1100111100110	0011000011001	0110010110012
110011001100110	1001101001102	0110010110013	0011000011000
000111100001111	0011110011110	1100001100001	1001011001012
101101001011010	0110100110102	1001011001013	1100001100000
011110000111100	1111001111000	0000110000111	0101100101102
110100101101001	1010011010012	0101100101103	0000110000110
000000011111111	0000001111111	1111110000002	1010100101013
001011111101000	0101111010001	1010000101112	1111010000103
010111011010001	1011100100011	0100011011102	0001001110113
011100111000110	1110010001101	0001101110012	0100111011003
100101110110100	0010111101003	1101000010110	1000010111101
101110010100011	0111001000113	1000110111000	1101100010011
110010110011010	1001010110103	0110101001010	0011111100001
111001010001101	1100100011013	0011011100100	0110001001111

■ **EXAMPLE 8.1.7.**

In Table 8.11 the first choice set contains the treatment combination 0000000000000. If the attribute levels are ordered from least preferred at level 0 to most preferred at level $\ell_q - 1$ then the treatment combination of all zeros will be dominated by the other treatment combinations in the choice set. We can use a different fraction for the starting design to avoid this problem (see Construction 2.3.9). First we choose a treatment combination that is not in F. For example, the treatment combination 1111110000000 is not in F and if we add it to F, using the modular arithmetic appropriate to each attribute, we will obtain a different fraction which is also an OMEP. This OMEP is displayed in the first column of Table 8.13. Neither the treatment combinations with low values, 0000000000000, nor the treatment combination with high values, 1111111111113, appears in this OMEP. Then by adding the same set of generators as we used in the previous example, we get the choice sets given in Table 8.13. These choice sets have the same efficiency and C_M matrix as before. Note that in no choice set does one treatment combination dominate another. □

We can use the same method if there are treatment combinations that are unrealistic. Suppose that in Example 8.1.5, the treatment combination with the 2-level attributes at the low level and the 4-level attributes at the high level (00033) is considered unrealistic and hence we do not want it appearing in the choice experiment. This treatment combination is in the starting design (see Table 8.8) and can be avoided either by using a different collapsing scheme for the first three 4-level attributes, such as $0 \to 0$, $1 \to 0$, $2 \to 1$ and $3 \to 1$, or by using Construction 2.3.9 to obtain a different starting design. After we have chosen the appropriate starting design, the set of generators should be chosen so that no unrealistic treatment combination results from the addition of the generators to the starting design.

Table 8.13 A Different $2^{12} \times 4//16$ and Choice Sets for $m = 3$

$F = 2^{12} \times 4//16$ + 1111110000000	$F + 1111111111111$	$F + 1010101010102$
1111110000000	0000001111111	0101011010102
1111000110111	0000111001002	0101101100013
1110101011012	0001010100103	0100000001110
1110011101103	0001100010010	0100110111001
1001110111013	0110001000100	0011011101111
1001000001102	0110111110013	0011101011000
1000101100001	0111010011112	0010000110103
1000011010110	0111100101001	0010110000012
0101111011101	1010000100012	1111010001003
0101001101010	1010110010101	1111100111112
0100100000113	1011011111000	1110001010011
0100010110002	1011101001113	1110111100100
0011111100112	1100000011003	1001010110010
0011001010003	1100110101110	1001100000101
0010100111100	1101011000011	1000001101002
0010010001011	1101101110102	1000111011113

8.1.3 More on Choosing Generators

When the number of levels increases, it is sometimes possible to obtain a more efficient choice experiment by using difference sets to choose sets of generators for the choice sets, as long as there is a difference set for the appropriate values of ℓ_q and m. In the following example we use the difference sets given in Tables 2.34 and 2.35.

■ EXAMPLE 8.1.8.

Let $k = 17$, $\ell_q = 2$ for $q = 1, \ldots, 16$, and $\ell_{17} = 13$. There are two possible starting designs, both available in Kuhfeld (2006): $2^{16} \times 13//52$; or $2^{16} \times 16//32$ in which the 16-level attribute will need to be collapsed to 13 levels, resulting in unequal replication of levels. In this example we first consider the choice sets constructed from the $2^{16} \times 13//52$ starting design for $m = 2, 3, 4, 5$ (see Table 8.14).

For $m = 2$ the only possible entry for the binary attributes in $\mathbf{g}_2$ is 1. For the 13-level attribute the entry in $\mathbf{g}_2$ can be any of the numbers 1 to 12 and the efficiency will be the same. Choice sets are constructed using the set of generators

$$G = (00000000000000000, 11111111111111111)$$

resulting in 52 choice sets with efficiency of 86.23%. For a higher efficiency two or more sets of generators are required. For example, if we use the sets of generators

$$G_1 = (00000000000000000, 11111111111111111)$$

and

$$G_2 = (00000000000000000, 11111111111111115)$$

the 104 choice sets have an efficiency of 96.84%.

For $m = 3$ an obvious set of generators to use is

$$G = (0000000000000000, 0101010101010111, 1010101010101010 2)$$

and the 52 choice sets are 93.10% efficient. While this efficiency is good, we can use the elements of either of the difference sets (0,1,4) and (0,2,7), given in Table 2.35 for $\ell = 13$, $m = 3$ and $\lambda = 1$, as the entries in the generators for the 13-level attribute to increase the efficiency. Thus we could use either of the sets of generators G_1 and G_2, where

$$G_1 = (00000000000000000, 01010101010101011, 10101010101010104)$$

and

$$G_2 = (00000000000000000, 01010101010101012, 10101010101010107).$$

Either would result in 52 choice sets that are 98.30% efficient. However, if we use both G_1 and G_2 the 104 choice sets are 100% efficient.

For $m = 4$ we can use the difference set (0,1,3,9) given in Table 2.34 for the entries in $\mathbf{g}_i$, $i = 1, 2, 3, 4$. The 52 choice sets are 100% efficient (see Table 8.14 for the sets of generators). On the other hand, if we use 0, 1, 2 and 3, which is not a difference set, for the entries in $\mathbf{g}_i$, $i = 1, 2, 3, 4$, for the 13-level attribute, the 52 choice sets are only 96.31% efficient.

For $m = 5$ there is the difference family (0,1,2,4,8), (0,1,3,6,12), and (0,2,5,6,10) in Table 2.35. Using just one difference set from the difference family for the entries in $\mathbf{g}_i$, $i = 1, \ldots, 5$, for the 13-level attribute, gives 52 choice sets that are 99.87%, 99.87%, and 99.65% efficient, respectively. However, if we use all three difference sets the 156 choice sets are 100% efficient.

Now we consider the choice sets constructed from the $2^{16} \times 16//32$. The 16-level attribute is collapsed to 13 levels by replacing all 13s with 0, all 14s with 1 and all 15s with 2. In Table 8.14 we give the efficiencies for $m = 2, 3, 4, 5$ using the same sets of generators as for the other starting design. While these designs all have only 32 choice sets, there is unequal replication of the levels of the 13-level attribute and the efficiencies are less than for the corresponding designs constructed from the $2^{16} \times 13//52$. For all of the choice experiments in this example, all pairs of main effects are uncorrelated. □

In the following two examples there is one attribute with a large number of levels. The OMEPs required for the starting designs are not already available and must be constructed. In both cases the number of choice sets is far more than a respondent can complete, and the choice sets need to be split into blocks of a manageable number of choice sets. While some practitioners may never use an attribute with such a large number of levels, we have been asked to construct choice experiments with large numbers of levels for one of the attributes on several occasions.

■ EXAMPLE 8.1.9.

Suppose there are $k = 8$ attributes with levels $\ell_1 = \ell_2 = 2$, $\ell_3 = \ell_4 = \ell_5 = \ell_6 = 4$, $\ell_7 = 8$, $\ell_8 = 36$, and $m = 3$. Since there is no starting design readily available, we will need to construct one. One way to do this is to start with the $4^8 \times 8//32$ shown in the first column of Table 8.15. We discard the first 4-level attribute and collapse the levels of the seventh and eighth 4-level attributes to 2 levels (0 and 1 become 0, 2 and 3 become 1) to obtain a $4^5 \times 2^2 \times 8//32$, which is shown in the second column of Table 8.15. We then create a 9-level attribute using Construction 2.3.6. We write down 9 copies of the

Table 8.14 Different Designs for $k = 17$, $\ell_q = 2$, $q = 1, \ldots, 16$, and $\ell_{17} = 13$ when $m = 2, 3, 4, 5$ for Main Effects Only

	$2^{16} \times 13//52$		$2^{16} \times 16//32$	
Generators	**# Choice Sets**	**Efficiency**	**# Choice Sets**	**Efficiency**
m = 2				
00000000000000000, 11111111111111111	52	86.23%	32	84.60%
m = 3				
00000000000000000, 01010101010101011, 10101010101010102	52	93.10%	32	91.68%
00000000000000000, 01010101010101011, 10101010101010104	52	98.30%	32	97.28%
m = 4				
00000000000000000, 10101010101010101, 01010101010101012, 11111111111111113	52	96.31%	32	95.16%
00000000000000000, 10101010101010101, 01010101010101013, 11111111111111119	52	100%	32	99.38%
m = 5				
00000000000000000, 10101010101010101, 01010101010101012, 11111111000000003, 00000000111111114	52	98.02%	32	97.11%
00000000000000000, 10101010101010101, 01010101010101012, 11111111000000004, 00000000111111118	52	99.87%	32	99.38%

$4^5 \times 2^2 \times 8//32$ OMEP, one above the other, and then adjoin a 9-level attribute with 32 copies of 0 then 32 copies of 1, and so on, up to 32 copies of 8, as shown in Table 8.16. The resulting array is a $9 \times 4^5 \times 2^2 \times 8//(32 \times 9 = 288)$ OMEP.

We will now create a 36-level attribute from the 9-level attribute and the first 4-level attribute by using contractive replacement (see Construction 2.3.5). The first two columns of the $9 \times 4^5 \times 2^2 \times 8//288$ form 8 copies of an OA[36, 4, 9, 2], which could be extended to form a tight OMEP. Thus replacing 00 with 0, 01 with 1, 02 with 2, and so on, results in a $36 \times 4^4 \times 2^2 \times 8//288$ OMEP. Reordering the columns, we get the $4^4 \times 2^2 \times 8 \times 36//288$ design in the columns labeled F in Appendix 8. A.1.

Now that we have the starting design, we can create the choice sets of size 3. By trying various sets of generators, the most efficient design in 288 choice sets we found is 93.73% efficient and can be obtained by adding a set of generators such as

$$G = (00000000, 11111011, 33330136).$$

There are many other combinations of $\mathbf{g}_2$ and $\mathbf{g}_3$ that result in a design with the same efficiency.

In order to calculate the C_M matrix for this design, we need to be able to calculate the rows in the B matrix for all of the attributes, including the 36-level attribute. The orthogonal polynomials for the 36-level attribute can be constructed using Kronecker products from the orthogonal polynomials for the 4-level and 9-level attributes. This method is illustrated in Example 6.3.3. □

■ EXAMPLE 8.1.10.

Suppose there are $k = 9$ attributes with levels $\ell_1 = \ell_2 = \ell_3 = \ell_4 = \ell_5 = 4$, $\ell_6 = 2$, $\ell_7 = \ell_8 = 8$, $\ell_9 = 24$, and $m = 3$. Again there is no starting design readily available, and we will need to construct one. First we take the $8^9//64$ from Sloane (2006b) and collapse the first 5 columns from 8 levels to 4 levels (0 and 1 become 0; 2 and 3 become 1; 4 and 5 become 2; 6 and 7 become 3). Column 6 is then collapsed from 8 levels to 2 levels (0, 1, 2 and 3 become 0; 4, 5, 6, and 7 become 1). We leave columns 7, 8, and 9 as they are, thus giving us a $4^5 \times 2 \times 8^3//64$ OMEP. This design is given in Table 8.17.

We will now create a 24-level attribute from the last 8-level attribute by using Construction 2.3.8. We write down 3 copies of the $4^5 \times 2 \times 8^3//64$ OMEP, changing the levels of column (attribute) 9 in two of the copies in order to create a column with 24 levels. In the first copy, the levels of column 9 remain unchanged (levels 0 to 7), in the 2nd copy add 8 to the levels of column 9 to get levels 8 to 15, and in the 3rd copy add 16 to the levels of column 9 to get levels 16 to 23; this results in a 24-level attribute in column 9. This design is now a $4^5 \times 2 \times 8^2 \times 24//(64 \times 3 = 192)$ OMEP, which is the required starting design (see the first column, F, in Appendix 8. A.2).

To create the choice sets we can add a set of generators such as

$$G = (000000000, 111111111, 33333033(10))$$

where (10) means that the number 10 is added (modulo 24) to the levels of the ninth attribute. This design has an efficiency of 96.32%, and the main effects are all pairwise uncorrelated. Various other sets of generators, such as

$$G = (000000000, 111111112, 333330335),$$

are just as good. To calculate the B matrix, and therefore the C_M matrix, we can calculate the orthogonal polynomials for the 24-level attribute from the polynomials for the 6-level and 4-level attributes (see Example 6.3.3). □

Table 8.15 $4^5 \times 2^2 \times 8//32$ and $4^8 \times 8//32$ OMEPs

$4^8 \times 8//32$	$4^5 \times 2^2 \times 8//32$
0 0 0 0 0 0 0 0 0	0 0 0 0 0 0 0 0 0
0 0 1 1 3 3 2 2 1	0 0 1 1 3 3 1 1 1
0 1 2 3 0 1 2 3 2	0 1 2 3 0 1 1 1 2
0 1 3 2 3 2 0 1 3	0 1 3 2 3 2 0 0 3
0 2 0 2 1 3 1 3 4	0 2 0 2 1 3 0 1 4
0 2 1 3 2 0 3 1 5	0 2 1 3 2 0 1 0 5
0 3 2 1 1 2 3 0 6	0 3 2 1 1 2 1 0 6
0 3 3 0 2 1 1 2 7	0 3 3 0 2 1 0 1 7
1 0 2 3 2 3 1 0 3	1 0 2 3 2 3 0 0 3
1 0 3 2 1 0 3 2 2	1 0 3 2 1 0 1 1 2
1 1 0 0 2 2 3 3 1	1 1 0 0 2 2 1 1 1
1 1 1 1 1 1 1 1 0	1 1 1 1 1 1 0 0 0
1 2 2 1 3 0 0 3 7	1 2 2 1 3 0 0 1 7
1 2 3 0 0 3 2 1 6	1 2 3 0 0 3 1 0 6
1 3 0 2 3 1 2 0 5	1 3 0 2 3 1 1 0 5
1 3 1 3 0 2 0 2 4	1 3 1 3 0 2 0 1 4
2 0 2 0 3 1 3 1 4	2 0 2 0 3 1 1 0 4
2 0 3 1 0 2 1 3 5	2 0 3 1 0 2 0 1 5
2 1 0 3 3 0 1 2 6	2 1 0 3 3 0 0 1 6
2 1 1 2 0 3 3 0 7	2 1 1 2 0 3 1 0 7
2 2 2 2 2 2 2 2 0	2 2 2 2 2 2 1 1 0
2 2 3 3 1 1 0 0 1	2 2 3 3 1 1 0 0 1
2 3 0 1 2 3 0 1 2	2 3 0 1 2 3 0 0 2
2 3 1 0 1 0 2 3 3	2 3 1 0 1 0 1 1 3
3 0 0 3 1 2 2 1 7	3 0 0 3 1 2 1 0 7
3 0 1 2 2 1 0 3 6	3 0 1 2 2 1 0 1 6
3 1 2 0 1 3 0 2 5	3 1 2 0 1 3 0 1 5
3 1 3 1 2 0 2 0 4	3 1 3 1 2 0 1 0 4
3 2 0 1 0 1 3 2 3	3 2 0 1 0 1 1 1 3
3 2 1 0 3 2 1 0 2	3 2 1 0 3 2 0 0 2
3 3 2 2 0 0 1 1 1	3 3 2 2 0 0 0 0 1
3 3 3 3 3 3 3 3 0	3 3 3 3 3 3 1 1 0

Table 8.16 $9 \times 4^5 \times 2^2 \times 8//288$ by Adding Another Attribute to the $4^5 \times 2^2 \times 8//32$

9-Level Attribute	$4^5 \times 2^2 \times 8$ Attributes
0 ⋮ 0	$4^5 \times 2^2 \times 8//32$
1 ⋮ 1	$4^5 \times 2^2 \times 8//32$
⋮	⋮
8 ⋮ 8	$4^5 \times 2^2 \times 8//32$

8.2 SMALL NEAR-OPTIMAL DESIGNS FOR MAIN EFFECTS PLUS TWO-FACTOR INTERACTIONS

In this chapter thus far we have been discussing techniques to construct choice experiments for main effects only which satisfy as many as possible of the seven desirable properties given at the beginning of this chapter. These designs are appropriate if all interactions can be assumed to be negligible. However, if there is no reason to believe that this is the case then a design that allows for the estimation of two-factor interactions, or even higher-order interactions, should be used.

In this section, we give some techniques for getting a starting design and choosing sets of generators to construct the choice sets. In Theorem 5.1.2, we give the maximum determinant of the C_{MT} matrix for binary attributes and hence we can calculate efficiencies, but for the situation in which there is at least one attribute with more than two levels we do not know the maximum possible determinant of the C_{MT} matrix. For this case we can still calculate $\det(C_{MT})$ for various designs (see Lemma 6.5.2) and choose the most appropriate design from those considered.

It is often the case, when wanting to estimate main effects and all or some of the two-factor interactions, that more than one set of generators needs to be added to an already large starting design. This results in a large number of choice sets, which will need to be split into blocks so that the respondents are presented with a manageable number of choice sets.

8.2.1 Getting a Starting Design

Designs of resolution 5 can be used to estimate main effects and all two-factor interactions although finding such a design can be challenging. If there are four or fewer attributes (so $k \leq 4$), then the only resolution 5 design is the complete factorial, independent of the number of levels of each of the attributes. If there are more than four attributes ($k > 4$),

Table 8.17 $4^5 \times 2 \times 8^3//64$ OMEP Obtained from $8^9//64$ by Collapsing the Levels of 6 Attributes

$8^9//64$		$4^5 \times 2 \times 8^3//64$	
000000000	404444444	000000000	202221444
011234567	412160735	000111567	201030735
022345671	421607352	011121671	210301352
033456712	436073521	011221712	213030521
044567123	440735216	022231123	220311216
055671234	457352160	022330234	223120160
066712345	463521607	033300345	231210607
077123456	475216073	033010456	232101073
101111111	505555555	000000111	202221555
110472653	516327014	000230653	203111014
124726530	523270146	012311530	211130146
137265304	532701463	013131304	211300463
142653047	547014632	021320047	223001632
156530472	550146327	023210472	220021327
165304726	561463270	032101726	230230270
173047265	574632701	031021265	232310701
202222222	606666666	101110222	303331666
214051376	615743102	102020376	302320102
220513764	627431025	110200764	313210025
235137640	634310257	112011640	312100257
241376405	643102574	120131405	321000574
253764051	651025743	121331051	320011743
267640513	660257431	133320513	330121431
276405137	672574310	133201137	331231310
303333333	707777777	101110333	303331777
317506241	713615420	103201241	301301420
325062417	726154203	112030417	313021203
330624175	731542036	110311175	310220036
346241750	745420361	123120750	322210361
352417506	754203615	121201506	322100615
364175062	762036154	132031062	331011154
371750624	770361542	130320624	330130542

then we can find a resolution 5 design that uses fewer treatment combinations than the complete factorial.

If only some two-factor interactions are to be estimated, under the assumption that the remaining two-factor and higher-order interactions are negligible, a resolution 4 fractional factorial design can be used, as long as the main effects and interactions of interest are orthogonal. Some fractional factorial designs of resolution 5 (strength 4) are given in Sloane (2006b); otherwise a fractional or complete factorial will need to be constructed. Some constructions for fractional factorial designs of resolution 5 are given in Section 2.2.1, and more constructions are given in Dey (1985) and Hedayat et al. (1999).

8.2.2 Designs for Two-Level Attributes

In Sloane (2006b) there are some designs for binary attributes with resolution at least 5. In Section 2.2.1 we give a construction for resolution 5 designs for 2^k factorial designs.

In Theorem 5.1.2 we give the maximum determinant of the C_{MT} matrix for binary attributes and the form of the optimal design. In Section 5.2.2 we give a construction for near-optimal designs.

■ **EXAMPLE 8.2.1.**
Let $k = 7$ and $\ell_q = 2$ for $q = 1, \ldots, 7$. From Table 2.10, we see that the smallest known 2^7 fractional factorial design of at least resolution 5 has $N = 64$ treatment combinations. Such a design can be found in Dey (1985) or Sloane (2006b) or by using the construction for 2-level fractions in Section 2.2.1. The design in Table 8.18 is from Dey (1985).

Theorem 5.1.2 states that the D-optimal design is given by

$$y_4 = \frac{m(m-1)}{2^7\binom{7}{4}}$$

and all other $y_i = 0$. For the optimal design for $m = 2$, choice sets with the difference vector containing only 4s are required. Hence we need to choose sets of generators so that there are four attributes different between any pair of treatment combinations in a choice set.

A method that results in near-optimal choice sets, for binary attributes for $m = 2$, is given in Section 5.2.2. The following sets of generators,

$$G_1 = (0000000, 0001111),$$

$$G_2 = (0000000, 0110011),$$

$$G_3 = (0000000, 1010101),$$

when added to the starting design in Table 8.19, result in choice sets that are 91.85% efficient with $\det(C_{MT}) = (1/384)^{12}(1/192)^{12}(1/128)^4$: Note that there is at least one 1 in the position corresponding to each attribute, and for any two attributes there is at least one $\mathbf{g}_2$ in which the corresponding positions have a 0 and a 1 entry. These conditions cannot be satisfied in less than three sets of generators. Any repeated choice sets were deleted, thereby reducing the number of choice sets from $64 \times 3 = 192$ to 96.

For $m = 3$ the sets of generators

$$G_1 = (0000000, 0001111, 0110011)$$

and

$$G_2 = (0000000, 1010101, 1101001)$$

give us difference vectors containing only 4s. The resulting design is 95.71% efficient.

Similarly, for $m = 4$, using

$$G_1 = (0000000, 0001111, 0110011, 1010101)$$

results in a design that is 98.98% efficient, and for $m = 5$ the design obtained using

$$G_1 = (0000000, 0001111, 0110011, 1010101, 1101001)$$

is 99.09% efficient.

In all of the above designs C_{MT}, and therefore C_{MT}^{-1}, are diagonal, which means that the main effects and two-factor interactions are all uncorrelated. □

In the previous example we have used a starting design of at least resolution 5. What happens if we use a starting design of resolution 3 or 4? In Sloane (2006b) there is a 2^7 resolution 3 design in 8 runs, in which the main effects and two-factor interactions are confounded, and a 2^8 resolution 4 design in 16 runs, in which the two-factor interactions are confounded with each other. If either of these designs is used as the starting design, then many more sets of generators are required to achieve a non-zero $\det(C_{MT})$. Moreover, the C_{MT}^{-1} matrix will not be diagonal and therefore some, if not all, pairs of effects will be correlated.

Table 8.18 Fractional Factorial of Resolution 7 for $\ell_q = 2, q = 1, \ldots, 7$

0000000	0011000	0111001	1011100
1000001	0010100	0110101	1011010
0100001	0010010	0110011	1010110
0010001	0001100	0101101	1001110
0001001	0001010	0101011	0111100
0000101	0000110	0100111	0111010
0000011	1110001	0011101	0110110
1100000	1101001	0011011	0101110
1010000	1100101	0010111	0011110
1001000	1100011	0001111	1111101
1000100	1011001	1111000	1111011
1000010	1010101	1110100	1110111
0110000	1010011	1110010	1101111
0101000	1001101	1101100	1011111
0100100	1001011	1101010	0111111
0100010	1000111	1100110	1111110

8.2.3 Designs for Attributes with More than Two Levels

At this stage there are not many fractional factorial designs available when there is at least one attribute with more than two levels. There are some available in Sloane (2006b), there are Constructions 2.2.2 and 2.3.2, and there are constructions in Section 2.3.2, Dey (1985) and Hedayat et al. (1999). Furthermore, if there is at least one attribute with more than two levels, we do not know the maximum $\det(C_{MT})$, other than for some small examples in the appendices of Chapter 6. However, we do have an expression for $\det(C_{MT})$ which allows

Table 8.19 Near-Optimal Choice Sets for $\ell_q = 2, q = 1, \ldots, 7$ and $m = 2$ for Main Effects and All Two-Factor Interactions

F	$F + 0001111$	F	$F + 0110011$	F	$F + 1010101$
0000000	0001111	0000000	0110011	0000000	1010101
1000001	1001110	1000001	1110010	1000001	0010100
0100001	0101110	0100001	0010010	0100001	1110100
0010001	0011110	0010001	0100010	0010001	1000100
0001001	0000110	0001001	0111010	0001001	1011100
0000101	0001010	0000101	0110110	0000101	1010000
0000011	0001100	0000011	0110000	0000011	1010110
1100000	1101111	1100000	1010011	1100000	0110101
1010000	1011111	1010000	1100011	1001000	0011101
1001000	1000111	1001000	1111011	1000010	0010111
1000100	1001011	1000100	1110111	0110000	1100101
1000010	1001101	1000010	1110001	0101000	1111101
0110000	0111111	0101000	0011011	0100100	1110001
0101000	0100111	0100100	0010111	0100010	1110111
0100100	0101011	0011000	0101011	0011000	1001101
0100010	0101101	0010100	0100111	0010010	1000111
0011000	0010111	0001100	0111111	0001100	1011001
0010100	0011011	0001010	0111001	0001010	1011111
0010010	0011101	0000110	0110101	0000110	1010011
1110001	1111110	1101001	1011010	1101001	0111100
1101001	1100110	1100101	1010110	1100011	0110110
1100101	1101010	1011001	1101010	1001011	0011110
1100011	1101100	1010101	1100110	0111001	1101100
1011001	1010110	1001101	1111110	0110011	1100110
1010101	1011010	1001011	1111000	0101101	1111000
1010011	1011100	1000111	1110100	0101011	1111110
0111001	0110110	0101101	0011110	0100111	1110010
0110101	0111010	0011101	0101110	0011011	1001110
0110011	0111100	0001111	0111100	0001111	1011010
1111000	1110111	1101100	1011111	1101010	0111111
1110100	1111011	1011100	1101111	0111010	1101111
1110010	1111101	1001110	1111101	0101110	1111011

us to compare different choice experiments. In this situation it is best to construct some large designs to try to find a very large $\det(C_{MT})$, then use this as a basis for determining the efficiency of smaller designs that can be used in practice.

EXAMPLE 8.2.2.

Suppose that $k = 5$, $\ell_1 = \ell_2 = \ell_3 = \ell_4 = 2$, and $\ell_5 = 4$. There is no resolution 5 design readily available, so we need to construct an OA$[32; 2^4, 4; 4]$ using Construction 2.3.2. The column $\mathbf{b}_1$ has 16 0s and 16 1s, the column $\mathbf{b}_2$ has 8 0s, 8 1s, 8 0s, then 8 1s, and so on up to the column $\mathbf{b}_5$, which alternates 0s and 1s. Then columns $\mathbf{b}_3$, $\mathbf{b}_4$, $\mathbf{b}_5$, and $\mathbf{b}_1 + \mathbf{b}_3 + \mathbf{b}_4 + \mathbf{b}_5$ represent the four columns for the 2-level attributes, where the addition is performed modulo 2. Columns $\mathbf{b}_1$, $\mathbf{b}_2$, and $\mathbf{b}_1 + \mathbf{b}_2$ become the one column for the four-level attribute, by replacing 000 with 0, 011 with 1, 101 with 2, and 110 with 3. This design is given in the first column of Table 8.20.

By looking at the complete factorial, and also the OA$[32; 2^4, 4; 4]$, plus generators $\mathbf{g}_2$ with at least 2 zeros, the largest $\det(C_{MT})$ we could find is $(1/160)^3(3/320)^{21}$. This was obtained by taking the OA$[32; 2^4, 4; 4]$ and adding the 10 sets of generators

$$G_1 = (00000, 00111),\ G_2 = (00000, 01012),\ G_3 = (00000, 01103),$$

$$G_4 = (00000, 01110),\ G_5 = (00000, 10011),\ G_6 = (00000, 10102),$$

$$G_7 = (00000, 10110),\ G_8 = (00000, 11003),\ G_9 = (00000, 11010),$$

$$\text{and } G_{10} = (00000, 11100),$$

resulting in 320 choice sets. However, if we just add the sets of generators G_1, G_6, and G_9, we get 96 choice sets which are 90.38% efficient relative to the largest $\det(C_{MT})$ found. This design is given in Table 8.20.

For $m = 3$ adding the sets of generators

$$G_1 = (00000, 00111, 10102) \text{ and } G_2 = (00000, 01103, 11010)$$

to the OA$[32; 2^4, 4; 4]$ gives us 64 choice sets that are 96.44% efficient relative to the largest $\det(C_{MT})$ we found. Similarly, for $m = 4$ the most efficient design in 32 choice sets was obtained by using the set of generators

$$G = (00000, 00111, 10102, 01110)$$

and for $m = 5$ using the set of generators

$$G = (00000, 00111, 10102, 01110, 11003).$$

These last two designs are 96.29% and 94.66% efficient respectively, relative in each case to the largest $\det(C_{MT})$ we could find. □

EXAMPLE 8.2.3.

Let $k = 3$, $\ell_1 = \ell_2 = 3$, and $\ell_3 = 5$. Since there are only three attributes, we need to use the complete factorial in 45 treatment combinations as the starting design (see the F columns in Table 8.21).

For $m = 2$ we do know the difference vectors for the optimal design and the maximum value of $\det(C_{MT})$, since this is one of the examples included in Appendix 6. A.2. From

Table 8.20 Choice Sets for $k = 5$, $\ell_1 = \ell_2 = \ell_3 = \ell_4 = 2$, $\ell_5 = 4$ when $m = 2$ for Main Effects and All Two-Factor Interactions

F	$F + 00111$	F	$F + 10102$	F	$F + 11010$
00000	00111	00000	10102	00000	11010
00110	00001	00110	10012	00110	11100
01010	01101	01010	11112	01010	10000
01100	01011	01100	11002	01100	10110
10010	10101	10010	00112	10010	01000
10100	10011	10100	00002	10100	01110
11000	11111	11000	01102	11000	00010
11110	11001	11110	01012	11110	00100
00001	00112	00001	10103	00001	11011
00111	00002	00111	10013	00111	11101
01011	01102	01011	11113	01011	10001
01101	01012	01101	11003	01101	10111
10011	10102	10011	00113	10011	01001
10101	10012	10101	00003	10101	01111
11001	11112	11001	01103	11001	00011
11111	11002	11111	01013	11111	00101
00012	00103	00012	10110	00012	11002
00102	00013	00102	10000	00102	11112
01002	01113	01002	11100	01002	10012
01112	01003	01112	11010	01112	10102
10002	10113	10002	00100	10002	01012
10112	10003	10112	00010	10112	01102
11012	11103	11012	01110	11012	00002
11102	11013	11102	01000	11102	00112
00013	00100	00013	10111	00013	11003
00103	00010	00103	10001	00103	11113
01003	01110	01003	11101	01003	10013
01113	01000	01113	11011	01113	10103
10003	10110	10003	00101	10003	01013
10113	10000	10113	00011	10113	01103
11013	11100	11013	01111	11013	00003
11103	11010	11103	01001	11103	00113

this table we see that all choice sets with difference vectors 011, 101 and 110 will give us the optimal design with

$$\det(C_{MT}) = (\tfrac{1}{100})^2(\tfrac{1}{100})^2(\tfrac{1}{90})^4(\tfrac{3}{200})^4(\tfrac{23}{1800})^8(\tfrac{23}{1800})^8.$$

This design will have 450 choice sets. To get an efficient design in a smaller number of choice sets, we still need to use the complete factorial as the starting design, but we can add the two sets of generators

$$G_1 = (000, 101) \text{ and } G_2 = (000, 110),$$

or various other sets of generators that have the difference vectors (101) and (110), such as

$$G_1 = (000, 104) \text{ and } G_2 = (000, 220),$$

$$\text{or } G_1 = (000, 104) \text{ and } G_2 = (000, 220).$$

Any of these sets of generators results in a design that is 94.80% efficient.

For larger choice set sizes we do not know the difference vectors for the optimal design or the maximum value of $\det(C_{MT})$. However, we can use the same idea and add sets of generators to the complete factorial so that there are two attributes different between any pair of treatment combinations in the choice sets. For $m = 3$ we use the set of generators

$$G_1 = (000, 101, 110),$$

(or indeed others such as $G_1 = (000, 013, 203)$ or $G_1 = (000, 022, 102)$, since they all have the difference vector (011,101,110)), to get 45 choice sets with efficiency 97.17% relative to the largest $\det(C_{MT})$ we could find.

Similarly, for $m = 4$ we add

$$G_1 = (000, 101, 110, 011)$$

to the complete factorial and for $m = 5$ we add

$$G_1 = (000, 101, 110, 011, 203).$$

Both of these two designs consist of 45 choice sets and are 98.51% and 99.14% efficient, respectively, relative in each case to the largest $\det(C_{MT})$ we could determine.

All of the designs constructed for this example have a C_{MT} matrix which is block diagonal. This means that C_{MT}^{-1} is also block diagonal and therefore the corresponding pairs of main effects are uncorrelated. □

8.2.4 Designs for Main Effects plus Some Two-Factor Interactions

When we believe that some of the two-factor interactions and all higher-order interactions are zero, then we can sometimes use a resolution 4 starting design as long as the two-factor interactions to be estimated are not confounded with each other. The construction of resolution 4 designs for 2-level attributes is discussed in Section 2.2.1. When at least one attribute has more than two levels then there are some resolution 4 (strength 3) designs in Sloane (2006b), and Dey (1985) and Hedayat et al. (1999) give constructions for resolution 4 designs. Alternatively we can construct a starting design from a resolution 3 OMEP in

Table 8.21 Choice Sets for $k = 3$, $\ell_1 = \ell_2 = 3$, and $\ell_3 = 5$ when $m = 3$ for Main Effects and All Two-Factor Interactions

F	$F + 101$	$F + 110$	F	$F + 101$	$F + 110$
000	101	110	113	214	223
001	102	111	114	210	224
002	103	112	120	221	200
003	104	113	121	222	201
004	100	114	122	223	202
010	111	120	123	224	203
011	112	121	124	220	204
012	113	122	200	001	010
013	114	123	201	002	011
014	110	124	202	003	012
020	121	100	203	004	013
021	122	101	204	000	014
022	123	102	210	011	020
023	124	103	211	012	021
024	120	104	212	013	022
100	201	210	213	014	023
101	202	211	214	010	024
102	203	212	220	021	000
103	204	213	221	022	001
104	200	214	222	023	002
110	211	220	223	024	003
111	212	221	224	020	004
112	213	222			

such a way that the interactions of interest are not confounded with each other or the main effects (see Construction 2.3.6).

In the first example below we are only interested in the two-factor interactions between a subset of the attributes, and in the following example we wish to estimate the two-factor interactions between one attribute and each of the other attributes.

■ EXAMPLE 8.2.4.

Let $k = 4$ and $\ell_1 = \ell_2 = \ell_3 = \ell_4 = 2$. Suppose that we are confident that two-factor interactions AC, BC and CD, and all higher-order interactions are negligible. We wish to estimate the main effects and $f = 3$ of the two-factor interactions AB, AD and BD. Then the 2^{4-1} design of resolution 4 in Table 2.7 can be used as the starting design. We can check that none of the effects to be estimated is aliased by checking the aliasing structure of this design (see Section 2.2.1).

We use C_{MT_f} to denote the information matrix for main effects and some two-factor interactions. The sets of generators are chosen so that there is a 1 entry in at least one $\mathbf{g}_2$ vector, and for each pair of attributes 1 and 2, 1 and 4, and 2 and 4, there is at least one $\mathbf{g}_2$ with a 0 entry and a 1 entry in the corresponding attribute positions. The largest $\det(C_{MT_f})$ can be obtained by adding the sets of generators

$$G_1 = (0000, 0111),\ G_2 = (0000, 1011) \text{ and } G_3 = (0000, 1110)$$

to the starting design resulting in 24 choice sets. However, by adding just two of these we can get the design in 16 choice sets in Table 8.22. This design is 95.26% efficient relative to the largest $\det(C_{MT_f})$ found.

For larger choice set sizes we will need only one set of generators to estimate all of the effects of interest in 8 choice sets.

$$G_1 = (0000, 0111, 1011) \text{ for } m = 3,$$

$$G_1 = (0000, 0111, 1011, 1110) \text{ for } m = 4,$$

$$G_1 = (0000, 0111, 1011, 1101, 1110) \text{ for } m = 5.$$

These designs are 100%, 97.67% and 98.67% efficient respectively, relative to the largest $\det(C_{MT_f})$ found for the particular value of m. □

Table 8.22 Choice Sets for $k = 4$ Binary Attributes when $m = 2$ for Main Effects and Interactions AB, AD, and BD

F	$F + 0111$	F	$F + 1011$
0000	0111	0000	1011
0011	0100	0011	1000
0101	0010	0101	1110
0110	0001	0110	1101
1001	1110	1001	0010
1010	1101	1010	0001
1100	1011	1100	0111
1111	1000	1111	0100

■ EXAMPLE 8.2.5.

Let $k = 6$, $\ell_1 = 3$, $\ell_2 = \ell_3 = \ell_4 = \ell_5 = 2$, and $\ell_6 = 4$. Suppose that only the two-factor interactions between the first attribute, and each of the other attributes, are to be estimated. In this example the $f = 5$ interactions AB, AC, AD, AE, and AF are to be estimated, and all other interactions are assumed to be zero. We can construct a starting design by taking a $2^4 \times 4$ resolution 3 design in 8 treatment combinations (available in Sloane (2006b)) and using Construction 2.3.6 to create the attribute with 3 levels. Three copies of the $2^4 \times 4//8$ OMEP were written down, one above the other. Then the 3-level attribute was adjoined, as the first column, with 8 copies of level 0, 8 copies of level 1 and 8 copies of level 2. The $3 \times 2^4 \times 4//16$ OMEP is shown in column F in Table 8.23.

When choosing the sets of generators to construct the choice sets, we need to make sure that each combination of entries in $\mathbf{g}_2$ for the first attribute and each of the other attributes contains a 0 entry and at least one non-zero entry. The sets of generators which we used to construct the choice sets are

$$G_1 = (000000, 111111) \text{ and } G_2 = (000000, 011111) \text{ for } m = 2,$$

$$G_1 = (000000, 111110) \text{ for } m = 3,$$

$$G_1 = (000000, 011111, 1000003, 111112) \text{ for } m = 4,$$

$$G_1 = (000000, 111111, 2000003, 011111, 100000) \text{ for } m = 5.$$

These designs are 92.07%, 98.68%, 99.39%, and 99.87% efficient, respectively, relative to the largest $\det(C_{MT_f})$ we could find in each case. In all of these designs all effects to be estimated are uncorrelated. The choice sets for $m = 3$ are given in Table 8.23. □

8.3 OTHER STRATEGIES FOR CONSTRUCTING CHOICE EXPERIMENTS

In this section we compare a number of common strategies for constructing discrete choice experiments. We discuss each of the strategies in detail for one example, then give summary tables for several designs. This is done for designs for which the main effects only are of interest, assuming all of the interactions are negligible, as well as for main effects plus some or all of the two-factor interactions, assuming the other interactions are negligible.

We consider five design strategies that have been routinely adopted in the past and are commonly found in the published literature on DCEs in marketing, transportation, and applied economics. We also construct a design using the techniques given in the first part of this chapter to use as a comparison. The strategies consist of the following methods to allocate the treatment combinations to the choice sets:

1. Random method 1;

2. Random method 2;

3. Satisfying the criteria in Huber and Zwerina (1996);

4. The L^{MA} Method;

5. The SAS macros;

6. The techniques used earlier in this chapter.

For the purposes of this comparison we assume that we have no prior information about the parameters to be estimated. In other words, we assume that the treatment combinations

Table 8.23 Choice Sets for $k = 6$ Attributes, $\ell_1 = 3$, $\ell_2 = \ell_3 = \ell_4 = \ell_5 = 2$, and $\ell_6 = 4$ when $m = 3$ for Main Effects and Interactions AB, AC, AD, AE, and AF

F	$F + 111110$	$F + 011113$
000000	111110	011113
011110	100000	000003
000111	111001	011000
011001	100111	000110
001012	110102	010101
010102	101012	001011
001103	110013	010012
010013	101103	001102
100000	211110	111113
111110	200000	100003
100111	211001	111000
111001	200111	100110
101012	210102	110101
110102	201012	101011
101103	210013	110012
110013	201103	101102
200000	011110	211113
211110	000000	200003
200111	011001	211000
211001	000111	200110
201012	010102	210101
210102	001012	201011
201103	010013	210012
210013	001103	201102

are equally attractive. This assumption only applies to three of the strategies: the Huber and Zwerina criteria of utility balance; the SAS macros, in which the β values are assumed to be zero; and the Street–Burgess method, in which the π values are all equal to 1. The other three strategies do not allow for the inclusion of prior information about the parameters.

We now discuss each strategy in detail when there are five attributes all with 4 levels and the choice sets are of size 2. For all of the strategies except the L^{MA} method, we construct choice experiments with 16 choice sets. For the L^{MA} method we cannot construct a design in less than 64 choice sets.

Strategy 1: Random Method 1 For this strategy we take an OMEP in t treatment combinations, where t is divisible by m, then randomly place the treatment combinations into the m options to create $N = t/m$ choice sets; see McKenzie et al. (2001) for a study in which this method was used to construct the choice sets. For the example in which there are five 4-level attributes and $m = 2$, we need a 4^5 OMEP in $16 \times m$ treatment combinations. So for $m = 2$ we take the first 5 columns of the $4^9//32$ OMEP from Sloane (2006b). The 32 treatment combinations were then randomly placed in the pairs resulting in a design with 16 choice sets. One such design is given in Table 8.24 with the C_M matrix displayed in Table 8.25. C_M^{-1} can then be calculated, and since this matrix has no off-diagonal block matrices consisting of all 0s, all pairs of main effects are correlated. This design is 44.44% efficient.

Table 8.24 Random Method 1 Choice Sets

Option 1	Option 2
2 1 0 2 3	0 0 3 1 1
0 1 2 2 0	0 3 0 1 1
3 1 1 1 3	3 2 2 3 1
0 2 1 2 0	2 2 3 2 3
2 3 2 3 0	1 2 0 3 2
3 1 1 3 1	2 3 2 1 2
3 3 3 2 2	1 3 1 2 1
3 3 3 0 0	0 1 2 0 2
1 1 3 3 2	3 0 0 2 2
3 0 0 0 0	3 2 2 1 3
2 0 1 3 0	1 0 2 2 1
0 3 0 3 3	1 0 2 0 3
2 2 3 0 1	1 1 3 1 0
0 0 3 3 3	0 2 1 0 2
2 1 0 0 1	1 2 0 1 0
1 3 1 0 3	2 0 1 1 2

No general comments can be made about designs that are constructed in this way. We constructed 100 different designs for $m = 2$ for the five 4-level attributes, and the efficiency ranged from 32.04% up to 59.61%, with an average of 45.08% for those that were non-zero. There were 12 with an efficiency of 0, which means that at least one main effect was not able to be estimated. In none of the designs constructed was any pair of main effects uncorrelated.

Strategy 2: Random Method 2 The second design strategy is similar to the first but uses m different OMEPs, the first OMEP to represent the treatment combinations that

Table 8.25 Random Method 1 Choice Sets: C_M Matrix

$$
\left[\begin{array}{ccc|ccc|ccc|ccc|ccc}
\frac{1}{2560} & 0 & \frac{-1}{5120} & \frac{-1}{81920} & \frac{3}{16384\sqrt{5}} & \frac{-1}{40960} & \frac{1}{81920} & \frac{\sqrt{5}}{16384} & \frac{1}{40960} & \frac{-1}{81920} & \frac{1}{16384\sqrt{5}} & \frac{-3}{20480} & \frac{3}{81920} & \frac{1}{4096\sqrt{5}} & \frac{1}{81920} \\
0 & \frac{3}{8192} & 0 & \frac{-1}{16384\sqrt{5}} & \frac{1}{16384} & \frac{-1}{8192\sqrt{5}} & \frac{-3}{16384\sqrt{5}} & \frac{1}{16384} & \frac{1}{4096\sqrt{5}} & \frac{3}{16384\sqrt{5}} & 0 & \frac{1}{16384\sqrt{5}} & \frac{-\sqrt{5}}{16384} & \frac{-1}{16384} & 0 \\
\frac{-1}{5120} & 0 & \frac{29}{40960} & \frac{-1}{40960} & \frac{-1}{4096\sqrt{5}} & \frac{-1}{20480} & \frac{1}{40960} & 0 & \frac{1}{20480} & \frac{-1}{40960} & \frac{-3}{16384\sqrt{5}} & \frac{1}{81920} & \frac{-1}{20480} & \frac{3}{16384\sqrt{5}} & \frac{7}{81920} \\
\hline
\frac{-1}{81920} & \frac{-1}{16384\sqrt{5}} & \frac{-1}{40960} & \frac{11}{20480} & \frac{1}{16384\sqrt{5}} & \frac{-7}{81920} & \frac{-3}{81920} & \frac{1}{8192\sqrt{5}} & \frac{-1}{81920} & 0 & \frac{1}{16384\sqrt{5}} & \frac{1}{16384} & \frac{1}{16384} & \frac{1}{8192\sqrt{5}} & \frac{1}{16384} \\
\frac{3}{16384\sqrt{5}} & \frac{1}{16384} & \frac{-1}{4096\sqrt{5}} & \frac{1}{16384\sqrt{5}} & \frac{1}{2048} & \frac{-3}{16384\sqrt{5}} & \frac{1}{8192\sqrt{5}} & \frac{3}{16384} & \frac{-1}{16384\sqrt{5}} & \frac{-3}{16384\sqrt{5}} & \frac{-1}{16384} & \frac{1}{4096\sqrt{5}} & \frac{-3}{8192\sqrt{5}} & \frac{1}{16384} & \frac{3}{16384\sqrt{5}} \\
\frac{-1}{40960} & \frac{-1}{8192\sqrt{5}} & \frac{-1}{20480} & \frac{-7}{81920} & \frac{-3}{16384\sqrt{5}} & \frac{23}{40960} & \frac{-1}{81920} & \frac{-1}{16384\sqrt{5}} & \frac{1}{10240} & \frac{1}{16384} & \frac{1}{8192\sqrt{5}} & \frac{-1}{16384} & \frac{1}{16384} & \frac{9}{16384\sqrt{5}} & 0 \\
\hline
\frac{1}{81920} & \frac{-3}{16384\sqrt{5}} & \frac{1}{40960} & \frac{-3}{81920} & \frac{1}{8192\sqrt{5}} & \frac{-1}{81920} & \frac{5}{8192} & \frac{-1}{16384\sqrt{5}} & \frac{-3}{16384} & \frac{3}{81920} & 0 & \frac{11}{81920} & \frac{1}{40960} & \frac{1}{8192\sqrt{5}} & \frac{1}{20480} \\
\frac{\sqrt{5}}{16384} & \frac{1}{16384} & 0 & \frac{1}{8192\sqrt{5}} & \frac{3}{16384} & \frac{-1}{16384\sqrt{5}} & \frac{-1}{16384\sqrt{5}} & \frac{1}{2048} & \frac{3}{16384\sqrt{5}} & \frac{1}{4096\sqrt{5}} & \frac{1}{16384} & \frac{3}{16384\sqrt{5}} & \frac{3}{16384\sqrt{5}} & 0 & \frac{1}{16384\sqrt{5}} \\
\frac{1}{40960} & \frac{1}{4096\sqrt{5}} & \frac{1}{20480} & \frac{-1}{81920} & \frac{-1}{16384\sqrt{5}} & \frac{1}{10240} & \frac{-3}{16384} & \frac{3}{16384\sqrt{5}} & \frac{1}{2048} & \frac{1}{81920} & \frac{\sqrt{5}}{16384} & \frac{3}{20480} & \frac{-11}{81920} & \frac{1}{4096\sqrt{5}} & \frac{3}{81920} \\
\hline
\frac{-1}{81920} & \frac{3}{16384\sqrt{5}} & \frac{-1}{40960} & 0 & \frac{-3}{16384\sqrt{5}} & \frac{1}{16384} & \frac{3}{81920} & \frac{1}{4096\sqrt{5}} & \frac{1}{81920} & \frac{17}{40960} & \frac{1}{8192\sqrt{5}} & \frac{1}{10240} & \frac{-3}{81920} & \frac{\sqrt{5}}{16384} & \frac{1}{20480} \\
\frac{1}{16384\sqrt{5}} & 0 & \frac{-3}{16384\sqrt{5}} & \frac{1}{16384\sqrt{5}} & \frac{-1}{16384} & \frac{1}{8192\sqrt{5}} & 0 & \frac{1}{16384} & \frac{\sqrt{5}}{16384} & \frac{1}{8192\sqrt{5}} & \frac{1}{2048} & \frac{-3}{8192\sqrt{5}} & \frac{-1}{4096\sqrt{5}} & \frac{-1}{16384} & \frac{7}{16384\sqrt{5}} \\
\frac{-3}{20480} & \frac{1}{16384\sqrt{5}} & \frac{1}{81920} & \frac{1}{16384} & \frac{1}{4096\sqrt{5}} & \frac{-1}{16384} & \frac{11}{81920} & \frac{3}{16384\sqrt{5}} & \frac{3}{20480} & \frac{1}{10240} & \frac{-3}{8192\sqrt{5}} & \frac{23}{40960} & \frac{-1}{81920} & 0 & \frac{-7}{81920} \\
\hline
\frac{3}{81920} & \frac{-\sqrt{5}}{16384} & \frac{-1}{20480} & \frac{1}{16384} & \frac{-3}{8192\sqrt{5}} & \frac{1}{16384} & \frac{1}{40960} & \frac{3}{16384\sqrt{5}} & \frac{-11}{81920} & \frac{-3}{81920} & \frac{-1}{4096\sqrt{5}} & \frac{-1}{81920} & \frac{21}{40960} & \frac{-1}{8192\sqrt{5}} & \frac{1}{20480} \\
\frac{1}{4096\sqrt{5}} & \frac{-1}{16384} & \frac{3}{16384\sqrt{5}} & \frac{1}{8192\sqrt{5}} & \frac{1}{16384} & \frac{9}{16384\sqrt{5}} & \frac{1}{8192\sqrt{5}} & 0 & \frac{1}{4096\sqrt{5}} & \frac{\sqrt{5}}{16384} & \frac{-1}{16384} & 0 & \frac{-1}{8192\sqrt{5}} & \frac{5}{8192} & \frac{-1}{4096\sqrt{5}} \\
\frac{1}{81920} & 0 & \frac{7}{81920} & \frac{1}{16384} & \frac{3}{16384\sqrt{5}} & 0 & \frac{1}{20480} & \frac{1}{16384\sqrt{5}} & \frac{3}{81920} & \frac{1}{20480} & \frac{7}{16384\sqrt{5}} & \frac{-7}{81920} & \frac{1}{20480} & \frac{-1}{4096\sqrt{5}} & \frac{3}{5120}
\end{array}\right]
$$

appear as the first option in the choice sets, the second OMEP to represent the treatment combinations that appear as the second option in the choice sets, and so on up to the mth OMEP to represent the treatment combinations in the last option (see Louviere et al. (2000), p. 114).

For the situation with five 4-level attributes and $m = 2$, one such design is shown for $m = 2$ in Table 8.26. The first OMEP is the $4^5//16$ which is shown in the first column of Table 8.8. The second OMEP is also a $4^5//16$ design obtained by adding the treatment combination 01231, using addition modulo 4, to the first OMEP. This method of obtaining another OMEP, when we assume that the 16 treatment combinations in the first OMEP are unwanted, is described in Construction 2.3.9. We then check that none of the treatment combinations which appear in the first OMEP are in the second OMEP. Note that each level of each attribute appears equally often in each option but this does not preclude the possibility that all pairs may have the same level of one, or more, attributes.

The design in Table 8.26 is 31.17% efficient and the C_M matrix for this design is given in Table 8.27. C_M^{-1} can then be calculated, and since this matrix has no off-diagonal block matrices consisting of all 0s, all pairs of main effects are correlated.

No general comments can be made about designs that are constructed in this way. We constructed 100 different designs for $m = 2$ and the efficiency ranging from 21.54% up to 55.59%, with an average of 35.46% for those that were non-zero. There were 53 with an efficiency of 0, which means that at least one main effect was not able to be estimated. In none of the designs constructed were the main effects uncorrelated.

Table 8.26 Random Method 2 Choice Sets

Option 1	Option 2
0 0 0 0 0	1 2 2 2 3
0 1 1 1 1	0 2 3 0 2
0 2 2 2 2	1 3 1 3 2
0 3 3 3 3	2 2 1 1 1
1 0 1 2 3	0 1 2 3 1
1 1 0 3 2	0 0 1 2 0
1 2 3 0 1	3 1 1 0 3
1 3 2 1 0	2 0 3 3 3
2 0 2 3 1	2 3 2 0 0
2 1 3 2 0	0 3 0 1 3
2 2 0 1 3	3 3 3 2 1
2 3 1 0 2	2 1 0 2 2
3 0 3 1 2	3 2 0 3 0
3 1 2 0 3	3 0 2 1 2
3 2 1 3 0	1 1 3 1 0
3 3 0 2 1	1 0 0 0 1

Strategy 3: Huber & Zwerina Criteria Huber and Zwerina (1996) give four criteria for efficient choice designs: level balance, orthogonality, utility balance, and minimal overlap (see Section 3.4.2). In this strategy we construct the choice sets by placing the treatment combinations from a fractional factorial design (an OMEP for main effects only) into the m options, so that the four criteria are satisfied. Orthogonality and level balance are satisfied by using an OMEP for the treatment combinations in option 1.

Table 8.27 Random Method 2 Choice Sets: C_M Matrix

$\frac{13}{40960}$	$\frac{-1}{16384\sqrt{5}}$	$\frac{-3}{81920}$	$\frac{1}{81920}$	$\frac{-3}{16384\sqrt{5}}$	$\frac{1}{40960}$	$\frac{-3}{81920}$	$\frac{-\sqrt{5}}{16384}$	$\frac{7}{40960}$	$\frac{3}{20480}$	$\frac{-1}{16384\sqrt{5}}$	$\frac{-11}{81920}$	$\frac{1}{40960}$	$\frac{-1}{16384\sqrt{5}}$	$\frac{-1}{81920}$
$\frac{-1}{16384\sqrt{5}}$	$\frac{5}{8192}$	$\frac{-7}{16384\sqrt{5}}$	$\frac{1}{4096\sqrt{5}}$	$\frac{3}{16384}$	$\frac{3}{16384\sqrt{5}}$	$\frac{-1}{16384\sqrt{5}}$	$\frac{-1}{16384}$	$\frac{-1}{8192\sqrt{5}}$	$\frac{3}{16384\sqrt{5}}$	$\frac{1}{16384}$	$\frac{-1}{4096\sqrt{5}}$	$\frac{-1}{8192\sqrt{5}}$	$\frac{1}{16384}$	$\frac{1}{16384\sqrt{5}}$
$\frac{-3}{81920}$	$\frac{-7}{16384\sqrt{5}}$	$\frac{17}{40960}$	$\frac{17}{81920}$	$\frac{1}{4096\sqrt{5}}$	$\frac{9}{81920}$	$\frac{1}{20480}$	0	$\frac{-1}{40960}$	$\frac{-1}{81920}$	$\frac{-1}{8192\sqrt{5}}$	$\frac{3}{81920}$	$\frac{1}{20480}$	$\frac{-1}{8192\sqrt{5}}$	$\frac{-1}{40960}$
$\frac{1}{81920}$	$\frac{1}{4096\sqrt{5}}$	$\frac{17}{81920}$	$\frac{13}{20480}$	0	$\frac{1}{20480}$	$\frac{-3}{81920}$	$\frac{-1}{4096\sqrt{5}}$	$\frac{-11}{81920}$	$\frac{-1}{81920}$	$\frac{3}{16384\sqrt{5}}$	$\frac{-1}{40960}$	$\frac{-1}{20480}$	$\frac{3}{16384\sqrt{5}}$	$\frac{-3}{81920}$
$\frac{-3}{16384\sqrt{5}}$	$\frac{3}{16384}$	$\frac{1}{4096\sqrt{5}}$	0	$\frac{5}{8192}$	0	$\frac{3}{16384\sqrt{5}}$	0	$\frac{11}{16384\sqrt{5}}$	$\frac{-1}{4096\sqrt{5}}$	0	$\frac{1}{8192\sqrt{5}}$	$\frac{1}{16384\sqrt{5}}$	0	$\frac{-3}{16384\sqrt{5}}$
$\frac{1}{40960}$	$\frac{3}{16384\sqrt{5}}$	$\frac{9}{81920}$	$\frac{1}{20480}$	0	$\frac{29}{40960}$	$\frac{1}{20480}$	$\frac{1}{8192\sqrt{5}}$	$\frac{-1}{40960}$	$\frac{3}{81920}$	$\frac{1}{16384\sqrt{5}}$	$\frac{3}{40960}$	$\frac{-3}{81920}$	$\frac{1}{16384\sqrt{5}}$	$\frac{1}{20480}$
$\frac{-3}{81920}$	$\frac{-1}{16384\sqrt{5}}$	$\frac{1}{20480}$	$\frac{-3}{81920}$	$\frac{3}{16384\sqrt{5}}$	$\frac{1}{20480}$	$\frac{13}{20480}$	$\frac{\sqrt{5}}{16384}$	$\frac{-1}{81920}$	$\frac{1}{81920}$	$\frac{-1}{8192\sqrt{5}}$	$\frac{-23}{81920}$	$\frac{-1}{40960}$	$\frac{-3}{8192\sqrt{5}}$	$\frac{-1}{20480}$
$\frac{-\sqrt{5}}{16384}$	$\frac{-1}{16384}$	0	$\frac{-1}{4096\sqrt{5}}$	0	$\frac{1}{8192\sqrt{5}}$	$\frac{\sqrt{5}}{16384}$	$\frac{1}{2048}$	$\frac{-\sqrt{5}}{16384}$	$\frac{1}{8192\sqrt{5}}$	$\frac{3}{16384}$	$\frac{-1}{16384\sqrt{5}}$	$\frac{3}{16384\sqrt{5}}$	$\frac{-1}{16384}$	$\frac{-1}{4096\sqrt{5}}$
$\frac{7}{40960}$	$\frac{-1}{8192\sqrt{5}}$	$\frac{-1}{40960}$	$\frac{-11}{81920}$	$\frac{11}{16384\sqrt{5}}$	$\frac{-1}{40960}$	$\frac{-1}{81920}$	$\frac{-\sqrt{5}}{16384}$	$\frac{19}{40960}$	$\frac{-3}{81920}$	$\frac{1}{16384\sqrt{5}}$	$\frac{-3}{40960}$	$\frac{1}{81920}$	$\frac{3}{16384\sqrt{5}}$	$\frac{1}{40960}$
$\frac{3}{20480}$	$\frac{3}{16384\sqrt{5}}$	$\frac{-1}{81920}$	$\frac{-1}{81920}$	$\frac{-1}{4096\sqrt{5}}$	$\frac{3}{81920}$	$\frac{1}{81920}$	$\frac{1}{8192\sqrt{5}}$	$\frac{-3}{81920}$	$\frac{11}{20480}$	$\frac{3}{16384\sqrt{5}}$	$\frac{-7}{81920}$	$\frac{1}{10240}$	$\frac{-3}{16384\sqrt{5}}$	$\frac{11}{81920}$
$\frac{-1}{16384\sqrt{5}}$	$\frac{1}{16384}$	$\frac{-1}{8192\sqrt{5}}$	$\frac{3}{16384\sqrt{5}}$	0	$\frac{1}{16384\sqrt{5}}$	$\frac{-1}{8192\sqrt{5}}$	$\frac{3}{16384}$	$\frac{1}{16384\sqrt{5}}$	$\frac{3}{16384\sqrt{5}}$	$\frac{3}{4096}$	$\frac{1}{16384\sqrt{5}}$	$\frac{1}{8192\sqrt{5}}$	$\frac{1}{16384}$	$\frac{-1}{16384\sqrt{5}}$
$\frac{-11}{81920}$	$\frac{-1}{4096\sqrt{5}}$	$\frac{3}{81920}$	$\frac{-1}{40960}$	$\frac{1}{8192\sqrt{5}}$	$\frac{3}{40960}$	$\frac{-23}{81920}$	$\frac{-1}{16384\sqrt{5}}$	$\frac{-3}{40960}$	$\frac{-7}{81920}$	$\frac{1}{16384\sqrt{5}}$	$\frac{23}{40960}$	$\frac{3}{40960}$	$\frac{-3}{8192\sqrt{5}}$	$\frac{1}{40960}$
$\frac{1}{40960}$	$\frac{-1}{8192\sqrt{5}}$	$\frac{1}{20480}$	$\frac{-1}{20480}$	$\frac{1}{16384\sqrt{5}}$	$\frac{-3}{81920}$	$\frac{-1}{40960}$	$\frac{3}{16384\sqrt{5}}$	$\frac{1}{81920}$	$\frac{1}{10240}$	$\frac{1}{8192\sqrt{5}}$	$\frac{3}{40960}$	$\frac{27}{40960}$	$\frac{1}{4096\sqrt{5}}$	$\frac{-1}{40960}$
$\frac{-1}{16384\sqrt{5}}$	$\frac{1}{16384}$	$\frac{-1}{8192\sqrt{5}}$	$\frac{3}{16384\sqrt{5}}$	0	$\frac{1}{16384\sqrt{5}}$	$\frac{-3}{8192\sqrt{5}}$	$\frac{-1}{16384}$	$\frac{3}{16384\sqrt{5}}$	$\frac{-3}{16384\sqrt{5}}$	$\frac{1}{16384}$	$\frac{-3}{8192\sqrt{5}}$	$\frac{1}{4096\sqrt{5}}$	$\frac{1}{2048}$	$\frac{-1}{8192\sqrt{5}}$
$\frac{-1}{81920}$	$\frac{1}{16384\sqrt{5}}$	$\frac{-1}{40960}$	$\frac{-3}{81920}$	$\frac{-3}{16384\sqrt{5}}$	$\frac{1}{20480}$	$\frac{-1}{20480}$	$\frac{-1}{4096\sqrt{5}}$	$\frac{1}{40960}$	$\frac{11}{81920}$	$\frac{-1}{16384\sqrt{5}}$	$\frac{1}{40960}$	$\frac{-1}{40960}$	$\frac{-1}{8192\sqrt{5}}$	$\frac{13}{40960}$

Minimal overlap is achieved when there are as many differences in the levels of an attribute as possible in a choice set. Under the null hypothesis, all treatment combinations have equal utility, which means we have utility balance. If this is not the case, then see Section 7.4. Some researchers have used these criteria to construct their choice sets (see, for example, Ryan et al. (2001)).

For $m = 2$, we take a $4^5//16$ OMEP from Sloane (2006b), and this gives the treatment combinations in option 1. We then use the same OMEP for the treatment combinations in option 2, pairing the treatment combinations in a choice set so that the minimal overlap criterion is satisfied, or the pairs come as close to it as possible. In effect, this means that for each attribute there should be the maximum number of different levels in the choice set. Each level appears either 0 or 1 times in each pair and, over the whole choice experiment, each option displays the possible levels of each attribute equally often. One set of pairs that results from this approach is given in Table 8.28. Unfortunately, for this example, it is not possible for any pairs to have no repeated levels of one attribute. To see this, consider the profile 00000. When this profile is paired with any of the other treatment combinations, one attribute will have a repeated level because every other profile contains one 0. However, it is possible to change the attribute that is repeated from choice set to choice set. For the design in Table 8.28 $\det(C_M) = 0$ which means that at least one of the main effects can not be estimated from this design.

It is possible to get good designs this way, but, as noted in Section 3.4.2, satisfying these criteria does not guarantee that the design is optimal and optimal designs do not necessarily satisfy all four criteria. The C_M matrix for the design in Table 8.28 is given in Table 8.29, and since $\det(C_M) = 0$ we cannot calculate C_M^{-1}.

Table 8.28 Choice Sets which Satisfy Huber & Zwerina Criteria

Option 1	Option 2
0 0 0 0 0	1 1 0 3 2
0 1 1 1 1	0 2 2 2 2
0 2 2 2 2	2 3 1 0 2
0 3 3 3 3	1 0 1 2 3
1 0 1 2 3	0 1 1 1 1
1 1 0 3 2	3 0 3 1 2
1 2 3 0 1	1 3 2 1 0
1 3 2 1 0	3 2 1 3 0
2 0 2 3 1	2 1 3 2 0
2 1 3 2 0	0 3 3 3 3
2 2 0 1 3	3 3 0 2 1
2 3 1 0 2	3 1 2 0 3
3 0 3 1 2	2 2 0 1 3
3 1 2 0 3	1 2 3 0 1
3 2 1 3 0	2 0 2 3 1
3 3 0 2 1	0 0 0 0 0

Strategy 4: L^{MA} Method The fourth strategy requires an OMEP for $m \times k$ attributes. The first k attributes become the treatment combinations in option 1, the next k attributes become the treatment combinations in option 2, and so on until the last k attributes which

Table 8.29 Choice Sets which Satisfy Huber & Zwerina Criteria: C_M Matrix

$$
\left[\begin{array}{ccc|ccc|ccc|ccc|ccc}
\frac{9}{20480} & \frac{3}{16384\sqrt{5}} & \frac{7}{81920} & \frac{-3}{81920} & 0 & \frac{9}{81920} & \frac{1}{81920} & \frac{-3}{16384\sqrt{5}} & \frac{3}{20480} & \frac{1}{20480} & \frac{-1}{4096\sqrt{5}} & \frac{-3}{20480} & \frac{1}{40960} & \frac{-1}{16384\sqrt{5}} & \frac{-1}{81920} \\
\frac{3}{16384\sqrt{5}} & \frac{3}{4096} & \frac{1}{16384\sqrt{5}} & 0 & 0 & 0 & \frac{7}{16384\sqrt{5}} & 0 & \frac{-1}{16384\sqrt{5}} & \frac{1}{16384\sqrt{5}} & \frac{1}{16384} & \frac{1}{8192\sqrt{5}} & \frac{-1}{8192\sqrt{5}} & \frac{1}{16384} & \frac{1}{16384\sqrt{5}} \\
\frac{7}{81920} & \frac{1}{16384\sqrt{5}} & \frac{17}{40960} & \frac{-1}{81920} & 0 & \frac{3}{81920} & \frac{1}{40960} & \frac{-1}{16384\sqrt{5}} & \frac{-1}{81920} & \frac{13}{81920} & \frac{-3}{16384\sqrt{5}} & \frac{-7}{40960} & \frac{1}{20480} & \frac{-1}{8192\sqrt{5}} & \frac{-1}{40960} \\
\hline
\frac{-3}{81920} & 0 & \frac{-1}{81920} & \frac{11}{20480} & 0 & \frac{1}{10240} & \frac{-1}{16384} & \frac{1}{4096\sqrt{5}} & \frac{1}{16384} & \frac{13}{81920} & \frac{1}{16384\sqrt{5}} & \frac{3}{40960} & \frac{1}{40960} & \frac{-3}{16384\sqrt{5}} & \frac{-1}{81920} \\
0 & 0 & 0 & 0 & \frac{3}{4096} & 0 & \frac{1}{4096\sqrt{5}} & \frac{-1}{8192} & \frac{-1}{8192\sqrt{5}} & \frac{-1}{4096\sqrt{5}} & 0 & \frac{1}{8192\sqrt{5}} & 0 & \frac{-1}{8192} & 0 \\
\frac{9}{81920} & 0 & \frac{3}{81920} & \frac{1}{10240} & 0 & \frac{7}{10240} & \frac{-1}{16384} & \frac{-1}{8192\sqrt{5}} & \frac{-1}{16384} & \frac{1}{81920} & \frac{-3}{16384\sqrt{5}} & \frac{3}{20480} & \frac{-3}{40960} & \frac{-1}{16384\sqrt{5}} & \frac{3}{81920} \\
\hline
\frac{1}{81920} & \frac{7}{16384\sqrt{5}} & \frac{1}{40960} & \frac{-1}{16384} & \frac{1}{4096\sqrt{5}} & \frac{-1}{16384} & \frac{3}{8192} & \frac{\sqrt{5}}{16384} & \frac{-1}{16384} & \frac{-1}{16384} & \frac{-3}{16384\sqrt{5}} & 0 & \frac{-1}{40960} & \frac{-1}{4096\sqrt{5}} & \frac{-1}{20480} \\
\frac{-3}{16384\sqrt{5}} & 0 & \frac{-1}{16384\sqrt{5}} & \frac{1}{4096\sqrt{5}} & \frac{-1}{8192} & \frac{-1}{8192\sqrt{5}} & \frac{\sqrt{5}}{16384} & \frac{1}{4096} & \frac{\sqrt{5}}{16384} & \frac{-1}{16384\sqrt{5}} & \frac{1}{16384} & \frac{-1}{8192\sqrt{5}} & \frac{-1}{8192\sqrt{5}} & \frac{-1}{16384} & \frac{1}{16384\sqrt{5}} \\
\frac{3}{20480} & \frac{-1}{16384\sqrt{5}} & \frac{-1}{81920} & \frac{1}{16384} & \frac{-1}{8192\sqrt{5}} & \frac{-1}{16384} & \frac{-1}{16384} & \frac{\sqrt{5}}{16384} & \frac{3}{4096} & \frac{-1}{8192} & \frac{1}{4096\sqrt{5}} & 0 & \frac{-7}{40960} & \frac{-3}{16384\sqrt{5}} & \frac{-3}{81920} \\
\hline
\frac{1}{20480} & \frac{1}{16384\sqrt{5}} & \frac{13}{81920} & \frac{13}{81920} & \frac{-1}{4096\sqrt{5}} & \frac{1}{81920} & \frac{-1}{16384} & \frac{-1}{16384\sqrt{5}} & \frac{-1}{8192} & \frac{1}{2560} & \frac{1}{8192\sqrt{5}} & \frac{-3}{40960} & \frac{3}{20480} & \frac{-\sqrt{5}}{16384} & \frac{-1}{81920} \\
\frac{-1}{4096\sqrt{5}} & \frac{1}{16384} & \frac{-3}{16384\sqrt{5}} & \frac{1}{16384\sqrt{5}} & 0 & \frac{-3}{16384\sqrt{5}} & \frac{-3}{16384\sqrt{5}} & \frac{1}{16384} & \frac{1}{4096\sqrt{5}} & \frac{1}{8192\sqrt{5}} & \frac{1}{2048} & \frac{1}{4096\sqrt{5}} & \frac{1}{4096\sqrt{5}} & \frac{-1}{16384} & \frac{3}{16384\sqrt{5}} \\
\frac{-3}{20480} & \frac{1}{8192\sqrt{5}} & \frac{-7}{40960} & \frac{3}{40960} & \frac{1}{8192\sqrt{5}} & \frac{3}{20480} & 0 & \frac{-1}{8192\sqrt{5}} & 0 & \frac{-3}{40960} & \frac{1}{4096\sqrt{5}} & \frac{3}{5120} & \frac{1}{20480} & 0 & \frac{1}{10240} \\
\hline
\frac{1}{40960} & \frac{-1}{8192\sqrt{5}} & \frac{1}{20480} & \frac{1}{40960} & 0 & \frac{-3}{40960} & \frac{-1}{40960} & \frac{-1}{8192\sqrt{5}} & \frac{-7}{40960} & \frac{3}{20480} & \frac{1}{4096\sqrt{5}} & \frac{1}{20480} & \frac{1}{2560} & \frac{1}{8192\sqrt{5}} & \frac{1}{20480} \\
\frac{-1}{16384\sqrt{5}} & \frac{1}{16384} & \frac{-1}{8192\sqrt{5}} & \frac{-3}{16384\sqrt{5}} & \frac{-1}{8192} & \frac{-1}{16384\sqrt{5}} & \frac{-1}{4096\sqrt{5}} & \frac{-1}{16384} & \frac{-3}{16384\sqrt{5}} & \frac{-\sqrt{5}}{16384} & \frac{-1}{16384} & 0 & \frac{1}{8192\sqrt{5}} & \frac{5}{8192} & \frac{-3}{8192\sqrt{5}} \\
\frac{-1}{81920} & \frac{1}{16384\sqrt{5}} & \frac{-1}{40960} & \frac{-1}{81920} & 0 & \frac{3}{81920} & \frac{-1}{20480} & \frac{1}{16384\sqrt{5}} & \frac{-3}{81920} & \frac{-1}{81920} & \frac{3}{16384\sqrt{5}} & \frac{1}{10240} & \frac{1}{20480} & \frac{-3}{8192\sqrt{5}} & \frac{19}{40960}
\end{array}\right]
$$

become the treatment combinations in the last option. This strategy is sometimes called an L^{MA} approach, and the details can be found in Louviere et al. (2000).

In the example where we have five 4-level attributes and $m = 2$, we need an OMEP with ten 4-level attributes. The smallest such design has 64 treatment combinations. For each treatment combination in the OMEP the first five attributes are used to represent the treatment combinations in the first option, and the final five attributes are used to represent the treatment combinations in the second option. So there are 64 pairs in the experiment in total.

The design, which is given in Table 8.30, is 75% efficient and the information matrix for the design is given by

$$C_M = \frac{1}{2048} I_{15}.$$

Since C_M, and therefore C_M^{-1}, are diagonal, the main effects of all the attributes are uncorrelated, although this is not always the case. The main disadvantage of this strategy is that it usually results in a lot more choice sets than the other strategies. Also pairs of main effects are often correlated and there can be repeated treatment combinations in the choice sets which must be deleted.

Strategy 5: SAS Macros This strategy uses the SAS macros (see Kuhfeld (2006)) to generate a starting OMEP and then construct choice sets using a search algorithm. An efficiency of the design constructed is provided by SAS, but it is not relative to the optimal design and there is no indication of how close to the best possible a design is. The user must nominate the number of treatment combinations in the candidate set from which the search algorithm selects treatment combinations for the choice sets. Bech (2003) has used this method to obtain the choice sets for his study.

For the example with five 4-level attributes with $m = 2$, we tried a number of different candidate sets, from which the choice sets were constructed. We have selected the choice sets with the highest efficiency (1.587), as calculated by SAS. The choice sets are displayed in Table 8.31 and we have calculated the efficiency of this design to be 94.49%.

The C_M^{-1} matrix for the design in Table 8.31 is given in Table 8.32. From this matrix we can see that the main effects of the first four attributes are correlated and only the main effects of the fifth attribute are independent of the main effects of the other attributes. Changing the contrasts does not affect the way the main effects are correlated.

It is worth noting that it is possible to get duplicate choice sets and also repeated treatment combinations within a choice set from the SAS macros.

Strategy 6: Street–Burgess Method This strategy uses the techniques described earlier in this chapter to construct the choice sets.

Again we use the $4^5//16$ from Sloane (2006b) as the starting design, which becomes the treatment combinations in option 1 (see the first column in Table 8.33). We then construct the other option by adding a set of generators to this starting design. For $m = 2$ we use $G = (00000, 11111)$, and this design is 94.49% efficient.

The design is given in Table 8.33. We use the third B_4 matrix given in Example 6.3.2, and the corresponding C_M matrix is given in Table 8.34. Since C_M, and therefore C_M^{-1}, are diagonal, the main effects of all the attributes are uncorrelated.

Table 8.30 L^{MA} Choice Sets

Option 1	Option 2	Option 1	Option 2
0 1 2 3 0	1 2 3 0 1	0 1 0 0 2	0 2 2 0 3
1 0 3 1 0	3 2 0 2 3	1 0 1 2 2	2 2 1 2 1
2 2 0 2 0	0 2 1 3 0	2 2 2 1 2	1 2 0 3 2
3 3 1 0 0	2 2 2 1 2	3 3 3 3 2	3 2 3 1 0
0 2 1 0 2	1 0 3 1 3	0 2 3 3 0	0 0 2 1 1
1 3 0 2 2	3 0 0 3 1	1 3 2 1 0	2 0 1 3 3
2 1 3 1 2	0 0 1 2 2	2 1 1 2 0	1 0 0 2 0
3 0 2 3 2	2 0 2 0 0	3 0 0 0 0	3 0 3 0 2
0 3 3 1 3	1 1 3 2 0	0 3 1 2 1	0 1 2 2 2
1 2 2 3 3	3 1 0 0 2	1 2 0 0 1	2 1 1 0 0
2 0 1 0 3	0 1 1 1 1	2 0 3 3 1	1 1 0 1 3
3 1 0 2 3	2 1 2 3 3	3 1 2 1 1	3 1 3 3 1
0 0 0 2 1	1 3 3 3 2	0 0 2 1 3	0 3 2 3 0
1 1 1 0 1	3 3 0 1 0	1 1 3 3 3	2 3 1 1 2
2 3 2 3 1	0 3 1 0 3	2 3 0 0 3	1 3 0 0 1
3 2 3 1 1	2 3 2 2 1	3 2 1 2 3	3 3 3 2 3
0 1 3 2 1	3 2 1 0 2	0 1 1 1 3	2 2 0 0 0
1 0 2 0 1	1 2 2 2 0	1 0 0 3 3	0 2 3 2 2
2 2 1 3 1	2 2 3 3 3	2 2 3 0 3	3 2 2 3 1
3 3 0 1 1	0 2 0 1 1	3 3 2 2 3	1 2 1 1 3
0 2 0 1 3	3 0 1 1 0	0 2 2 2 1	2 0 0 1 2
1 3 1 3 3	1 0 2 3 2	1 3 3 0 1	0 0 3 3 0
2 1 2 0 3	2 0 3 2 1	2 1 0 3 1	3 0 2 2 3
3 0 3 2 3	0 0 0 0 3	3 0 1 1 1	1 0 1 0 1
0 3 2 0 2	3 1 1 2 3	0 3 0 3 0	2 1 0 2 1
1 2 3 2 2	1 1 2 0 1	1 2 1 1 0	0 1 3 0 3
2 0 0 1 2	2 1 3 1 2	2 0 2 2 0	3 1 2 1 0
3 1 1 3 2	0 1 0 3 0	3 1 3 0 0	1 1 1 3 2
0 0 1 3 0	3 3 1 3 1	0 0 3 0 2	2 3 0 3 3
1 1 0 1 0	1 3 2 1 3	1 1 2 2 2	0 3 3 1 1
2 3 3 2 0	2 3 3 0 0	2 3 1 1 2	3 3 2 0 2
3 2 2 0 0	0 3 0 2 2	3 2 0 3 2	1 3 1 2 0

Table 8.31 Choice Sets from SAS Macros

Option 1	Option 2
3 1 0 2 0	1 3 2 0 3
2 0 1 3 3	1 3 2 0 0
3 0 2 1 1	1 2 0 3 2
2 2 2 2 3	3 3 3 3 2
1 0 3 2 1	2 3 0 1 0
1 2 0 3 1	0 3 1 2 2
0 1 2 3 0	3 2 1 0 1
1 1 1 1 3	0 0 0 0 2
2 3 0 1 1	3 2 1 0 0
0 2 3 1 3	2 0 1 3 0
0 3 1 2 1	2 1 3 0 2
0 1 2 3 1	1 0 3 2 0
3 3 3 3 3	1 1 1 1 2
0 0 0 0 3	2 2 2 2 2
2 1 3 0 1	3 0 2 1 2
3 1 0 2 3	0 2 3 1 0

Table 8.32 Choice Sets from SAS Macros: C_M^{-1} Matrix

1792	0	0	256	0	0	$\frac{768}{5}$	0	$\frac{-1024}{5}$	$\frac{-768}{5}$	0	$\frac{1024}{5}$	0	0	0
0	1536	0	0	−512	0	0	0	0	0	0	0	0	0	0
0	0	1792	0	0	256	$\frac{-1024}{5}$	0	$\frac{-768}{5}$	$\frac{1024}{5}$	0	$\frac{768}{5}$	0	0	0
256	0	0	1792	0	0	$\frac{-768}{5}$	0	$\frac{1024}{5}$	$\frac{768}{5}$	0	$\frac{-1024}{5}$	0	0	0
0	−512	0	0	1536	0	0	0	0	0	0	0	0	0	0
0	0	256	0	0	1792	$\frac{1024}{5}$	0	$\frac{768}{5}$	$\frac{-1024}{5}$	0	$\frac{-768}{5}$	0	0	0
$\frac{768}{5}$	0	$\frac{-1024}{5}$	$\frac{-768}{5}$	0	$\frac{1024}{5}$	1792	0	0	256	0	0	0	0	0
0	0	0	0	0	0	0	1536	0	0	−512	0	0	0	0
$\frac{-1024}{5}$	0	$\frac{-768}{5}$	$\frac{1024}{5}$	0	$\frac{768}{5}$	0	0	1792	0	0	256	0	0	0
$\frac{-768}{5}$	0	$\frac{1024}{5}$	$\frac{768}{5}$	0	$\frac{-1024}{5}$	256	0	0	1792	0	0	0	0	0
0	0	0	0	0	0	0	−512	0	0	1536	0	0	0	0
$\frac{1024}{5}$	0	$\frac{768}{5}$	$\frac{-1024}{5}$	0	$\frac{-768}{5}$	0	0	256	0	0	1792	0	0	0
0	0	0	0	0	0	0	0	0	0	0	0	$\frac{9216}{5}$	0	$\frac{-2048}{5}$
0	0	0	0	0	0	0	0	0	0	0	0	0	2048	0
0	0	0	0	0	0	0	0	0	0	0	0	$\frac{-2048}{5}$	0	$\frac{6144}{5}$

Table 8.33 Street–Burgess Choice Sets

Option 1	Option 2
0 0 0 0 0	1 1 1 1 1
0 1 1 1 1	1 2 2 2 2
1 3 3 3 3	0 2 2 2 2
1 0 0 0 0	0 3 3 3 3
2 1 2 3 0	1 0 1 2 3
1 1 0 3 2	2 2 1 0 3
1 2 3 0 1	2 3 0 1 2
2 0 3 2 1	1 3 2 1 0
2 0 2 3 1	3 1 3 0 2
3 2 0 3 1	2 1 3 2 0
3 3 1 2 0	2 2 0 1 3
2 3 1 0 2	3 0 2 1 3
0 1 0 2 3	3 0 3 1 2
0 2 3 1 0	3 1 2 0 3
3 2 1 3 0	0 3 2 0 1
3 3 0 2 1	0 0 1 3 2

Table 8.34 Street–Burgess Choice Sets: C_M Matrix

$$
\left[\begin{array}{ccc|ccc|ccc|ccc|ccc}
\frac{1}{2048} & 0 & 0 & 0 & 0 & 0 & 0 & 0 & 0 & 0 & 0 & 0 & 0 & 0 & 0 \\
0 & \frac{1}{1024} & 0 & 0 & 0 & 0 & 0 & 0 & 0 & 0 & 0 & 0 & 0 & 0 & 0 \\
0 & 0 & \frac{1}{2048} & 0 & 0 & 0 & 0 & 0 & 0 & 0 & 0 & 0 & 0 & 0 & 0 \\
\hline
0 & 0 & 0 & \frac{1}{2048} & 0 & 0 & 0 & 0 & 0 & 0 & 0 & 0 & 0 & 0 & 0 \\
0 & 0 & 0 & 0 & \frac{1}{1024} & 0 & 0 & 0 & 0 & 0 & 0 & 0 & 0 & 0 & 0 \\
0 & 0 & 0 & 0 & 0 & \frac{1}{2048} & 0 & 0 & 0 & 0 & 0 & 0 & 0 & 0 & 0 \\
\hline
0 & 0 & 0 & 0 & 0 & 0 & \frac{1}{2048} & 0 & 0 & 0 & 0 & 0 & 0 & 0 & 0 \\
0 & 0 & 0 & 0 & 0 & 0 & 0 & \frac{1}{1024} & 0 & 0 & 0 & 0 & 0 & 0 & 0 \\
0 & 0 & 0 & 0 & 0 & 0 & 0 & 0 & \frac{1}{2048} & 0 & 0 & 0 & 0 & 0 & 0 \\
\hline
0 & 0 & 0 & 0 & 0 & 0 & 0 & 0 & 0 & \frac{1}{2048} & 0 & 0 & 0 & 0 & 0 \\
0 & 0 & 0 & 0 & 0 & 0 & 0 & 0 & 0 & 0 & \frac{1}{1024} & 0 & 0 & 0 & 0 \\
0 & 0 & 0 & 0 & 0 & 0 & 0 & 0 & 0 & 0 & 0 & \frac{1}{2048} & 0 & 0 & 0 \\
\hline
0 & 0 & 0 & 0 & 0 & 0 & 0 & 0 & 0 & 0 & 0 & 0 & \frac{1}{2048} & 0 & 0 \\
0 & 0 & 0 & 0 & 0 & 0 & 0 & 0 & 0 & 0 & 0 & 0 & 0 & \frac{1}{1024} & 0 \\
0 & 0 & 0 & 0 & 0 & 0 & 0 & 0 & 0 & 0 & 0 & 0 & 0 & 0 & \frac{1}{2048}
\end{array}\right]
$$

8.4 COMPARISON OF STRATEGIES

In this section we compare the different strategies for constructing choice sets for main effects only, then for main effects plus all two-factor interactions, and finally for main effects plus some two-factor interactions. We do this for different design specifications of attributes and levels, for choice set sizes from two to five, and for the different effects to be estimated. These design specifications were chosen to reflect a variety of numbers of attributes and the levels of these attributes. For the first five strategies, any choice set which contained a particular treatment combination more than once, or was a duplicate of another choice set, was deleted before calculating the efficiency of the design. These duplicates do not occur when using the Street–Burgess method.

For each design specification, the number of choice sets was chosen to be the minimum number of choice sets from which the effects of interest could be estimated independently, by any of the construction methods. It is possible to construct designs with non-zero efficiencies in fewer choice sets for all of the strategies, but the effects are correlated.

We first compare designs, for estimating main effects only, constructed using the six strategies in the previous section. The results are shown in Table 8.35. All six strategies are considered for the 4^5 design only, as the results for the two random strategies were similar for the other designs considered. The table shows the number of choice sets for each design constructed, as well as the efficiency relative to the optimal design (see Theorem 6.3.1), and an asterisk indicates that all the main effects are uncorrelated. For the two random strategies the efficiencies were averaged over 100 designs with non-zero efficiencies.

As can be seen in Table 8.35, the two random strategies do not produce efficient designs, and the main effects are almost always correlated. It is possible to construct optimal designs by satisfying the Huber & Zwerina criteria, but it is also possible to create designs that cannot estimate all of the main effects or designs in which the main effects are correlated, even though the designs satisfy the four criteria. The main problems with the L^{MA} method are that the number of choice sets is larger than necessary and the efficiency is not as high as other methods. The designs produced by the SAS macros are, as expected from a search algorithm, highly efficient and sometimes optimal. However, these designs rarely provide uncorrelated estimates of the main effects, unlike the Street–Burgess designs which are highly efficient with uncorrelated estimates of the main effects.

We now consider designs for main effects and two-factor interactions. We compare designs that have been constructed using the L^{MA} method, the SAS macros, and the Street–Burgess method, as these are the only strategies of the six that routinely produce designs that allow for the estimation of the main effects plus interactions. These results are displayed in Table 8.36; an asterisk denotes a choice experiment in which the main effects and two-factor interactions are uncorrelated. There are five different design specifications; the first three are for designs that estimate main effects plus all two-factor interactions, and the final two are for estimating main effects and only some of the two-factor interactions. For the 2^4 design only the two-factor interactions involving the first, second, and fourth attributes are to be estimated — that is, AB, AD, and BD. For the $3 \times 2^4 \times 4$ design the two-factor interactions between the first attribute and each of the other attributes are to be estimated — that is, AB, AC, AD AE, and AF.

For the 2^7 design we know the maximum value of $\det(C_{MT})$ (see Theorem 5.1.2), so it is possible to calculate the efficiency relative to the optimal design for any choice experiment we construct. However, if at least one of the attributes has more than two levels, then we do not know the maximum value of $\det(C_{MT})$. In these cases the efficiencies in Table 8.36, which are in italics, are calculated relative to the largest $\det(C_{MT})$ we could find.

Table 8.35 Comparison of Construction Methods for Main Effects Only

Design	Construction Method	Choice Set Size			
		m = 2	m = 3	m = 4	m = 5
4^5	Random 1	45.08% 16 ch sets	68.96% 16 ch sets	72.48% 16 ch sets	82.35% 16 ch sets
	Random 2	35.46% 16 ch sets	65.22% 16 ch sets	70.86% 16 ch sets	80.78% 16 ch sets
	H & Z Criteria	0 16 ch sets	71.82% 16 ch sets	76.70% 16 ch sets	83.97% 16 ch sets
	L^{MA}	75.00%* 64 ch sets	75.00%* 64 ch sets	75.00%* 64 ch sets	83.33%* 128 ch sets
	SAS Macros	94.49% 16 ch sets	100%* 16 ch sets	100%* 16 ch sets	99.98% 16 ch sets
	S & B	94.49%* 16 ch sets	100%* 16 ch sets	100%* 16 ch sets	100%* 16 ch sets
$2 \times 3 \times 6$	H & Z Criteria	0 18 ch sets	99.39% 12 ch sets	99.39% 18 ch sets	100%* 36 ch sets
	L^{MA}	73.94%* 36 ch sets	80.12% 70 ch sets	85.04% 70 ch sets	86.36% 68 ch sets
	SAS Macros	99.39% 18 ch sets	99.71% 12 ch sets	99.92% 18 ch sets	99.995% 36 ch sets
	S & B	90.56%* 18 ch sets	99.39%* 12 ch sets	99.78%* 18 ch sets	100%* 36 ch sets
$2^6 \times 4^6$	H & Z Criteria	63.97% 32 ch sets	83.70% 32 ch sets	85.19% 32 ch sets	95.27% 32 ch sets
	L^{MA}	67.77%* 64 ch sets	75.00%* 96 ch sets	78.20% 124 ch sets	85.44% 124 ch sets
	SAS Macros	87.63% 32 ch sets	98.95% 32 ch sets	99.33% 32 ch sets	99.54% 32 ch sets
	S & B	95.84%* 32 ch sets	100%* 32 ch sets	100%* 32 ch sets	100%* 32 ch sets
$2^{16} \times 13$	SAS Macros	79.57% 52 ch sets	97.72% 52 ch sets	95.50% 52 ch sets	99.24% 52 ch sets
	S & B	86.23%* 52 ch sets	98.30%* 52 ch sets	100%* 52 ch sets	99.87%* 52 ch sets

An asterisk (*) indicates that all the effects are uncorrelated.

Table 8.36 Comparison of Construction Methods for Main Effects and Two-Factor Interactions

Design & Interactions	Construction Method	Choice Set Size m = 2	m = 3	m = 4	m = 5
2^7 All 2fi's	L^{MA} Resoln. 5	88.36% 253 ch sets	88.64% 1002 ch sets	82.10% 2043 ch sets	— —
	L^{MA} Resoln. 3	0 26 ch sets	65.69% 26 ch sets	82.10% 51 ch sets	80.05% 34 ch sets
	SAS Macros	97.89% 96 ch sets	99.60% 128 ch sets	98.92% 64 ch sets	99.15% 64 ch sets
	S & B	91.85%* 96 ch sets	95.71%* 128 ch sets	98.98%* 64 ch sets	99.09%* 64 ch sets
$2^4 \times 4$ All 2fi's	L^{MA} Resoln. 3	_59.47%_ 32 ch sets	_76.02%_ 32 ch sets	0 64 ch sets	0 64 ch sets
	SAS Macros	_96.65%_ 96 ch sets	_99.38%_ 64 ch sets	_98.59%_ 32 ch sets	_99.02%_ 32 ch sets
	S & B	_90.38%*_ 96 ch sets	_96.44%*_ 64 ch sets	_96.29%*_ 32 ch sets	_94.66%*_ 32 ch sets
$3 \times 3 \times 5$ All 2fi's	SAS Macros	_98.15%_ 90 ch sets	_98.01%_ 45 ch sets	_98.42%_ 45 ch sets	_99.45%_ 45 ch sets
	S & B	_94.80%*_ 90 ch sets	_97.17%*_ 45 ch sets	_98.51%*_ 45 ch sets	_99.14%*_ 45 ch sets
2^4 AB,AD,BD	SAS Macros	_97.79%_ 16 ch sets	_100%_ 8 ch sets	_96.76%_ 6 ch sets	_100%_ 7 ch sets
	S & B	_95.26%*_ 16 ch sets	_100%*_ 8 ch sets	_97.67%*_ 8 ch sets	_98.67%*_ 8 ch sets
$3 \times 2^4 \times 4$ AB,AC,AD,	SAS Macros	_98.23%_ 47 ch sets	_98.34%_ 24 ch sets	_99.64%_ 24 ch sets	_99.63%_ 24 ch sets
AE,AF	S & B	_92.07%*_ 48 ch sets	_98.68%*_ 24 ch sets	_99.39%*_ 24 ch sets	_99.87%*_ 24 ch sets

An asterisk (*) indicates that all the effects are uncorrelated; numbers in italics indicate that the efficiency is relative to the largest $\det(C_{MT})$ found.

The details of the method used for each of the Street–Burgess designs are given in Section 8.2.2.

As we have seen, the Street–Burgess designs are not necessarily the most efficient, nor are they in the smallest possible number of choice sets. However, these designs are highly efficient and can always independently estimate the effects of interest.

8.5 REFERENCES AND COMMENTS

A comparison of the 6 strategies for one example, with five 4-level attributes and $m = 2$, has appeared in Street et al. (2005).

Appendix

8. A.1 NEAR-OPTIMAL CHOICE SETS FOR $\ell_1 = \ell_2 = 2$, $\ell_3 = \ell_4 = \ell_5 = \ell_6 = 4$, $\ell_7 = 8$, $\ell_8 = 36$, AND $m = 3$ FOR MAIN EFFECTS ONLY

F	$F + 11111011$	$F + 33330136$	F	$F + 11111011$	$F + 33330136$
00000000	11111011	33330136	11111101	22220112	00001037
22220002	33331013	11110138	33331103	00000114	22221039
21030013	32101024	10320149	30121112	01230123	23011048
03210011	10321022	32100147	12301110	23010121	01231046
13120023	20231034	02010159	02031122	13100133	31321058
31300021	02011032	20230157	20211120	31320131	13101056
32110030	03221041	21000166	23001131	30110142	12331067
10330032	21001043	03220168	01221133	12330144	30111069
23311042	30020053	12201178	32200143	03311154	21130079
01131040	12200051	30021176	10020141	21131152	03310077
02321051	13030062	31211107	13230150	20301161	02120006
20101053	31210064	13031109	31010152	02121163	20300008
30231061	01300072	23121117	21320160	32031171	10210016
12011063	23120074	01301119	03100162	10211173	32030018
11201072	22310003	00131128	00310173	11021104	33200029
33021070	00130001	22311126	22130171	33201102	11020027
00000004	11111015	333301310	11111105	22220116	000010311
22220006	33331017	111101312	33331107	00000118	222210313
21030017	32101028	103201413	30121116	01230127	230110412
03210015	10321026	321001411	12301114	23010125	012310410
13120027	20231038	020101513	02031126	13100137	313210512
31300025	02011036	202301511	20211124	31320135	131010510
32110034	03221045	210001610	23001135	30110146	123310611
10330036	21001047	032201612	01221137	12330148	301110613
23311046	30020057	122011712	32200147	03311158	211300713
01131044	12200055	300211710	10020145	21131156	033100711
02321055	13030066	312111011	13230154	20301165	021200010
20101057	31210068	130311013	31010156	02121167	203000012
30231065	01300076	231211111	21320164	32031175	102100110
12011067	23120078	013011113	03100166	10211177	320300112
11201076	22310007	001311212	00310177	11021108	332000213
33021074	00130005	223111210	22130175	33201106	110200211
00000008	11111019	333301314	11111109	222201110	000010315
222200010	333310111	111101316	333311011	000001112	222210317
210300111	321010212	103201417	301211110	012301211	230110416
03210019	103210210	321001415	12301118	23010129	012310414
131200211	202310312	020101517	020311210	131001311	313210516
31300029	020110310	202301515	20211128	31320139	131010514
32110038	03221049	210001614	23001139	301101410	123310615
103300310	210010411	032201616	012211311	123301412	301110617
233110410	300200511	122011716	322001411	033111512	211300717
01131048	12200059	300211714	10020149	211311510	033100715
02321059	130300610	312111015	13230158	20301169	021200014
201010511	312100612	130311017	310101510	021211611	203000016
30231069	013000710	231211115	21320168	32031179	102100114
120110611	231200712	013011117	031001610	102111711	320300116
112010710	223100011	001311216	003101711	110211012	332000217
33021078	00130009	223111214	22130179	332011010	110200215
000000012	111110113	333301318	111111013	222201114	000010319
222200014	333310115	111101320	333311015	000001116	222210321
210300115	321010216	103201421	301211114	012301215	230110420
032100113	103210214	321001419	123011112	230101213	012310418
131200215	202310316	020101521	020311214	131001315	313210520
313000213	020110314	202301519	202111212	313201313	131010518
321100312	032210413	210001618	230011313	301101414	123310619
103300314	210010415	032201620	012211315	123301416	301110621
233110414	300200515	122011720	322001415	033111516	211300721
011310412	122000513	300211718	100201413	211311514	033100719
023210513	130300614	312111019	132301512	203011613	021200018
201010515	312100616	130311021	310101514	021211615	203000020
302310613	013000714	231211119	213201612	320311713	102100118
120110615	231200716	013011121	031001614	102111715	320300120
112010714	223100015	001311220	003101715	110211016	332000221
330210712	001300013	223111218	221301713	332011014	110200219
000000016	111110117	333301322	111111017	222201118	000010323
222200018	333310119	111101324	333311019	000001120	222210325
210300119	321010220	103201425	301211118	012301219	230110424
032100117	103210218	321001423	123011116	230101217	012310422
131200219	202310320	020101525	020311218	131001319	313210524
313000217	020110318	202301523	202111216	313201317	131010522
321100316	032210417	210001622	230011317	301101418	123310623
103300318	210010419	032201624	012211319	123301420	301110625

F	$F + 11111011$	$F + 33330136$	F	$F + 11111011$	$F + 33330136$
2 3 3 1 1 0 4 18	3 0 0 2 0 0 5 19	1 2 2 0 1 1 7 24	3 2 2 0 0 1 4 19	0 3 3 1 1 1 5 20	2 1 1 3 0 0 7 25
0 1 1 3 1 0 4 16	1 2 2 0 0 0 5 17	3 0 0 2 1 1 7 22	1 0 0 2 0 1 4 17	2 1 1 3 1 1 5 18	0 3 3 1 0 0 7 23
0 2 3 2 1 0 5 17	1 3 0 3 0 0 6 18	3 1 2 1 1 1 0 23	1 3 2 3 0 1 5 16	2 0 3 0 1 1 6 17	0 2 1 2 0 0 0 22
2 0 1 0 1 0 5 19	3 1 2 1 0 0 6 20	1 3 0 3 1 1 0 25	3 1 0 1 0 1 5 18	0 2 1 2 1 1 6 19	2 0 3 0 0 0 0 24
3 0 2 3 1 0 6 17	0 1 3 0 0 0 7 18	2 3 1 2 1 1 1 23	2 1 3 2 0 1 6 16	3 2 0 3 1 1 7 17	1 0 2 1 0 0 1 22
1 2 0 1 1 0 6 19	2 3 1 2 0 0 7 20	0 1 3 0 1 1 1 25	0 3 1 0 0 1 6 18	1 0 2 1 1 1 7 19	3 2 0 3 0 0 1 24
1 1 2 0 1 0 7 18	2 2 3 1 0 0 0 19	0 0 1 3 1 1 2 24	0 0 3 1 0 1 7 19	1 1 0 2 1 1 0 20	3 3 2 0 0 0 2 25
3 3 0 2 1 0 7 16	0 0 1 3 0 0 0 17	2 2 3 1 1 1 2 22	2 2 1 3 0 1 7 17	3 3 2 0 1 1 0 18	1 1 0 2 0 0 2 23
0 0 0 0 0 0 0 20	1 1 1 1 1 0 1 21	3 3 3 3 0 1 3 26	1 1 1 1 1 1 0 21	2 2 2 2 0 1 1 22	0 0 0 0 1 0 3 27
2 2 2 2 0 0 0 22	3 3 3 3 1 0 1 23	1 1 1 1 0 1 3 28	3 3 3 3 1 1 0 23	0 0 0 0 0 1 1 24	2 2 2 2 1 0 3 29
2 1 0 3 0 0 1 23	3 2 1 0 1 0 2 24	1 0 3 2 0 1 4 29	3 0 1 2 1 1 1 22	0 1 2 3 0 1 2 23	2 3 0 1 1 0 4 28
0 3 2 1 0 0 1 21	1 0 3 2 1 0 2 22	3 2 1 0 0 1 4 27	1 2 3 0 1 1 1 20	2 3 0 1 0 1 2 21	0 1 2 3 1 0 4 26
1 3 1 2 0 0 2 23	2 0 2 3 1 0 3 24	0 2 0 1 0 1 5 29	0 2 0 3 1 1 2 22	1 3 1 0 0 1 3 23	3 1 3 2 1 0 5 28
3 1 3 0 0 0 2 21	0 2 0 1 1 0 3 22	2 0 2 3 0 1 5 27	2 0 2 1 1 1 2 20	3 1 3 2 0 1 3 21	1 3 1 0 1 0 5 26
3 2 1 1 0 0 3 20	0 3 2 2 1 0 4 21	2 1 0 0 0 1 6 26	2 3 0 0 1 1 3 21	3 0 1 1 0 1 4 22	1 2 3 3 1 0 6 27
1 0 3 3 0 0 3 22	2 1 0 0 1 0 4 23	0 3 2 2 0 1 6 28	0 1 2 2 1 1 3 23	1 2 3 3 0 1 4 24	3 0 1 1 1 0 6 29
2 3 3 1 1 0 4 22	3 0 0 2 0 0 5 23	1 2 2 0 1 1 7 28	3 2 2 0 0 1 4 23	0 3 3 1 1 1 5 24	2 1 1 3 0 0 7 29
0 1 1 3 1 0 4 20	1 2 2 0 0 0 5 21	3 0 0 2 1 1 7 26	1 0 0 2 0 1 4 21	2 1 1 3 1 1 5 22	0 3 3 1 0 0 7 27
0 2 3 2 1 0 5 21	1 3 0 3 0 0 6 22	3 1 2 1 1 1 0 27	1 3 2 3 0 1 5 20	2 0 3 0 1 1 6 21	0 2 1 2 0 0 0 26
2 0 1 0 1 0 5 23	3 1 2 1 0 0 6 24	1 3 0 3 1 1 0 29	3 1 0 1 0 1 5 22	0 2 1 2 1 1 6 23	2 0 3 0 0 0 0 28
3 0 2 3 1 0 6 21	0 1 3 0 0 0 7 22	2 3 1 2 1 1 1 27	2 1 3 2 0 1 6 20	3 2 0 3 1 1 7 21	1 0 2 1 0 0 1 26
1 2 0 1 1 0 6 23	2 3 1 2 0 0 7 24	0 1 3 0 1 1 1 29	0 3 1 0 0 1 6 22	1 0 2 1 1 1 7 23	3 2 0 3 0 0 1 28
1 1 2 0 1 0 7 22	2 2 3 1 0 0 0 23	0 0 1 3 1 1 2 28	0 0 3 1 0 1 7 23	1 1 0 2 1 1 0 24	3 3 2 0 0 0 2 29
3 3 0 2 1 0 7 20	0 0 1 3 0 0 0 21	2 2 3 1 1 1 2 26	2 2 1 3 0 1 7 21	3 3 2 0 1 1 0 22	1 1 0 2 0 0 2 27
0 0 0 0 0 0 0 24	1 1 1 1 1 0 1 25	3 3 3 3 0 1 3 30	1 1 1 1 1 1 0 25	2 2 2 2 0 1 1 26	0 0 0 0 1 0 3 31
2 2 2 2 0 0 0 26	3 3 3 3 1 0 1 27	1 1 1 1 0 1 3 32	3 3 3 3 1 1 0 27	0 0 0 0 0 1 1 28	2 2 2 2 1 0 3 33
2 1 0 3 0 0 1 27	3 2 1 0 1 0 2 28	1 0 3 2 0 1 4 33	3 0 1 2 1 1 1 26	0 1 2 3 0 1 2 27	2 3 0 1 1 0 4 32
0 3 2 1 0 0 1 25	1 0 3 2 1 0 2 26	3 2 1 0 0 1 4 31	1 2 3 0 1 1 1 24	2 3 0 1 0 1 2 25	0 1 2 3 1 0 4 30
1 3 1 2 0 0 2 27	2 0 2 3 1 0 3 28	0 2 0 1 0 1 5 33	0 2 0 3 1 1 2 26	1 3 1 0 0 1 3 27	3 1 3 2 1 0 5 32
3 1 3 0 0 0 2 25	0 2 0 1 1 0 3 26	2 0 2 3 0 1 5 31	2 0 2 1 1 1 2 24	3 1 3 2 0 1 3 25	1 3 1 0 1 0 5 30
3 2 1 1 0 0 3 24	0 3 2 2 1 0 4 25	2 1 0 0 0 1 6 30	2 3 0 0 1 1 3 25	3 0 1 1 0 1 4 26	1 2 3 3 1 0 6 31
1 0 3 3 0 0 3 26	2 1 0 0 1 0 4 27	0 3 2 2 0 1 6 32	0 1 2 2 1 1 3 27	1 2 3 3 0 1 4 28	3 0 1 1 1 0 6 33
2 3 3 1 1 0 4 26	3 0 0 2 0 0 5 27	1 2 2 0 1 1 7 32	3 2 2 0 0 1 4 27	0 3 3 1 1 1 5 28	2 1 1 3 0 0 7 33
0 1 1 3 1 0 4 24	1 2 2 0 0 0 5 25	3 0 0 2 1 1 7 30	1 0 0 2 0 1 4 25	2 1 1 3 1 1 5 26	0 3 3 1 0 0 7 31
0 2 3 2 1 0 5 25	1 3 0 3 0 0 6 26	3 1 2 1 1 1 0 31	1 3 2 3 0 1 5 24	2 0 3 0 1 1 6 25	0 2 1 2 0 0 0 30
2 0 1 0 1 0 5 27	3 1 2 1 0 0 6 28	1 3 0 3 1 1 0 33	3 1 0 1 0 1 5 26	0 2 1 2 1 1 6 27	2 0 3 0 0 0 0 32
3 0 2 3 1 0 6 25	0 1 3 0 0 0 7 26	2 3 1 2 1 1 1 31	2 1 3 2 0 1 6 24	3 2 0 3 1 1 7 25	1 0 2 1 0 0 1 30
1 2 0 1 1 0 6 27	2 3 1 2 0 0 7 28	0 1 3 0 1 1 1 33	0 3 1 0 0 1 6 26	1 0 2 1 1 1 7 27	3 2 0 3 0 0 1 32
1 1 2 0 1 0 7 26	2 2 3 1 0 0 0 27	0 0 1 3 1 1 2 32	0 0 3 1 0 1 7 27	1 1 0 2 1 1 0 28	3 3 2 0 0 0 2 33
3 3 0 2 1 0 7 24	0 0 1 3 0 0 0 25	2 2 3 1 1 1 2 30	2 2 1 3 0 1 7 25	3 3 2 0 1 1 0 26	1 1 0 2 0 0 2 31
0 0 0 0 0 0 0 28	1 1 1 1 1 0 1 29	3 3 3 3 0 1 3 34	1 1 1 1 1 1 0 29	2 2 2 2 0 1 1 30	0 0 0 0 1 0 3 35
2 2 2 2 0 0 0 30	3 3 3 3 1 0 1 31	1 1 1 1 0 1 3 0	3 3 3 3 1 1 0 31	0 0 0 0 0 1 1 32	2 2 2 2 1 0 3 1
2 1 0 3 0 0 1 31	3 2 1 0 1 0 2 32	1 0 3 2 0 1 4 1	3 0 1 2 1 1 1 30	0 1 2 3 0 1 2 31	2 3 0 1 1 0 4 0
0 3 2 1 0 0 1 29	1 0 3 2 1 0 2 30	3 2 1 0 0 1 4 35	1 2 3 0 1 1 1 28	2 3 0 1 0 1 2 29	0 1 2 3 1 0 4 34
1 3 1 2 0 0 2 31	2 0 2 3 1 0 3 32	0 2 0 1 0 1 5 1	0 2 0 3 1 1 2 30	1 3 1 0 0 1 3 31	3 1 3 2 1 0 5 0
3 1 3 0 0 0 2 29	0 2 0 1 1 0 3 30	2 0 2 3 0 1 5 35	2 0 2 1 1 1 2 28	3 1 3 2 0 1 3 29	1 3 1 0 1 0 5 34
3 2 1 1 0 0 3 28	0 3 2 2 1 0 4 29	2 1 0 0 0 1 6 34	2 3 0 0 1 1 3 29	3 0 1 1 0 1 4 30	1 2 3 3 1 0 6 35
1 0 3 3 0 0 3 30	2 1 0 0 1 0 4 31	0 3 2 2 0 1 6 0	0 1 2 2 1 1 3 31	1 2 3 3 0 1 4 32	3 0 1 1 1 0 6 1
2 3 3 1 1 0 4 30	3 0 0 2 0 0 5 31	1 2 2 0 1 1 7 0	3 2 2 0 0 1 4 31	0 3 3 1 1 1 5 32	2 1 1 3 0 0 7 1
0 1 1 3 1 0 4 28	1 2 2 0 0 0 5 29	3 0 0 2 1 1 7 34	1 0 0 2 0 1 4 29	2 1 1 3 1 1 5 30	0 3 3 1 0 0 7 35
0 2 3 2 1 0 5 29	1 3 0 3 0 0 6 30	3 1 2 1 1 1 0 35	1 3 2 3 0 1 5 28	2 0 3 0 1 1 6 29	0 2 1 2 0 0 0 34
2 0 1 0 1 0 5 31	3 1 2 1 0 0 6 32	1 3 0 3 1 1 0 1	3 1 0 1 0 1 5 30	0 2 1 2 1 1 6 31	2 0 3 0 0 0 0 0
3 0 2 3 1 0 6 29	0 1 3 0 0 0 7 30	2 3 1 2 1 1 1 35	2 1 3 2 0 1 6 28	3 2 0 3 1 1 7 29	1 0 2 1 0 0 1 34
1 2 0 1 1 0 6 31	2 3 1 2 0 0 7 32	0 1 3 0 1 1 1 1	0 3 1 0 0 1 6 30	1 0 2 1 1 1 7 31	3 2 0 3 0 0 1 0
1 1 2 0 1 0 7 30	2 2 3 1 0 0 0 31	0 0 1 3 1 1 2 0	0 0 3 1 0 1 7 31	1 1 0 2 1 1 0 32	3 3 2 0 0 0 2 1
3 3 0 2 1 0 7 28	0 0 1 3 0 0 0 29	2 2 3 1 1 1 2 34	2 2 1 3 0 1 7 29	3 3 2 0 1 1 0 30	1 1 0 2 0 0 2 35
0 0 0 0 0 0 0 32	1 1 1 1 1 0 1 33	3 3 3 3 0 1 3 2	1 1 1 1 1 1 0 33	2 2 2 2 0 1 1 34	0 0 0 0 1 0 3 3
2 2 2 2 0 0 0 34	3 3 3 3 1 0 1 35	1 1 1 1 0 1 3 4	3 3 3 3 1 1 0 35	0 0 0 0 0 1 1 0	2 2 2 2 1 0 3 5
2 1 0 3 0 0 1 35	3 2 1 0 1 0 2 0	1 0 3 2 0 1 4 5	3 0 1 2 1 1 1 34	0 1 2 3 0 1 2 35	2 3 0 1 1 0 4 4
0 3 2 1 0 0 1 33	1 0 3 2 1 0 2 34	3 2 1 0 0 1 4 3	1 2 3 0 1 1 1 32	2 3 0 1 0 1 2 33	0 1 2 3 1 0 4 2
1 3 1 2 0 0 2 35	2 0 2 3 1 0 3 0	0 2 0 1 0 1 5 5	0 2 0 3 1 1 2 34	1 3 1 0 0 1 3 35	3 1 3 2 1 0 5 4
3 1 3 0 0 0 2 33	0 2 0 1 1 0 3 34	2 0 2 3 0 1 5 3	2 0 2 1 1 1 2 32	3 1 3 2 0 1 3 33	1 3 1 0 1 0 5 2
3 2 1 1 0 0 3 32	0 3 2 2 1 0 4 33	2 1 0 0 0 1 6 2	2 3 0 0 1 1 3 33	3 0 1 1 0 1 4 34	1 2 3 3 1 0 6 3
1 0 3 3 0 0 3 34	2 1 0 0 1 0 4 35	0 3 2 2 0 1 6 4	0 1 2 2 1 1 3 35	1 2 3 3 0 1 4 0	3 0 1 1 1 0 6 5
2 3 3 1 1 0 4 34	3 0 0 2 0 0 5 35	1 2 2 0 1 1 7 4	3 2 2 0 0 1 4 35	0 3 3 1 1 1 5 0	2 1 1 3 0 0 7 5
0 1 1 3 1 0 4 32	1 2 2 0 0 0 5 33	3 0 0 2 1 1 7 2	1 0 0 2 0 1 4 33	2 1 1 3 1 1 5 34	0 3 3 1 0 0 7 3
0 2 3 2 1 0 5 33	1 3 0 3 0 0 6 34	3 1 2 1 1 1 0 3	1 3 2 3 0 1 5 32	2 0 3 0 1 1 6 33	0 2 1 2 0 0 0 2
2 0 1 0 1 0 5 35	3 1 2 1 0 0 6 0	1 3 0 3 1 1 0 5	3 1 0 1 0 1 5 34	0 2 1 2 1 1 6 35	2 0 3 0 0 0 0 4
3 0 2 3 1 0 6 33	0 1 3 0 0 0 7 34	2 3 1 2 1 1 1 3	2 1 3 2 0 1 6 32	3 2 0 3 1 1 7 33	1 0 2 1 0 0 1 2
1 2 0 1 1 0 6 35	2 3 1 2 0 0 7 0	0 1 3 0 1 1 1 5	0 3 1 0 0 1 6 34	1 0 2 1 1 1 7 35	3 2 0 3 0 0 1 4
1 1 2 0 1 0 7 34	2 2 3 1 0 0 0 35	0 0 1 3 1 1 2 4	0 0 3 1 0 1 7 35	1 1 0 2 1 1 0 0	3 3 2 0 0 0 2 5
3 3 0 2 1 0 7 32	0 0 1 3 0 0 0 33	2 2 3 1 1 1 2 2	2 2 1 3 0 1 7 33	3 3 2 0 1 1 0 34	1 1 0 2 0 0 2 3

8. A.2 NEAR-OPTIMAL CHOICE SETS FOR $\ell_1 = \ell_2 = \ell_3 = \ell_4 = \ell_5 = 4$, $\ell_6 = 2$, $\ell_7 = \ell_8 = 8$, $\ell_9 = 24$, AND $m = 3$ FOR MAIN EFFECTS ONLY

F	F + 1 1 1 1 1 1 1 1 1	F + 3 3 3 3 3 0 3 3 10
0 0 0 0 0 0 0 0 0	1 1 1 1 1 1 1 1 1	3 3 3 3 3 0 3 3 10
0 0 0 1 1 1 5 6 7	1 1 1 2 2 0 6 7 8	3 3 3 0 0 1 0 1 17
0 1 1 1 2 1 6 7 1	1 2 2 2 3 0 7 0 2	3 0 0 0 1 1 1 2 11
0 1 1 2 2 1 7 1 2	1 2 2 3 3 0 0 2 3	3 0 0 1 1 1 2 4 12
0 2 2 2 3 1 1 2 3	1 3 3 3 0 0 2 3 4	3 1 1 1 2 1 4 5 13
0 2 2 3 3 0 2 3 4	1 3 3 0 0 1 3 4 5	3 1 1 2 2 0 5 6 14
0 3 3 3 0 0 3 4 5	1 0 0 0 1 1 4 5 6	3 2 2 2 3 0 6 7 15
0 3 3 0 1 0 4 5 6	1 0 0 1 2 1 5 6 7	3 2 2 3 0 0 7 0 16
0 0 0 0 0 0 1 1 1	1 1 1 1 1 1 2 2 2	3 3 3 3 3 0 4 4 11
0 0 0 2 3 0 6 5 3	1 1 1 3 0 1 7 6 4	3 3 3 1 2 0 1 0 13
0 1 2 3 1 1 5 3 0	1 2 3 0 2 0 6 4 1	3 0 1 2 0 1 0 6 10
0 1 3 1 3 1 3 0 4	1 2 0 2 0 0 4 1 5	3 0 2 0 2 1 6 3 14
0 2 1 3 2 0 0 4 7	1 3 2 0 3 1 1 5 8	3 1 0 2 1 0 3 7 17
0 2 3 2 1 0 4 7 2	1 3 0 3 2 1 5 0 3	3 1 2 1 0 0 7 2 12
0 3 2 1 0 1 7 2 6	1 0 3 2 1 0 0 3 7	3 2 1 0 3 1 2 5 16
0 3 1 0 2 1 2 6 5	1 0 2 1 3 0 3 7 6	3 2 0 3 1 1 5 1 15
1 0 1 1 1 0 2 2 2	2 1 2 2 2 1 3 3 3	0 3 0 0 0 0 5 5 12
1 0 2 0 2 0 3 7 6	2 1 3 1 3 1 4 0 7	0 3 1 3 1 0 6 2 16
1 1 0 2 0 0 7 6 4	2 2 1 3 1 1 0 7 5	0 0 3 1 3 0 2 1 14
1 1 2 0 1 1 6 4 0	2 2 3 1 2 0 7 5 1	0 0 1 3 0 1 1 7 10
1 2 0 1 3 1 4 0 5	2 3 1 2 0 0 5 1 6	0 1 3 0 2 1 7 3 15
1 2 1 3 3 1 0 5 1	2 3 2 0 0 0 1 6 2	0 1 0 2 2 1 3 0 11
1 3 3 3 2 0 5 1 3	2 0 0 0 3 1 6 2 4	0 2 2 2 1 0 0 4 13
1 3 3 2 0 1 1 3 7	2 0 0 3 1 0 2 4 8	0 2 2 1 3 1 4 6 17
1 0 1 1 1 0 3 3 3	2 1 2 2 2 1 4 4 4	0 3 0 0 0 0 6 6 13
1 0 3 2 0 1 2 4 1	2 1 0 3 1 0 3 5 2	0 3 2 1 3 1 5 7 11
1 1 2 0 3 0 4 1 7	2 2 3 1 0 1 5 2 8	0 0 1 3 2 0 7 4 17
1 1 0 3 1 1 1 7 5	2 2 1 0 2 0 2 0 6	0 0 3 2 0 1 4 2 15
1 2 3 1 2 0 7 5 0	2 3 0 2 3 1 0 6 1	0 1 2 0 1 0 2 0 10
1 2 1 2 0 1 5 0 6	2 3 2 3 1 0 6 1 7	0 1 0 1 3 1 0 3 16
1 3 2 0 3 1 0 6 2	2 0 3 1 0 0 1 7 3	0 2 1 3 2 1 3 1 12
1 3 0 3 2 0 6 2 4	2 0 1 0 3 1 7 3 5	0 2 3 2 1 0 1 5 14
2 0 2 2 2 1 4 4 4	3 1 3 3 3 0 5 5 5	1 3 1 1 1 1 7 7 14
2 0 1 0 3 0 7 3 5	3 1 2 1 0 1 0 4 6	1 3 0 3 2 0 2 6 15
2 1 0 3 0 1 3 5 2	3 2 1 0 1 0 4 6 3	1 0 3 2 3 1 6 0 12
2 1 3 0 3 0 5 2 1	3 2 0 1 0 1 6 3 2	1 0 2 3 2 0 0 5 11
2 2 0 3 1 1 2 1 6	3 3 1 0 2 0 3 2 7	1 1 3 2 0 1 5 4 16
2 2 3 1 2 0 1 6 0	3 3 0 2 3 1 2 7 1	1 1 2 0 1 0 4 1 10
2 3 1 2 1 0 6 0 7	3 0 2 3 2 1 7 1 8	1 2 0 1 0 0 1 3 17
2 3 2 1 0 1 0 7 3	3 0 3 2 1 0 1 0 4	1 2 1 0 3 1 3 2 13
2 0 2 2 2 1 5 5 5	3 1 3 3 3 0 6 6 6	1 3 1 1 1 1 0 0 15
2 0 3 1 1 1 0 1 4	3 1 0 2 2 0 1 2 5	1 3 2 0 0 1 3 4 14
2 1 1 1 3 0 1 4 6	3 2 2 2 0 1 2 5 7	1 0 0 0 2 0 4 7 16
2 1 1 3 0 0 4 6 3	3 2 2 0 1 1 5 7 4	1 0 0 2 3 0 7 1 13
2 2 3 0 0 1 6 3 2	3 3 0 1 1 0 7 4 3	1 1 2 3 3 1 1 6 12
2 2 0 0 2 1 3 2 7	3 3 1 1 3 0 4 3 8	1 1 3 3 1 1 6 5 17

F	F + 1 1 1 1 1 1 1 1 1	F + 3 3 3 3 0 3 3 10
2 3 0 2 3 0 2 7 0	3 0 1 3 0 1 3 0 1	1 2 3 1 2 0 5 2 10
2 3 2 3 1 0 7 0 1	3 0 3 0 2 1 0 1 2	1 2 1 2 0 0 2 3 11
3 0 3 3 3 1 6 6 6	0 1 0 0 0 0 7 7 7	2 3 2 2 2 1 1 1 16
3 0 2 3 2 0 1 0 2	0 1 3 0 3 1 2 1 3	2 3 1 2 1 0 4 3 12
3 1 3 2 1 0 0 2 5	0 2 0 3 2 1 1 3 6	2 0 2 1 0 0 3 5 15
3 1 2 1 0 0 2 5 7	0 2 3 2 1 1 3 6 8	2 0 1 0 3 0 5 0 17
3 2 1 0 0 0 5 7 4	0 3 2 1 1 1 6 0 5	2 1 0 3 3 0 0 2 14
3 2 0 0 1 1 7 4 3	0 3 1 1 2 0 0 5 4	2 1 3 3 0 1 2 7 13
3 3 0 1 2 1 4 3 1	0 0 1 2 3 0 5 4 2	2 2 3 0 1 1 7 6 11
3 3 1 2 3 1 3 1 0	0 0 2 3 0 0 4 2 1	2 2 0 1 2 1 6 4 10
3 0 3 3 3 1 7 7 7	0 1 0 0 0 0 0 0 8	2 3 2 2 2 1 2 2 17
3 0 1 3 0 1 4 2 0	0 1 2 0 1 0 5 3 1	2 3 0 2 3 1 7 5 10
3 1 3 0 2 1 2 0 3	0 2 0 1 3 0 3 1 4	2 0 2 3 1 1 5 3 13
3 1 0 2 2 0 0 3 6	0 2 1 3 3 1 1 4 7	2 0 3 1 1 0 3 6 16
3 2 2 2 1 0 3 6 1	0 3 3 3 2 1 4 7 2	2 1 1 1 0 0 6 1 11
3 2 2 1 0 0 6 1 5	0 3 3 2 1 1 7 2 6	2 1 1 0 3 0 1 4 15
3 3 1 0 1 1 1 5 4	0 0 2 1 2 0 2 6 5	2 2 0 3 0 1 4 0 14
3 3 0 1 3 0 5 4 2	0 0 1 2 0 1 6 5 3	2 2 3 0 2 0 0 7 12
0 0 0 0 0 0 0 0 8	1 1 1 1 1 1 1 1 9	3 3 3 3 3 0 3 3 18
0 0 0 1 1 1 5 6 15	1 1 1 2 2 0 6 7 16	3 3 3 0 0 1 0 1 1
0 1 1 1 2 1 6 7 9	1 2 2 2 3 0 7 0 10	3 0 0 0 1 1 1 2 19
0 1 1 2 2 1 7 1 10	1 2 2 3 3 0 0 2 11	3 0 0 1 1 1 2 4 20
0 2 2 2 3 1 1 2 11	1 3 3 3 0 0 2 3 12	3 1 1 1 2 1 4 5 21
0 2 2 3 3 0 2 3 12	1 3 3 0 0 1 3 4 13	3 1 1 2 2 0 5 6 22
0 3 3 3 0 0 3 4 13	1 0 0 0 1 1 4 5 14	3 2 2 2 3 0 6 7 23
0 3 3 0 1 0 4 5 14	1 0 0 1 2 1 5 6 15	3 2 2 3 0 0 7 0 0
0 0 0 0 0 0 1 1 9	1 1 1 1 1 1 2 2 10	3 3 3 3 3 0 4 4 19
0 0 0 2 3 0 6 5 11	1 1 1 3 0 1 7 6 12	3 3 3 1 2 0 1 0 21
0 1 2 3 1 1 5 3 8	1 2 3 0 2 0 6 4 9	3 0 1 2 0 1 0 6 18
0 1 3 1 3 1 3 0 12	1 2 0 2 0 0 4 1 13	3 0 2 0 2 1 6 3 22
0 2 1 3 2 0 0 4 15	1 3 2 0 3 1 1 5 16	3 1 0 2 1 0 3 7 1
0 2 3 2 1 0 4 7 10	1 3 0 3 2 1 5 0 11	3 1 2 1 0 0 7 2 20
0 3 2 1 0 1 7 2 14	1 0 3 2 1 0 0 3 15	3 2 1 0 3 1 2 5 0
0 3 1 0 2 1 2 6 13	1 0 2 1 3 0 3 7 14	3 2 0 3 1 1 5 1 23
1 0 1 1 1 0 2 2 10	2 1 2 2 2 1 3 3 11	0 3 0 0 0 0 5 5 20
1 0 2 0 2 0 3 7 14	2 1 3 1 3 1 4 0 15	0 3 1 3 1 0 6 2 0
1 1 0 2 0 0 7 6 12	2 2 1 3 1 1 0 7 13	0 0 3 1 3 0 2 1 22
1 1 2 0 1 1 6 4 8	2 2 3 1 2 0 7 5 9	0 0 1 3 0 1 1 7 18
1 2 0 1 3 1 4 0 13	2 3 1 2 0 0 5 1 14	0 1 3 0 2 1 7 3 23
1 2 1 3 3 1 0 5 9	2 3 2 0 0 0 1 6 10	0 1 0 2 2 1 3 0 19
1 3 3 3 2 0 5 1 11	2 0 0 0 3 1 6 2 12	0 2 2 2 1 0 0 4 21
1 3 3 2 0 1 1 3 15	2 0 0 3 1 0 2 4 16	0 2 2 1 3 1 4 6 1
1 0 1 1 1 0 3 3 11	2 1 2 2 2 1 4 4 12	0 3 0 0 0 0 6 6 21
1 0 3 2 0 1 2 4 9	2 1 0 3 1 0 3 5 10	0 3 2 1 3 1 5 7 19
1 1 2 0 3 0 4 1 15	2 2 3 1 0 1 5 2 16	0 0 1 3 2 0 7 4 1
1 1 0 3 1 1 1 7 13	2 2 1 0 2 0 2 0 14	0 0 3 2 0 1 4 2 23
1 2 3 1 2 0 7 5 8	2 3 0 2 3 1 0 6 9	0 1 2 0 1 0 2 0 18
1 2 1 2 0 1 5 0 14	2 3 2 3 1 0 6 1 15	0 1 0 1 3 1 0 3 0
1 3 2 0 3 1 0 6 10	2 0 3 1 0 0 1 7 11	0 2 1 3 2 1 3 1 20
1 3 0 3 2 0 6 2 12	2 0 1 0 3 1 7 3 13	0 2 3 2 1 0 1 5 22

F	F + 1 1 1 1 1 1 1 1 1	F + 3 3 3 3 0 3 3 10
2 0 2 2 2 1 4 4 12	3 1 3 3 3 0 5 5 13	1 3 1 1 1 1 7 7 22
2 0 1 0 3 0 7 3 13	3 1 2 1 0 1 0 4 14	1 3 0 3 2 0 2 6 23
2 1 0 3 0 1 3 5 10	3 2 1 0 1 0 4 6 11	1 0 3 2 3 1 6 0 20
2 1 3 0 3 0 5 2 9	3 2 0 1 0 1 6 3 10	1 0 2 3 2 0 0 5 19
2 2 0 3 1 1 2 1 14	3 3 1 0 2 0 3 2 15	1 1 3 2 0 1 5 4 0
2 2 3 1 2 0 1 6 8	3 3 0 2 3 1 2 7 9	1 1 2 0 1 0 4 1 18
2 3 1 2 1 0 6 0 15	3 0 2 3 2 1 7 1 16	1 2 0 1 0 0 1 3 1
2 3 2 1 0 1 0 7 11	3 0 3 2 1 0 1 0 12	1 2 1 0 3 1 3 2 21
2 0 2 2 2 1 5 5 13	3 1 3 3 3 0 6 6 14	1 3 1 1 1 1 0 0 23
2 0 3 1 1 1 0 1 12	3 1 0 2 2 0 1 2 13	1 3 2 0 0 1 3 4 22
2 1 1 1 3 0 1 4 14	3 2 2 2 0 1 2 5 15	1 0 0 0 2 0 4 7 0
2 1 1 3 0 0 4 6 11	3 2 2 0 1 1 5 7 12	1 0 0 2 3 0 7 1 21
2 2 3 0 0 1 6 3 10	3 3 0 1 1 0 7 4 11	1 1 2 3 3 1 1 6 20
2 2 0 0 2 1 3 2 15	3 3 1 1 3 0 4 3 16	1 1 3 3 1 1 6 5 1
2 3 0 2 3 0 2 7 8	3 0 1 3 0 1 3 0 9	1 2 3 1 2 0 5 2 18
2 3 2 3 1 0 7 0 9	3 0 3 0 2 1 0 1 10	1 2 1 2 0 0 2 3 19
3 0 3 3 3 1 6 6 14	0 1 0 0 0 0 7 7 15	2 3 2 2 2 1 1 1 0
3 0 2 3 2 0 1 0 10	0 1 3 0 3 1 2 1 11	2 3 1 2 1 0 4 3 20
3 1 3 2 1 0 0 2 13	0 2 0 3 2 1 1 3 14	2 0 2 1 0 0 3 5 23
3 1 2 1 0 0 2 5 15	0 2 3 2 1 1 3 6 16	2 0 1 0 3 0 5 0 1
3 2 1 0 0 0 5 7 12	0 3 2 1 1 1 6 0 13	2 1 0 3 3 0 0 2 22
3 2 0 0 1 1 7 4 11	0 3 1 1 2 0 0 5 12	2 1 3 3 0 1 2 7 21
3 3 0 1 2 1 4 3 9	0 0 1 2 3 0 5 4 10	2 2 3 0 1 1 7 6 19
3 3 1 2 3 1 3 1 8	0 0 2 3 0 0 4 2 9	2 2 0 1 2 1 6 4 18
3 0 3 3 3 1 7 7 15	0 1 0 0 0 0 0 0 16	2 3 2 2 2 1 2 2 1
3 0 1 3 0 1 4 2 8	0 1 2 0 1 0 5 3 9	2 3 0 2 3 1 7 5 18
3 1 3 0 2 1 2 0 11	0 2 0 1 3 0 3 1 12	2 0 2 3 1 1 5 3 21
3 1 0 2 2 0 0 3 14	0 2 1 3 3 1 1 4 15	2 0 3 1 1 0 3 6 0
3 2 2 2 1 0 3 6 9	0 3 3 3 2 1 4 7 10	2 1 1 1 0 0 6 1 19
3 2 2 1 0 0 6 1 13	0 3 3 2 1 1 7 2 14	2 1 1 0 3 0 1 4 23
3 3 1 0 1 1 1 5 12	0 0 2 1 2 0 2 6 13	2 2 0 3 0 1 4 0 22
3 3 0 1 3 0 5 4 10	0 0 1 2 0 1 6 5 11	2 2 3 0 2 0 0 7 20
0 0 0 0 0 0 0 0 16	1 1 1 1 1 1 1 1 17	3 3 3 3 3 0 3 3 2
0 0 0 1 1 1 5 6 23	1 1 1 2 2 0 6 7 0	3 3 3 0 0 1 0 1 9
0 1 1 1 2 1 6 7 17	1 2 2 2 3 0 7 0 18	3 0 0 0 1 1 1 2 3
0 1 1 2 2 1 7 1 18	1 2 2 3 3 0 0 2 19	3 0 0 1 1 1 2 4 4
0 2 2 2 3 1 1 2 19	1 3 3 3 0 0 2 3 20	3 1 1 1 2 1 4 5 5
0 2 2 3 3 0 2 3 20	1 3 3 0 0 1 3 4 21	3 1 1 2 2 0 5 6 6
0 3 3 3 0 0 3 4 21	1 0 0 0 1 1 4 5 22	3 2 2 2 3 0 6 7 7
0 3 3 0 1 0 4 5 22	1 0 0 1 2 1 5 6 23	3 2 2 3 0 0 7 0 8
0 0 0 0 0 1 1 1 17	1 1 1 1 1 1 2 2 18	3 3 3 3 3 0 4 4 3
0 0 0 2 3 0 6 5 19	1 1 1 3 0 1 7 6 20	3 3 3 1 2 0 1 0 5
0 1 2 3 1 1 5 3 16	1 2 3 0 2 0 6 4 17	3 0 1 2 0 1 0 6 2
0 1 3 1 3 1 3 0 20	1 2 0 2 0 0 4 1 21	3 0 2 0 2 1 6 3 6
0 2 1 3 2 0 0 4 23	1 3 2 0 3 1 1 5 0	3 1 0 2 1 0 3 7 9
0 2 3 2 1 0 4 7 18	1 3 0 3 2 1 5 0 19	3 1 2 1 0 0 7 2 4
0 3 2 1 0 1 7 2 22	1 0 3 2 1 0 0 3 23	3 2 1 0 3 1 2 5 8
0 3 1 0 2 1 2 6 21	1 0 2 1 3 0 3 7 22	3 2 0 3 1 1 5 1 7

F	F + 1 1 1 1 1 1 1 1 1	F + 3 3 3 3 0 3 3 3 10
1 0 1 1 1 0 2 2 18	2 1 2 2 2 1 3 3 19	0 3 0 0 0 0 5 5 4
1 0 2 0 2 0 3 7 22	2 1 3 1 3 1 4 0 23	0 3 1 3 1 0 6 2 8
1 1 0 2 0 0 7 6 20	2 2 1 3 1 1 0 7 21	0 0 3 1 3 0 2 1 6
1 1 2 0 1 1 6 4 16	2 2 3 1 2 0 7 5 17	0 0 1 3 0 1 1 7 2
1 2 0 1 3 1 4 0 21	2 3 1 2 0 0 5 1 22	0 1 3 0 2 1 7 3 7
1 2 1 3 3 1 0 5 17	2 3 2 0 0 0 1 6 18	0 1 0 2 2 1 3 0 3
1 3 3 3 2 0 5 1 19	2 0 0 0 3 1 6 2 20	0 2 2 2 1 0 0 4 5
1 3 3 2 0 1 1 3 23	2 0 0 3 1 0 2 4 0	0 2 2 1 3 1 4 6 9
1 0 1 1 1 0 3 3 19	2 1 2 2 2 1 4 4 20	0 3 0 0 0 0 6 6 5
1 0 3 2 0 1 2 4 17	2 1 0 3 1 0 3 5 18	0 3 2 1 3 1 5 7 3
1 1 2 0 3 0 4 1 23	2 2 3 1 0 1 5 2 0	0 0 1 3 2 0 7 4 9
1 1 0 3 1 1 1 7 21	2 2 1 0 2 0 2 0 22	0 0 3 2 0 1 4 2 7
1 2 3 1 2 0 7 5 16	2 3 0 2 3 1 0 6 17	0 1 2 0 1 0 2 0 2
1 2 1 2 0 1 5 0 22	2 3 2 3 1 0 6 1 23	0 1 0 1 3 1 0 3 8
1 3 2 0 3 1 0 6 18	2 0 3 1 0 0 1 7 19	0 2 1 3 2 1 3 1 4
1 3 0 3 2 0 6 2 20	2 0 1 0 3 1 7 3 21	0 2 3 2 1 0 1 5 6
2 0 2 2 2 1 4 4 20	3 1 3 3 3 0 5 5 21	1 3 1 1 1 1 7 7 6
2 0 1 0 3 0 7 3 21	3 1 2 1 0 1 0 4 22	1 3 0 3 2 0 2 6 7
2 1 0 3 0 1 3 5 18	3 2 1 0 1 0 4 6 19	1 0 3 2 3 1 6 0 4
2 1 3 0 3 0 5 2 17	3 2 0 1 0 1 6 3 18	1 0 2 3 2 0 0 5 3
2 2 0 3 1 1 2 1 22	3 3 1 0 2 0 3 2 23	1 1 3 2 0 1 5 4 8
2 2 3 1 2 0 1 6 16	3 3 0 2 3 1 2 7 17	1 1 2 0 1 0 4 1 2
2 3 1 2 1 0 6 0 23	3 0 2 3 2 1 7 1 0	1 2 0 1 0 0 1 3 9
2 3 2 1 0 1 0 7 19	3 0 3 2 1 0 1 0 20	1 2 1 0 3 1 3 2 5
2 0 2 2 2 1 5 5 21	3 1 3 3 3 0 6 6 22	1 3 1 1 1 1 0 0 7
2 0 3 1 1 1 0 1 20	3 1 0 2 2 0 1 2 21	1 3 2 0 0 1 3 4 6
2 1 1 1 3 0 1 4 22	3 2 2 2 0 1 2 5 23	1 0 0 0 2 0 4 7 8
2 1 1 3 0 0 4 6 19	3 2 2 0 1 1 5 7 20	1 0 0 2 3 0 7 1 5
2 2 3 0 0 1 6 3 18	3 3 0 1 1 0 7 4 19	1 1 2 3 3 1 1 6 4
2 2 0 0 2 1 3 2 23	3 3 1 1 3 0 4 3 0	1 1 3 3 1 1 6 5 9
2 3 0 2 3 0 2 7 16	3 0 1 3 0 1 3 0 17	1 2 3 1 2 0 5 2 2
2 3 2 3 1 0 7 0 17	3 0 3 0 2 1 0 1 18	1 2 1 2 0 0 2 3 3
3 0 3 3 3 1 6 6 22	0 1 0 0 0 0 7 7 23	2 3 2 2 2 1 1 1 8
3 0 2 3 2 0 1 0 18	0 1 3 0 3 1 2 1 19	2 3 1 2 1 0 4 3 4
3 1 3 2 1 0 0 2 21	0 2 0 3 2 1 1 3 22	2 0 2 1 0 0 3 5 7
3 1 2 1 0 0 2 5 23	0 2 3 2 1 1 3 6 0	2 0 1 0 3 0 5 0 9
3 2 1 0 0 0 5 7 20	0 3 2 1 1 1 6 0 21	2 1 0 3 3 0 0 2 6
3 2 0 0 1 1 7 4 19	0 3 1 1 2 0 0 5 20	2 1 3 3 0 1 2 7 5
3 3 0 1 2 1 4 3 17	0 0 1 2 3 0 5 4 18	2 2 3 0 1 1 7 6 3
3 3 1 2 3 1 3 1 16	0 0 2 3 0 0 4 2 17	2 2 0 1 2 1 6 4 2
3 0 3 3 3 1 7 7 23	0 1 0 0 0 0 0 0 0	2 3 2 2 2 1 2 2 9
3 0 1 3 0 1 4 2 16	0 1 2 0 1 0 5 3 17	2 3 0 2 3 1 7 5 2
3 1 3 0 2 1 2 0 19	0 2 0 1 3 0 3 1 20	2 0 2 3 1 1 5 3 5
3 1 0 2 2 0 0 3 22	0 2 1 3 3 1 1 4 23	2 0 3 1 1 0 3 6 8
3 2 2 2 1 0 3 6 17	0 3 3 3 2 1 4 7 18	2 1 1 1 0 0 6 1 3
3 2 2 1 0 0 6 1 21	0 3 3 2 1 1 7 2 22	2 1 1 0 3 0 1 4 7
3 3 1 0 1 1 1 5 20	0 0 2 1 2 0 2 6 21	2 2 0 3 0 1 4 0 6
3 3 0 1 3 0 5 4 18	0 0 1 2 0 1 6 5 19	2 2 3 0 2 0 0 7 4

Bibliography

Abel, J. R. and Greig, M. (2006). BIBDs with small block size. In Colbourn, C. and Dinitz, J., editors, *Handbook of Combinatorial Designs, Second Edition*, pages 71–79. Chapman & Hall/CRC Press, Boca Raton, FL.

Abelson, R. M. and Bradley, R. A. (1954). A 2 × 2 factorial with paired comparisons. *Biometrics*, 10:487–502.

Addelman, S. (1962). Orthogonal main-effect plans for asymmetrical factorial experiments. *Technometrics*, 4:21–46.

Addelman, S. (1972). Recent developments in the design of factorial experiments. *Journal of the American Statistical Association*, 67:103–111.

Agresti, A. (2002). *Categorical Data Analysis*. Wiley, New York.

Arentze, T., Borgers, A., Timmermans, H., and Mistro, R. D. (2003). Transport stated choice responses: Effects of task complexity, presentation format and literacy. *Transportation Research*, 39E:229–244.

Atkinson, A. C. and Donev, A. N. (1992). *Optimum Experimental Designs*. Clarendon Press, Oxford, U.K.

Bech, M. (2003). County council politicians' choice of hospital payment scheme: A discrete choice study. *Applied Health Economics and Health Policy*, 2:225–232.

Beder, J. H. (2004). On the definition of effects in fractional factorial designs. *Utilitas Mathematica*, 66:47–60.

Bose, R. C. (1947). Mathematical theory of the symmetrical factorial design. *Sankhya*, 8:107–66.

Bose, R. C. and Kishen, K. (1940). On the problem of confounding in the general symmetrical factorial design. *Sankhya*, 5:21–36.

Bradley, R. A. (1955). Rank analysis of incomplete block designs: III. Some large-sample results on estimation and power for a method of paired comparisons. *Biometrika*, 42:450–470.

Bradley, R. A. (1985). Paired comparisons. In Kotz, S. and Johnson, N., editors, *Encyclopedia of Statistical Sciences*, volume 6, pages 555–560. Wiley, New York.

Bradley, R. A. and El-Helbawy, A. T. (1976). Treatment contrasts in paired comparions: Basic procedures with applications to factorials. *Biometrika*, 63:255–262.

Bradley, R. A. and Gart, J. J. (1962). The asymptotic properties of ML estimators when sampling from associated populations. *Biometrika*, 49:205–214.

Bradley, R. A. and Terry, M. E. (1952). Rank analysis of incomplete block designs. I. The method of paired comparisons. *Biometrika*, 39:324–345.

Brazell, J. D. and Louviere, J. J. (1995). Length effects in conjoint choice experiments and surveys: An explanation based on cumulative cognitive burden. INFORMS Marketing Science Conference, Sydney, Australia.

Bryan, S. and Dolan, P. (2004). Discrete choice experiments in health economics. *The European Journal of Health Economics*, 5:199–202.

Bunch, D. S., Louviere, J. J., and Anderson, D. (1996). A comparison of experimental design strategies for multinomial logit models: The case of generic attributes. Technical report, University of California, Davis. Available at http://faculty.gsm.ucdavis.edu/~bunch/.

Burgess, L. and Street, D. J. (2003). Optimal designs for 2^k choice experiments. *Communications in Statistics - Theory and Methods*, 32:2185–2206.

Burgess, L. and Street, D. J. (2005). Optimal designs for asymmetric choice experiments. *Journal of Statistical Inference*, 134:288–301.

Burgess, L. and Street, D. J. (2006). The optimal size of choice sets in choice experiments. *Statistics*, 40:507–515.

Bush, K. A. (1952). Orthogonal arrays of index unity. *Annals of Mathematical Statistics*, 23:426–34.

Carlsson, F. (2003). The demand for intercity public transport: The case of business passengers. *Applied Economics*, 35:41–50.

Carlsson, F. and Martinsson, P. (2003). Design techniques for stated preference methods in health economics. *Health Economics*, 12:281–294.

Cattin, P. and Wittink, D. R. (1982). Commercial use of conjoint analysis: A survey. *Journal of Marketing*, 46:44–53.

Chakraborty, G., Ettenson, R., and Gaeth, G. (1994). How consumers choose health insurance. *Journal of Health Care Marketing*, 14:21–33.

Cramer, H. (1946). *Mathematical Methods of Statistics*. Princeton University Press, Princeton, NJ.

Critchlow, D. E. and Fligner, M. A. (1991). Paired comparison, triple comparison, and ranking experiments as generalized linear models, and their implementation on GLIM. *Psychometrika*, 56:517–533.

David, H. A. (1988). *The Method of Paired Comparisons*. Griffin Press, Oxford, U.K.

Davidson, R. R. (1970). On extending the Bradley–Terry model to accommodate ties in paired comparison experiments. *Journal of the American Statistical Association*, 65:317–328.

DeShazo, J. R. and Fermo, G. (2002). Designing choice sets for stated preference methods: The effects of complexity on choice consistency. *Journal of Environmental Economics and Management*, 44:123–143.

Dey, A. (1985). *Orthogonal Fractional Factorial Designs*. Wiley, New York.

Dhar, R. (1997). Consumer preference for a no-choice option. *Journal of Consumer Research*, 24:215–231.

Dillman, D. A. and Bowker, D. K. (2001). *Mail and Internet Surveys: The Tailored Design Method*. Wiley, New York.

Dykstra, O. (1956). A note on the rank analysis of incomplete block designs applications beyond the scope of existing tables. *Biometrics*, 12:301–306.

El-Helbawy, A., Ahmed, E., and Alharbey, A. H. (1994). Optimal designs for asymmetrical factorial paired comparison experiments. *Communications in Statistics - Simulation*, 23:663–681.

El-Helbawy, A. T. and Ahmed, E. A. (1984). Optimal design results for 2^n factorial paired comparison experiments. *Communications in Statistics. Theory and Methods*, 13:2827–2845.

El-Helbawy, A. T. and Bradley, R. A. (1977). Treatment contrasts in paired comparisons: Convergence of a basic iterative scheme for estimation. *Communications in Statistics: Planning and Inference*, A6:197–207.

El-Helbawy, A. T. and Bradley, R. A. (1978). Treatment contrasts in paired comparisons: Large-sample results, applications and some optimal designs. *Journal of the American Statistical Association*, 73:831–839.

EuroQoL, g. (2006). What is EQ-5D? Technical report, EuroQoL. Available at http://gs1.q4matics.com/EuroqolPublishWeb/.

Fechner, G. T. (1860). *Elemente der Psychophysik*. Breitkopf und Härtel, Leipzig.

Finney, D. J. (1945). The factorial replication of factorial arrangements. *Annals of Eugenics*, 12:291–301.

Fisher, R. A. (1935). *The Design of Experiments*. Macmillan, Hampshire, U.K.

Fisher, R. A. (1945). A system of confounding for factors with more than two alternatives giving completely orthogonal cubes and higher powers. *Annals of Eugenics*, 12:283–290.

Flynn, T. N., Louviere, J. J., Peters, T. J., and Coast, J. (2007). Best-worst scaling: What it can do for health care research and how to do it. *Journal of Health Economics*, 26:171–189.

Ford Jr., L. R. (1957). Solution of a ranking problem from binary comparisons. *American Mathematics Monthly*, 64:28–33.

Gerard, K., Shanahan, M., and Louviere, J. (2003). Using stated preference discrete choice modelling to inform health care decision-making: A pilot study of breast screening participation. *Applied Economics*, 35:1073–1085.

Grasshoff, U., Grossmann, H., Holling, H., and Schwabe, R. (2004). Optimal designs for main effects in linear paired comparison models. *Journal of Statistical Planning and Inference*, 126:361–376.

Green, P. (1974). On the design of choice experiments involving multifactor alternatives. *Journal of Consumer Research*, 1:61–68.

Haaijer, R., Kamakura, W., and Wedel, M. (2001). The 'no-choice' alternative in conjoint choice experiments. *International Journal of Market Research*, 43:93–106.

Hall, J., Kenny, P., King, M., Louviere, J., Viney, R., and Yeoh, A. (2002). Using stated preference discrete choice modelling to evaluate the introduction of varicella vaccination. *Health Economics*, 11:457–465.

Hanley, N., Mourato, S., and Wright, R. E. (2001). Choice modeling approaches: A superior alternative for environmental valuation? *Journal of Economic Surveys*, 15:435–462.

Hartmann, A. and Sattler, H. (2002). Commercial use of conjoint analysis in Austria, Germany and Switzerland. Technical report, University of Hamburg. Research papers on marketing and retailing.

Hedayat, A., Sloane, N. J. A., and Stufken, J. (1999). *Orthogonal Arrays: Theory and Applications*. Springer, New York.

Hensher, D. (1994). Stated preference analysis of travel choices: The state of practice. *Transportation*, 21:107–133.

Holling, H., Melles, T., and Reiners, W. (1998). How many attributes should be used in a paired comparison task?: An empirical examination using a new validation approach. Technical report, Westfälische Wilhelms-Universität Münster, Germany. Available at http://3mfuture.com/reiners/articles/how_many_attributes_should_be_used_in_a_paired_comparison_task.htm.

Huber, J. and Zwerina, K. (1996). The importance of utility balance in efficient choice designs. *Journal of Marketing Research*, 33:307–317.

Iyengar, S., Jiang, W., and Huberman, G. (2004). How much choice is too much: Determinants of individual contributions in 401K retirement plans. In Mitchell, O. and Utkus,

S., editors, *Pension Design and Structure: New Lessons from Behavioral Finance*, pages 83–95. Oxford University Press, Oxford.

Iyengar, S. S. and Lepper, M. R. (2000). When choice is demotivating: Can one desire too much of a good thing? *Journal of Personality and Social Psychology*, 79:995–1006.

John, J. A. and Dean, A. M. (1975). Single replicate factorial experiments in generalized cyclic designs: I. Symmetrical arrangements. *Journal of the Royal Statistical Society*, B 37:345–360.

Jungnickel, D., Pott, A., and Smith, K. W. (2006). Difference sets. In Colbourn, C. and Dinitz, J., editors, *Handbook of Combinatorial Designs, Second Edition*, pages 419–435. Chapman & Hall/CRC Press, Boca Raton, FL.

Kanninen, B. (2002). Optimal designs for multinomial choice experiments. *Journal of Marketing Research*, 39:214–227.

Kemperman, A. D. A. M., Borgers, A. W. J., Oppewal, H., and Timmermans, H. J. P. (2000). Consumer choice of theme parks: A conjoint choice model of seasonality effects and variety seeking behavior. *Leisure Sciences*, 22:1–18.

Kessels, R., Goos, P., and Vandebroek, M. (2006). A comparison of criteria to design efficient choice experiments. *Journal of Marketing Research*, 43:409–419.

Kuehl, R. (1999). *Design of Experiments: Statistical Principles of Research Design and Analysis*. Duxbury Press, Pacific Grove, MA.

Kuhfeld, W. F. (2006). Orthogonal arrays. Technical report, SAS Institute. Available at http://support.sas.com/techsup/technote/ts723.html.

Long, J. S. and Freese, J. (2006). *Regression Models for Categorical Dependent Variables Using Stata*. Stata Press, College Station, TX.

Longworth, L., Ratcliffe, J., and Boulton, M. (2001). Investigating women's preferences for intrapartum care: Home versus hospital births. *Health and Social Care in the Community*, 9:404–413.

Louviere, J., Carson, R. T., Burgess, L., Street, D. J., and Marley, T. S. (2007). Sequential preference questions factors influencing completion rates using an online panel. Technical report, University of Technology, Sydney. available at http://www.business.uts.edu.au/censoc/papers/index.html.

Louviere, J., Hensher, D., and Swait, J. (2000). *Stated Choice Methods: Analysis and Application*. Cambridge University Press, Cambridge.

Luce, R. D. (1959). *Individual Choice Behavior*. Wiley, New York.

MacKay, D. B. (1988). Thurstone's theory of comparative judgment. In Kotz, S. and Johnson, N., editors, *Encyclopedia of Statistical Sciences*, volume 9, pages 237–241. Wiley, New York.

Maddala, T., Phillips, K., and Johnson, F. (2002). Measuring preferences for health care interventions using conjoint analysis: An application to HIV testing. *Health Service Research*, 37:1681–1705.

Mason, R. L., Gunst, R. F., and Hess, J. L. (2003). *Statistical Design and Analysis of Experiments, with Applications to Engineering and Science*. Wiley, New York.

Mathon, R. and Rosa, A. (2006). 2-(v, k, λ) designs of small order. In Colbourn, C. and Dinitz, J., editors, *Handbook of Combinatorial Designs, Second Edition*, pages 25–58. Chapman & Hall/CRC Press, Boca Raton, FL.

McCormick, E. J. and Bachus, J. A. (1952). Paired comparison ratings: 1. The effect on ratings of reductions in the number of pairs. *Journal of Applied Psychology*, 36:123–127.

McCormick, E. J. and Roberts, W. K. (1952). Paired comparison ratings: 2. The reliability of ratings based on partial pairings. *Journal of Applied Psychology*, 36:188–192.

McKenzie, L., Cairns, J., and Osman, L. (2001). Symptom-based outcome measures for asthma: The use of discrete choice methods to assess patient preferences. *Health Policy*, 57:193–204.

Montgomery, D. C. (2001). *Design and Analysis of Experiments*. Wiley, New York.

Moore, B. (2001). personal communication.

Narula, S. C. (1978). Orthogonal polynomial regression for unequal spacing and frequencies. *Journal of Quality Technology*, 10:170–179.

Pendergrass, R. N. and Bradley, R. A. (1960). Ranking in triple comparisons. In *Contributions to probability and statistics*, pages 331–51. Stanford University Press, Stanford, CA.

Raghavarao, D. (1971). *Constructions and Combinatorial Problems in Design of Experiments*. Wiley, New York.

Raktoe, B. L., Hedayat, A., and Federer, W. T. (1981). *Factorial Designs*. Wiley, New York.

Roberts, I. (2000). personal communication.

Ryan, M. (1999). Using conjoint analysis to take account of patient preferences and go beyond health outcomes: An application to *in vitro* fertilisation. *Social Science and Medicine*, 48:535–546.

Ryan, M., Bate, A., Eastmond, C. J., and Ludbrook, A. (2001). Use of discrete choice experiments to elicit preferences. *Quality in Health Care*, 10:i55–i60.

Ryan, M. and Farrar, S. (2000). Using conjoint analysis to elicit preferences for health care. *British Medical Journal*, 320:1530–1533.

Ryan, M. and Gerard, K. (2003). Using discrete choice experiments to value health care programmes: Current practice and future research reflections. *Applied Health Economics and Health Policy*, 2:55–64.

Ryan, M. and Hughes, J. (1997). Using conjoint analysis to assess women's preferences for miscarriage management. *Health Economics*, 6:261–273.

Ryan, M., McIntosh, E., Dean, T., and Old, P. (2000). Trade-offs between location and waiting times in the provision of health care: The case of elective surgery on the Isle of Wight. *Journal of Public Health Medicine*, 22:202–210.

Sandor, Z. and Wedel, M. (2001). Designing conjoint choice experiments using managers' prior beliefs. *Journal of Marketing Research*, 38:430–444.

Scarpa, R., Willis, K., Acutt, M., and Ferrini, S. (2004). Monte Carlo simulation evidence on the effect of the status-quo in choice experiment models. *EAERE conference Budapest*.

Scholz, F. W. (1985). Maximum likelihood estimation. In Kotz, S. and Johnson, N., editors, *Encyclopedia of Statistical Sciences*, volume 5, pages 340–351. Wiley, New York.

Schwartz, B., Ward, A., Monterosso, J., Lyubomirsky, S., White, K., and Lehman, D. R. (2002). Maximizing versus satisficing: Happiness is a matter of choice. *Journal of Personality and Social Psychology*, 83:1178–1197.

Scott, A. (2002). Identifying and analysing dominant preferences in discrete choice experiments: An application in health care. *Journal of Economic Psychology*, 23:383–398.

Severin, V. (2000). *Comparing Statistical Efficiency and Respondent Efficiency in Choice Experiments*. PhD thesis, University of Sydney.

Sloane, N. J. A. (2006a). A library of Hadamard matrices. Technical report, AT&T Shannon Lab. Available at http://www.research.att.com/~njas/hadamard/.

Sloane, N. J. A. (2006b). A library of orthogonal arrays. Technical report, AT&T Shannon Lab. Available at http://www.research.att.com/~njas/oadir/.

Street, D. J., Bunch, D., and Moore, B. (2001). Optimal designs for 2^k paired comparison experiments. *Communications in Statistics - Theory and Methods*, 30:2149–2171.

Street, D. J. and Burgess, L. (2004a). Optimal and near-optimal pairs for the estimation of effects in 2-level choice experiments. *Journal of Statistical Planning and Inference*, 118:185–199.

Street, D. J. and Burgess, L. (2004b). Optimal stated preference choice experiments when all choice sets contain a specific option. *Statistical Methodology*, 1:37–45.

Street, D. J., Burgess, L., and Louviere, J. (2005). Quick and easy choice sets: Constructing optimal and nearly optimal stated choice experiments. *International Journal of Research Marketing*, 22:459–470.

Swait, J. and Adamowicz, W. (1996). The effect of choice environment and task demands on consumer behavior: Discriminating between contribution and confusion. Technical report, University of Alberta. Available at http://econpapers.repec.org/paper/wopalresp/9609.htm.

Tayyaran, M. R., Khan, A. M., and Anderson, D. A. (2003). Impact of telecommuting and intelligent transportation systems on residential location choice. *Transportation Planning and Technology*, 26:171 193.

Thompson, L. (2005). An S manual to accompany Agresti's Categorical Data Analysis. Technical report, University of Houston-Clear Lake. Available at https://home.comcast.net/~lthompson221/#CDA.

Thorndike, E. L. (1910). Handwriting. *Teachers College Record*, 11:1–93.

Thurstone, L. L. (1927). The method of paired comparisons for social values. *Journal of Abnormal and Social Psychology*, 21:384–400.

Timmermans, H. J. P., Borgers, A., Waerden, P. v. d., and Berenos, M. (2006). Order effects in stated-choice experiments: Study of transport mode choice decisions. Technical report, Transportation Research Board Annual Meeting 2006 Paper #06-0504. Abstract available at http://pubsindex.trb.org/document/view/default.asp?lbid=776375.

Train, K. E. (2003). *Discrete Choice Methods with Simulation*. Cambridge University Press, Cambridge.

van Berkum, E. E. M. (1987a). Optimal paired comparison designs for factorial and quadratic models. *Journal of Statistical Planning and Inference*, 15:265–278.

van Berkum, E. E. M. (1987b). *Optimal paired comparison designs for factorial experiments*. PhD thesis, Centrum voor Wiskunde en Informatica. CWI Tract **31**.

Venables, W. N. and Ripley, B. D. (2003). *Modern Applied Statistics with S*. Springer, New York.

Walker, B., Marsh, A., Wardman, M., and Niner, P. (2002). Modelling tenantsŠ choices in the public rented sector: A stated preference approach. *Urban Studies*, 39:665–688.

Wittink, D. R. and Cattin, P. (1989). Commercial use of conjoint analysis: An update. *Journal of Marketing*, 53:91–96.

Wittink, D. R., Vriens, M., and Burthenne, W. (1994). Commercial use of conjoint analysis in Europe: Results and critical reflections. *International Journal of Research in Marketing*, 11:41–52.

Yates, F. (1935). Complex experiments (with discussion). *Journal of the Royal Statistical Society*, Suppl 2:181–247.

Zermelo, E. (1929). Die berechnung der turnier-ergebnisse als ein maximumproblem der wahrscheinlichkeitsrechnung. *Mathematische Zeitschrift*, 29:436–60.

Zwerina, K., Huber, J., and Kuhfeld, W. F. (1996). A general method for constructing efficient choice designs. Technical report, SAS Institute. Available at http://support.sas.com/techsup/tnote/tnote_stat.html.

INDEX

$a_{\mathbf{v}_j}$, 142, 172
B_a, 72
B_f, 228
B_h, 72
B_{ℓ_q}, 180
B_n, 228
B_r, 72
C, 97, 100, 105, 147, 152, 180
C_{MT_f}, 278
C_c, 234
C_{cn}, 237
C_f, 228
C_{MT}, 240
C_n, 228
$C_{\mathrm{opt}.M.m=2}$, 234
$C_{\mathrm{opt}.M.\mathrm{any}\ m}$, 237
$C_{\mathrm{opt}.MT}$, 240
C_s, 242
$c_{\mathbf{v}_j}$, 142, 172
$D_{\mathbf{d}}$, 174
$D_{k,v}$, 97
Eff_A, 86, 117
Eff_D, 86, 112, 151, 157, 188
I_p, 72
i_v, 97
$i_{\mathbf{v}_j}$, 142, 172
L, 167
ℓ_q, 167
m, 167
M_{i_q}, 174
m_s, 242
$n_{i_1,i_2,\ldots,i_m}$, 79
$n^{(s)}_{i_1,i_2}$, 242
N_s, 242
$x_{\mathbf{v}_j;\mathbf{d}}$, 172
$x_{\mathbf{v}_j;i}$, 142
Λ, 69, 80, 97, 105, 118, 144, 152, 174
Λ_{cn}, 236
Λ_f, 228
Λ_n, 228
Λ_s, 242

A-efficiency, 86
A-optimal design, 84
adding another factor, 265
affine plane, 24
affirmation bias, 12
alias structure, 29
aliased effects, 29
alternative specific attributes, 11
associated distribution, 67
asymmetric factorial design, 16
attribute, 3, 15
    alternative specific, 11
availability designs, 11

balance
    level, 89
    utility, 89
balanced incomplete block design, 53

The Construction of Optimal Stated Choice Experiments. By D. J. Street and L. Burgess

symmetric, 53
bias
affirmation, 12
policy response, 12
rationalization, 12
BIBD, 53, 244
binary response experiment, 2–5, 233–234
block diagonal, 76, 210
Bradley–Terry model, 60–79
information matrix, 68
likelihood function, 61
maximum likelihood estimation, 62–65
ML estimates
convergence, 65–67

choice experiments
binary attributes
pairs, 95–135
various choice set sizes, 242
choice models, 58–60
choice probabilities, 59–60
choice set, 2
choice set size
optimal, 237–243
main effects and two-factor interactions, binary attributes, 240
main effects only, 237–240
class of competing designs, 95, 138
cognitive complexity, 12
collapsed attributes, 257
collapsing levels, 46, 258
common base and none options, 236–237
common base option, 2, 8–9, 234–236
comparison, 19, 20
independent, 19, 20
orthogonal, 20
complete factorial design, 16–25
confounded effects, 29
connected design, 66, 78
contractive replacement, 47, 260, 267
contrast, 20
geometric, 24
independent, 20
orthogonal, 20
orthogonal polynomial, 21–22
polynomial, 25
contrast matrix, 72, 105
binary attributes, 99, 144
interaction effects, 100
main effects, 100, 180
main effects and two-factor interaction effects, 198
normalized, 180

D-efficiency, 86
D-optimal design, 84
any m
asymmetric attributes, 187
binary attributes, 149, 154
asymmetric attributes
any m, 187
binary attributes
any m, 149, 154
paired comparisons, 103, 107
paired comparisons
binary attributes, 103, 107
D-optimal forced choice design
binary attributes
any m, 159–164
defining contrasts, 27, 33
defining effects, 29
defining equations, 27, 33
independent, 27
design
A-efficiency, 86
A-optimal, 84
connected, 66
D-efficiency, 86
D-optimal, 84
DCE
construction strategy, 279–287
strategy comparison, 291–293
E-optimal, 84
efficiency, 85
near-optimal, 159–164
orthogonal, 89
shifted, 89
difference family, 55
difference set, 54, 264
difference vector
choice set of size m, asymmetric attributes, 169–173
choice set of size m, binary attributes, 138–143
discrete choice experiment, 2
distribution
associated, 67
dominating option, 3, 10, 133, 164, 261

E-optimal design, 84
effects
aliased, 29
confounded, 29
defining, 29
interaction, 17–19, 23–25
main, 16–17, 19–20, 25
simple, 16
efficiency, 85
A-, 86
D-, 86
expansive replacement, 47, 260
experiment
paired comparison, 60

factor, 15
level, 15
qualitative, 19
quantitative, 19, 20

factorial design, 15
2^k, 16–19
2^k regular fractional, 27–33
3^k, 19–24
3^k regular fractional, 33–37
alias structure, 29
asymmetric, 16, 24–25
collapsing levels, 46
complete, 16–25
contractive replacement, 47
expansive replacement, 47
fractional, 27
interaction effects, 17–19, 23–25, 29
higher-order, 18
two-factor, 17
irregular fractional, 27, 41–52
resolution 3, 41
resolution 5, 42
main effects, 16–17, 19–20, 25
prime-power levels regular fractional, 39–41
resolution 3, 39
resolution 5, 40
regular fractional, 27–41
generator vector, 31, 37
resolution 3, 29, 31, 43, 251–267
resolution 4, 29, 271, 276–279
resolution 5, 30, 31, 163, 271–276
symmetric, 16
factorial designs
tables, 55
finite field, 37–39
irreducible polynomial, 38
primitive element, 39
Fisher information matrix, 67
foldover, 78
foldover treatment, 98
forced choice, 95
forced choice experiment, 2, 5–7, 57
optimal, 137, 167
forced choice stated preference experiment
optimal, 137, 167
fractional factorial design, 27
D-optimal design, binary attributes
any m, 159–164
as starting design, 249
irregular, 41–52

Galois field, 37–39
generator
paired comparison design, 122
generator vector, 37
generators
choosing, 264–267
estimable set, 129
fractional factorial designs, 31
main effects
asymmetric attributes, 191
small, optimal choice experiments, 160, 163

Hadamard matrix, 55
higher-order interaction effect, 18

IIA, 83
independence from irrelevant alternatives, 83
independent comparison, 20
independent contrast, 20
independent defining equations, 27
indicator variable, 97
information matrix, 97
Bradley–Terry model, 68
derivation, 174–180
derivation of, 97–99, 118–119, 143–147
Fisher, 67
general form, 176
general form, binary attributes, 144
main effect of attribute q, 184
main effects, 102, 148, 180, 184
main effects and two-factor interactions, 105, 152, 201
interaction, 17–19, 23–24
higher-order, 18, 24
linear × linear, 23
linear × quadratic, 23
quadratic × quadratic, 23
three-factor, 18
two-factor, 17
interaction effects, 17–19, 23–25
definition by restriction, 29
irreducible polynomial, 38
irregular fractional factorial design, 27, 41–52
resolution 3, 41
resolution 5, 42

labeled options, 11
level, 3, 15
level balance, 89
lexicographic order, 97

main effects, 16–17, 19–20, 25
correlated, 257
variance-covariance matrix, 74
main effects plan
orthogonal, 42
matrix
contrast, 72
merit, 61
minimum overlap, 89
MNL model, 58, 79–83
model
Bradley–Terry, 60–79
GEV, 58, 83
main effects and two-factor interactions, 105–117, 121–133, 152–159, 163–164, 197–210
main effects only, 100–105, 119–121, 147–151, 160–162, 180–189
mixed logit, 58
MNL, 58, 79–83

probit, 58
multiplicative inverse, 38

near-optimal design, 159–164
main effects only, 251–267
main effects plus some two-factor interactions
non-binary attributes, 276–279
main effects plus two-factor interactions, 269–279
binary attributes, 271–272
non-binary attributes, 272–276
none option, 2, 7–8, 228–233
null hypothesis
usual, 98, 144, 174

OMEP, 42
optimal choice set size, 237–243
optimal choice sets, 137, 167
asymmetric attributes
$m \geq 2$, 167–210
binary attributes
$m \geq 2$, 137–165
pairs, 95–117
optimality
A-, 84
D-, 84
E-, 84
options
labeled, 11
orthogonal array, 42
adding one more factor, 49
asymmetric, 42
collapsing levels, 46
construction, 43–44, 46, 47
contractive replacement, 47
expansive replacement, 47
index, 42
juxtaposing two OAs, 50
number of constraints, 42
number of levels, 42
parent design, 55
recursive construction, 49, 50
saturated, 47
strength, 42
symmetric, 43–44
table of, 55
resolution 3, 55
tight, 47, 261
orthogonal comparison, 20
orthogonal contrast, 20
orthogonal main effects plan, 42
orthogonal polynomial, 21–22
orthogonality, 89
orthonormal matrix, 72

paired comparison design
binary attributes, 95–135
A-optimal, 105, 113
D-optimal, 103
D-optimal, 107
fractional factorial, 118–119
paired comparison experiment, 60
partial profiles, 243–244
pencil, 24, 35
policy response bias, 12
polynomial
irreducible, 38
primitive element, 39
principal fraction, 28
principal minor, 74
prior information, 245–246
pseudo-factor, 39

qualitative factor, 19
quantitative factor, 19

rationalization bias, 12
regular fractional factorial
2^k resolution 3, 31
2^k resolution 5, 33
3^k resolution 3, 37
3^k resolution 5, 37
regular fractional factorial design, 27–41
2^k, 27–33
3^k, 33–37
generator vector, 31, 37
prime-power levels, 39–41
resolution 3, 39
resolution 5, 40
resolution 3 factorial design, 29, 31, 251–267
resolution 4 factorial design, 29, 271, 276–279
resolution 5 factorial design, 30, 31, 271–276
ring of polynomials, 39

SBIBD, 53
shifted design, 89
simple effects, 16
starting design, 249
obtaining, 256–263, 269–271
stated choice experiment, 2
stated preference choice experiment, 2
symmetric balanced incomplete block design, 53
symmetric factorial design, 16

tight orthogonal array, 261
treatment combinations, 15
standard order, 97
unrealistic, 51
Yates standard order, 97
two-factor interaction effect, 17

unrealistic treatment combination, 9–10, 51, 262
utility, 58
utility balance, 89

variance-covariance matrix, 84
main effects, 74
various choice set sizes, 242

Yates standard order, 97

WILEY SERIES IN PROBABILITY AND STATISTICS
ESTABLISHED BY WALTER A. SHEWHART AND SAMUEL S. WILKS

BECHHOFER, SANTNER, and GOLDSMAN · Design and Analysis of Experiments for Statistical Selection, Screening, and Multiple Comparisons
BELSLEY · Conditioning Diagnostics: Collinearity and Weak Data in Regression
† BELSLEY, KUH, and WELSCH · Regression Diagnostics: Identifying Influential Data and Sources of Collinearity
BENDAT and PIERSOL · Random Data: Analysis and Measurement Procedures, *Third Edition*
BERRY, CHALONER, and GEWEKE · Bayesian Analysis in Statistics and Econometrics: Essays in Honor of Arnold Zellner
BERNARDO and SMITH · Bayesian Theory
BHAT and MILLER · Elements of Applied Stochastic Processes, *Third Edition*
BHATTACHARYA and WAYMIRE · Stochastic Processes with Applications
BILLINGSLEY · Convergence of Probability Measures, *Second Edition*
BILLINGSLEY · Probability and Measure, *Third Edition*
BIRKES and DODGE · Alternative Methods of Regression
BLISCHKE AND MURTHY (editors) · Case Studies in Reliability and Maintenance
BLISCHKE AND MURTHY · Reliability: Modeling, Prediction, and Optimization
BLOOMFIELD · Fourier Analysis of Time Series: An Introduction, *Second Edition*
BOLLEN · Structural Equations with Latent Variables
BOLLEN and CURRAN · Latent Curve Models: A Structural Equation Perspective
BOROVKOV · Ergodicity and Stability of Stochastic Processes
BOULEAU · Numerical Methods for Stochastic Processes
BOX · Bayesian Inference in Statistical Analysis
BOX · R. A. Fisher, the Life of a Scientist
BOX and DRAPER · Response Surfaces, Mixtures, and Ridge Analyses, *Second Edition*
* BOX and DRAPER · Evolutionary Operation: A Statistical Method for Process Improvement
BOX and FRIENDS · Improving Almost Anything, *Revised Edition*
BOX, HUNTER, and HUNTER · Statistics for Experimenters: Design, Innovation, and Discovery, *Second Editon*
BOX and LUCEÑO · Statistical Control by Monitoring and Feedback Adjustment
BRANDIMARTE · Numerical Methods in Finance: A MATLAB-Based Introduction
BROWN and HOLLANDER · Statistics: A Biomedical Introduction
BRUNNER, DOMHOF, and LANGER · Nonparametric Analysis of Longitudinal Data in Factorial Experiments
BUCKLEW · Large Deviation Techniques in Decision, Simulation, and Estimation
CAIROLI and DALANG · Sequential Stochastic Optimization
CASTILLO, HADI, BALAKRISHNAN, and SARABIA · Extreme Value and Related Models with Applications in Engineering and Science
CHAN · Time Series: Applications to Finance
CHARALAMBIDES · Combinatorial Methods in Discrete Distributions
CHATTERJEE and HADI · Regression Analysis by Example, *Fourth Edition*
CHATTERJEE and HADI · Sensitivity Analysis in Linear Regression
CHERNICK · Bootstrap Methods: A Practitioner's Guide
CHERNICK and FRIIS · Introductory Biostatistics for the Health Sciences
CHILÈS and DELFINER · Geostatistics: Modeling Spatial Uncertainty
CHOW and LIU · Design and Analysis of Clinical Trials: Concepts and Methodologies, *Second Edition*
CLARKE and DISNEY · Probability and Random Processes: A First Course with Applications, *Second Edition*
* COCHRAN and COX · Experimental Designs, *Second Edition*
CONGDON · Applied Bayesian Modelling
CONGDON · Bayesian Models for Categorical Data
CONGDON · Bayesian Statistical Modelling

*Now available in a lower priced paperback edition in the Wiley Classics Library.
†Now available in a lower priced paperback edition in the Wiley–Interscience Paperback Series.

CONOVER · Practical Nonparametric Statistics, *Third Edition*
COOK · Regression Graphics
COOK and WEISBERG · Applied Regression Including Computing and Graphics
COOK and WEISBERG · An Introduction to Regression Graphics
CORNELL · Experiments with Mixtures, Designs, Models, and the Analysis of Mixture Data, *Third Edition*
COVER and THOMAS · Elements of Information Theory
COX · A Handbook of Introductory Statistical Methods
* COX · Planning of Experiments
CRESSIE · Statistics for Spatial Data, *Revised Edition*
CSÖRGŐ and HORVÁTH · Limit Theorems in Change Point Analysis
DANIEL · Applications of Statistics to Industrial Experimentation
DANIEL · Biostatistics: A Foundation for Analysis in the Health Sciences, *Eighth Edition*
* DANIEL · Fitting Equations to Data: Computer Analysis of Multifactor Data, *Second Edition*
DASU and JOHNSON · Exploratory Data Mining and Data Cleaning
DAVID and NAGARAJA · Order Statistics, *Third Edition*
* DEGROOT, FIENBERG, and KADANE · Statistics and the Law
DEL CASTILLO · Statistical Process Adjustment for Quality Control
DeMARIS · Regression with Social Data: Modeling Continuous and Limited Response Variables
DEMIDENKO · Mixed Models: Theory and Applications
DENISON, HOLMES, MALLICK and SMITH · Bayesian Methods for Nonlinear Classification and Regression
DETTE and STUDDEN · The Theory of Canonical Moments with Applications in Statistics, Probability, and Analysis
DEY and MUKERJEE · Fractional Factorial Plans
DILLON and GOLDSTEIN · Multivariate Analysis: Methods and Applications
DODGE · Alternative Methods of Regression
* DODGE and ROMIG · Sampling Inspection Tables, *Second Edition*
* DOOB · Stochastic Processes
DOWDY, WEARDEN, and CHILKO · Statistics for Research, *Third Edition*
DRAPER and SMITH · Applied Regression Analysis, *Third Edition*
DRYDEN and MARDIA · Statistical Shape Analysis
DUDEWICZ and MISHRA · Modern Mathematical Statistics
DUNN and CLARK · Basic Statistics: A Primer for the Biomedical Sciences, *Third Edition*
DUPUIS and ELLIS · A Weak Convergence Approach to the Theory of Large Deviations
EDLER and KITSOS · Recent Advances in Quantitative Methods in Cancer and Human Health Risk Assessment
* ELANDT-JOHNSON and JOHNSON · Survival Models and Data Analysis
ENDERS · Applied Econometric Time Series
† ETHIER and KURTZ · Markov Processes: Characterization and Convergence
EVANS, HASTINGS, and PEACOCK · Statistical Distributions, *Third Edition*
FELLER · An Introduction to Probability Theory and Its Applications, Volume I, *Third Edition*, Revised; Volume II, *Second Edition*
FISHER and VAN BELLE · Biostatistics: A Methodology for the Health Sciences
FITZMAURICE, LAIRD, and WARE · Applied Longitudinal Analysis
* FLEISS · The Design and Analysis of Clinical Experiments
FLEISS · Statistical Methods for Rates and Proportions, *Third Edition*
† FLEMING and HARRINGTON · Counting Processes and Survival Analysis
FULLER · Introduction to Statistical Time Series, *Second Edition*
† FULLER · Measurement Error Models

*Now available in a lower priced paperback edition in the Wiley Classics Library.
†Now available in a lower priced paperback edition in the Wiley–Interscience Paperback Series.

GALLANT · Nonlinear Statistical Models
GEISSER · Modes of Parametric Statistical Inference
GELMAN and MENG · Applied Bayesian Modeling and Causal Inference from Incomplete-Data Perspectives
GEWEKE · Contemporary Bayesian Econometrics and Statistics
GHOSH, MUKHOPADHYAY, and SEN · Sequential Estimation
GIESBRECHT and GUMPERTZ · Planning, Construction, and Statistical Analysis of Comparative Experiments
GIFI · Nonlinear Multivariate Analysis
GIVENS and HOETING · Computational Statistics
GLASSERMAN and YAO · Monotone Structure in Discrete-Event Systems
GNANADESIKAN · Methods for Statistical Data Analysis of Multivariate Observations, *Second Edition*
GOLDSTEIN and LEWIS · Assessment: Problems, Development, and Statistical Issues
GREENWOOD and NIKULIN · A Guide to Chi-Squared Testing
GROSS and HARRIS · Fundamentals of Queueing Theory, *Third Edition*
* HAHN and SHAPIRO · Statistical Models in Engineering
HAHN and MEEKER · Statistical Intervals: A Guide for Practitioners
HALD · A History of Probability and Statistics and their Applications Before 1750
HALD · A History of Mathematical Statistics from 1750 to 1930
† HAMPEL · Robust Statistics: The Approach Based on Influence Functions
HANNAN and DEISTLER · The Statistical Theory of Linear Systems
HEIBERGER · Computation for the Analysis of Designed Experiments
HEDAYAT and SINHA · Design and Inference in Finite Population Sampling
HEDEKER and GIBBONS · Longitudinal Data Analysis
HELLER · MACSYMA for Statisticians
HINKELMANN and KEMPTHORNE · Design and Analysis of Experiments, Volume 1: Introduction to Experimental Design
HINKELMANN and KEMPTHORNE · Design and Analysis of Experiments, Volume 2: Advanced Experimental Design
HOAGLIN, MOSTELLER, and TUKEY · Exploratory Approach to Analysis of Variance
* HOAGLIN, MOSTELLER, and TUKEY · Exploring Data Tables, Trends and Shapes
* HOAGLIN, MOSTELLER, and TUKEY · Understanding Robust and Exploratory Data Analysis
HOCHBERG and TAMHANE · Multiple Comparison Procedures
HOCKING · Methods and Applications of Linear Models: Regression and the Analysis of Variance, *Second Edition*
HOEL · Introduction to Mathematical Statistics, *Fifth Edition*
HOGG and KLUGMAN · Loss Distributions
HOLLANDER and WOLFE · Nonparametric Statistical Methods, *Second Edition*
HOSMER and LEMESHOW · Applied Logistic Regression, *Second Edition*
HOSMER and LEMESHOW · Applied Survival Analysis: Regression Modeling of Time to Event Data
† HUBER · Robust Statistics
HUBERTY · Applied Discriminant Analysis
HUBERTY and OLEJNIK · Applied MANOVA and Discriminant Analysis, *Second Edition*
HUNT and KENNEDY · Financial Derivatives in Theory and Practice, *Revised Edition*
HUSKOVA, BERAN, and DUPAC · Collected Works of Jaroslav Hajek—with Commentary
HUZURBAZAR · Flowgraph Models for Multistate Time-to-Event Data
IMAN and CONOVER · A Modern Approach to Statistics

*Now available in a lower priced paperback edition in the Wiley Classics Library.
†Now available in a lower priced paperback edition in the Wiley–Interscience Paperback Series.

† JACKSON · A User's Guide to Principle Components
JOHN · Statistical Methods in Engineering and Quality Assurance
JOHNSON · Multivariate Statistical Simulation
JOHNSON and BALAKRISHNAN · Advances in the Theory and Practice of Statistics: A Volume in Honor of Samuel Kotz
JOHNSON and BHATTACHARYYA · Statistics: Principles and Methods, *Fifth Edition*
JOHNSON and KOTZ · Distributions in Statistics
JOHNSON and KOTZ (editors) · Leading Personalities in Statistical Sciences: From the Seventeenth Century to the Present
JOHNSON, KOTZ, and BALAKRISHNAN · Continuous Univariate Distributions, Volume 1, *Second Edition*
JOHNSON, KOTZ, and BALAKRISHNAN · Continuous Univariate Distributions, Volume 2, *Second Edition*
JOHNSON, KOTZ, and BALAKRISHNAN · Discrete Multivariate Distributions
JOHNSON, KEMP, and KOTZ · Univariate Discrete Distributions, *Third Edition*
JUDGE, GRIFFITHS, HILL, LÜTKEPOHL, and LEE · The Theory and Practice of Econometrics, *Second Edition*
JUREČKOVÁ and SEN · Robust Statistical Procedures: Aymptotics and Interrelations
JUREK and MASON · Operator-Limit Distributions in Probability Theory
KADANE · Bayesian Methods and Ethics in a Clinical Trial Design
KADANE AND SCHUM · A Probabilistic Analysis of the Sacco and Vanzetti Evidence
KALBFLEISCH and PRENTICE · The Statistical Analysis of Failure Time Data, *Second Edition*
KARIYA and KURATA · Generalized Least Squares
KASS and VOS · Geometrical Foundations of Asymptotic Inference
† KAUFMAN and ROUSSEEUW · Finding Groups in Data: An Introduction to Cluster Analysis
KEDEM and FOKIANOS · Regression Models for Time Series Analysis
KENDALL, BARDEN, CARNE, and LE · Shape and Shape Theory
KHURI · Advanced Calculus with Applications in Statistics, *Second Edition*
KHURI, MATHEW, and SINHA · Statistical Tests for Mixed Linear Models
KLEIBER and KOTZ · Statistical Size Distributions in Economics and Actuarial Sciences
KLUGMAN, PANJER, and WILLMOT · Loss Models: From Data to Decisions, *Second Edition*
KLUGMAN, PANJER, and WILLMOT · Solutions Manual to Accompany Loss Models: From Data to Decisions, *Second Edition*
KOTZ, BALAKRISHNAN, and JOHNSON · Continuous Multivariate Distributions, Volume 1, *Second Edition*
KOVALENKO, KUZNETZOV, and PEGG · Mathematical Theory of Reliability of Time-Dependent Systems with Practical Applications
LACHIN · Biostatistical Methods: The Assessment of Relative Risks
LAD · Operational Subjective Statistical Methods: A Mathematical, Philosophical, and Historical Introduction
LAMPERTI · Probability: A Survey of the Mathematical Theory, *Second Edition*
LANGE, RYAN, BILLARD, BRILLINGER, CONQUEST, and GREENHOUSE · Case Studies in Biometry
LARSON · Introduction to Probability Theory and Statistical Inference, *Third Edition*
LAWLESS · Statistical Models and Methods for Lifetime Data, *Second Edition*
LAWSON · Statistical Methods in Spatial Epidemiology
LE · Applied Categorical Data Analysis
LE · Applied Survival Analysis
LEE and WANG · Statistical Methods for Survival Data Analysis, *Third Edition*
LEPAGE and BILLARD · Exploring the Limits of Bootstrap

*Now available in a lower priced paperback edition in the Wiley Classics Library.
†Now available in a lower priced paperback edition in the Wiley–Interscience Paperback Series.

LEYLAND and GOLDSTEIN (editors) · Multilevel Modelling of Health Statistics
LIAO · Statistical Group Comparison
LINDVALL · Lectures on the Coupling Method
LIN · Introductory Stochastic Analysis for Finance and Insurance
LINHART and ZUCCHINI · Model Selection
LITTLE and RUBIN · Statistical Analysis with Missing Data, *Second Edition*
LLOYD · The Statistical Analysis of Categorical Data
LOWEN and TEICH · Fractal-Based Point Processes
MAGNUS and NEUDECKER · Matrix Differential Calculus with Applications in Statistics and Econometrics, *Revised Edition*
MALLER and ZHOU · Survival Analysis with Long Term Survivors
MALLOWS · Design, Data, and Analysis by Some Friends of Cuthbert Daniel
MANN, SCHAFER, and SINGPURWALLA · Methods for Statistical Analysis of Reliability and Life Data
MANTON, WOODBURY, and TOLLEY · Statistical Applications Using Fuzzy Sets
MARCHETTE · Random Graphs for Statistical Pattern Recognition
MARDIA and JUPP · Directional Statistics
MASON, GUNST, and HESS · Statistical Design and Analysis of Experiments with Applications to Engineering and Science, *Second Edition*
McCULLOCH and SEARLE · Generalized, Linear, and Mixed Models
McFADDEN · Management of Data in Clinical Trials
* McLACHLAN · Discriminant Analysis and Statistical Pattern Recognition
McLACHLAN, DO, and AMBROISE · Analyzing Microarray Gene Expression Data
McLACHLAN and KRISHNAN · The EM Algorithm and Extensions
McLACHLAN and PEEL · Finite Mixture Models
McNEIL · Epidemiological Research Methods
MEEKER and ESCOBAR · Statistical Methods for Reliability Data
MEERSCHAERT and SCHEFFLER · Limit Distributions for Sums of Independent Random Vectors: Heavy Tails in Theory and Practice
MICKEY, DUNN, and CLARK · Applied Statistics: Analysis of Variance and Regression, *Third Edition*
* MILLER · Survival Analysis, *Second Edition*
MONTGOMERY, PECK, and VINING · Introduction to Linear Regression Analysis, *Fourth Edition*
MORGENTHALER and TUKEY · Configural Polysampling: A Route to Practical Robustness
MUIRHEAD · Aspects of Multivariate Statistical Theory
MULLER and STOYAN · Comparison Methods for Stochastic Models and Risks
MURRAY · X-STAT 2.0 Statistical Experimentation, Design Data Analysis, and Nonlinear Optimization
MURTHY, XIE, and JIANG · Weibull Models
MYERS and MONTGOMERY · Response Surface Methodology: Process and Product Optimization Using Designed Experiments, *Second Edition*
MYERS, MONTGOMERY, and VINING · Generalized Linear Models. With Applications in Engineering and the Sciences
† NELSON · Accelerated Testing, Statistical Models, Test Plans, and Data Analyses
† NELSON · Applied Life Data Analysis
NEWMAN · Biostatistical Methods in Epidemiology
OCHI · Applied Probability and Stochastic Processes in Engineering and Physical Sciences
OKABE, BOOTS, SUGIHARA, and CHIU · Spatial Tesselations: Concepts and Applications of Voronoi Diagrams, *Second Edition*
OLIVER and SMITH · Influence Diagrams, Belief Nets and Decision Analysis

*Now available in a lower priced paperback edition in the Wiley Classics Library.
†Now available in a lower priced paperback edition in the Wiley–Interscience Paperback Series.

PALTA · Quantitative Methods in Population Health: Extensions of Ordinary Regressions
PANJER · Operational Risk: Modeling and Analytics
PANKRATZ · Forecasting with Dynamic Regression Models
PANKRATZ · Forecasting with Univariate Box-Jenkins Models: Concepts and Cases
* PARZEN · Modern Probability Theory and Its Applications
PEÑA, TIAO, and TSAY · A Course in Time Series Analysis
PIANTADOSI · Clinical Trials: A Methodologic Perspective
PORT · Theoretical Probability for Applications
POURAHMADI · Foundations of Time Series Analysis and Prediction Theory
PRESS · Bayesian Statistics: Principles, Models, and Applications
PRESS · Subjective and Objective Bayesian Statistics, *Second Edition*
PRESS and TANUR · The Subjectivity of Scientists and the Bayesian Approach
PUKELSHEIM · Optimal Experimental Design
PURI, VILAPLANA, and WERTZ · New Perspectives in Theoretical and Applied Statistics
† PUTERMAN · Markov Decision Processes: Discrete Stochastic Dynamic Programming
QIU · Image Processing and Jump Regression Analysis
* RAO · Linear Statistical Inference and Its Applications, *Second Edition*
RAUSAND and HØYLAND · System Reliability Theory: Models, Statistical Methods, and Applications, *Second Edition*
RENCHER · Linear Models in Statistics
RENCHER · Methods of Multivariate Analysis, *Second Edition*
RENCHER · Multivariate Statistical Inference with Applications
* RIPLEY · Spatial Statistics
* RIPLEY · Stochastic Simulation
ROBINSON · Practical Strategies for Experimenting
ROHATGI and SALEH · An Introduction to Probability and Statistics, *Second Edition*
ROLSKI, SCHMIDLI, SCHMIDT, and TEUGELS · Stochastic Processes for Insurance and Finance
ROSENBERGER and LACHIN · Randomization in Clinical Trials: Theory and Practice
ROSS · Introduction to Probability and Statistics for Engineers and Scientists
ROSSI, ALLENBY, and McCULLOCH · Bayesian Statistics and Marketing
† ROUSSEEUW and LEROY · Robust Regression and Outlier Detection
* RUBIN · Multiple Imputation for Nonresponse in Surveys
RUBINSTEIN · Simulation and the Monte Carlo Method
RUBINSTEIN and MELAMED · Modern Simulation and Modeling
RYAN · Modern Experimental Design
RYAN · Modern Regression Methods
RYAN · Statistical Methods for Quality Improvement, *Second Edition*
SALEH · Theory of Preliminary Test and Stein-Type Estimation with Applications
* SCHEFFE · The Analysis of Variance
SCHIMEK · Smoothing and Regression: Approaches, Computation, and Application
SCHOTT · Matrix Analysis for Statistics, *Second Edition*
SCHOUTENS · Levy Processes in Finance: Pricing Financial Derivatives
SCHUSS · Theory and Applications of Stochastic Differential Equations
SCOTT · Multivariate Density Estimation: Theory, Practice, and Visualization
† SEARLE · Linear Models for Unbalanced Data
† SEARLE · Matrix Algebra Useful for Statistics
† SEARLE, CASELLA, and McCULLOCH · Variance Components
SEARLE and WILLETT · Matrix Algebra for Applied Economics
SEBER and LEE · Linear Regression Analysis, *Second Edition*
† SEBER · Multivariate Observations
† SEBER and WILD · Nonlinear Regression
SENNOTT · Stochastic Dynamic Programming and the Control of Queueing Systems

*Now available in a lower priced paperback edition in the Wiley Classics Library.
†Now available in a lower priced paperback edition in the Wiley–Interscience Paperback Series.

* SERFLING · Approximation Theorems of Mathematical Statistics
SHAFER and VOVK · Probability and Finance: It's Only a Game!
SILVAPULLE and SEN · Constrained Statistical Inference: Inequality, Order, and Shape Restrictions
SMALL and McLEISH · Hilbert Space Methods in Probability and Statistical Inference
SRIVASTAVA · Methods of Multivariate Statistics
STAPLETON · Linear Statistical Models
STAUDTE and SHEATHER · Robust Estimation and Testing
STOYAN, KENDALL, and MECKE · Stochastic Geometry and Its Applications, *Second Edition*
STOYAN and STOYAN · Fractals, Random Shapes and Point Fields: Methods of Geometrical Statistics
STREET and BURGESS · The Construction of Optimal Stated Choice Experiments: Theory and Methods
STYAN · The Collected Papers of T. W. Anderson: 1943–1985
SUTTON, ABRAMS, JONES, SHELDON, and SONG · Methods for Meta-Analysis in Medical Research
TAKEZAWA · Introduction to Nonparametric Regression
TANAKA · Time Series Analysis: Nonstationary and Noninvertible Distribution Theory
THOMPSON · Empirical Model Building
THOMPSON · Sampling, *Second Edition*
THOMPSON · Simulation: A Modeler's Approach
THOMPSON and SEBER · Adaptive Sampling
THOMPSON, WILLIAMS, and FINDLAY · Models for Investors in Real World Markets
TIAO, BISGAARD, HILL, PEÑA, and STIGLER (editors) · Box on Quality and Discovery: with Design, Control, and Robustness
TIERNEY · LISP-STAT: An Object-Oriented Environment for Statistical Computing and Dynamic Graphics
TSAY · Analysis of Financial Time Series, *Second Edition*
UPTON and FINGLETON · Spatial Data Analysis by Example, Volume II: Categorical and Directional Data
VAN BELLE · Statistical Rules of Thumb
VAN BELLE, FISHER, HEAGERTY, and LUMLEY · Biostatistics: A Methodology for the Health Sciences, *Second Edition*
VESTRUP · The Theory of Measures and Integration
VIDAKOVIC · Statistical Modeling by Wavelets
VIDAKOVIC and KVAM · Nonparametric Statistics with Applications to Science and Engineering
VINOD and REAGLE · Preparing for the Worst: Incorporating Downside Risk in Stock Market Investments
WALLER and GOTWAY · Applied Spatial Statistics for Public Health Data
WEERAHANDI · Generalized Inference in Repeated Measures: Exact Methods in MANOVA and Mixed Models
WEISBERG · Applied Linear Regression, *Third Edition*
WELSH · Aspects of Statistical Inference
WESTFALL and YOUNG · Resampling-Based Multiple Testing: Examples and Methods for *p*-Value Adjustment
WHITTAKER · Graphical Models in Applied Multivariate Statistics
WINKER · Optimization Heuristics in Economics: Applications of Threshold Accepting
WONNACOTT and WONNACOTT · Econometrics, *Second Edition*
WOODING · Planning Pharmaceutical Clinical Trials: Basic Statistical Principles
WOODWORTH · Biostatistics: A Bayesian Introduction
WOOLSON and CLARKE · Statistical Methods for the Analysis of Biomedical Data, *Second Edition*

*Now available in a lower priced paperback edition in the Wiley Classics Library.
†Now available in a lower priced paperback edition in the Wiley–Interscience Paperback Series.

WU and HAMADA · Experiments: Planning, Analysis, and Parameter Design Optimization
WU and ZHANG · Nonparametric Regression Methods for Longitudinal Data Analysis
YANG · The Construction Theory of Denumerable Markov Processes
YOUNG, VALERO-MORA, and FRIENDLY · Visual Statistics: Seeing Data with Dynamic Interactive Graphics
ZELTERMAN · Discrete Distributions—Applications in the Health Sciences
* ZELLNER · An Introduction to Bayesian Inference in Econometrics
ZHOU, OBUCHOWSKI, and McCLISH · Statistical Methods in Diagnostic Medicine

Lightning Source UK Ltd.
Milton Keynes UK
UKOW042036120412

190597UK00002B/5/P

9 780470 053324